The Student's Guide to Social Neuroscience

The Student's Guide to Social Neuroscience

Jamie Ward

Psychology Press
Taylor & Francis Group

HOVE AND NEW YORK

First published 2012
by Psychology Press
27 Church Road, Hove, East Sussex, BN3 2FA

Simultaneously published in the USA and Canada
by Psychology Press
711 Third Avenue, New York, NY 10017

*Psychology Press is an imprint of the Taylor & Francis Group,
an Informa business*

British Library Cataloguing in Publication Data
A catalogue record for this book is available from the British Library

Library of Congress Cataloging in Publication Data
Ward, Jamie.
  The student's guide to social neuroscience / Jamie Ward.
    p. cm.
Includes bibliographical references and index.
ISBN 978–1–84872–004–6 (hardcover) – ISBN 978–1–84872–005–3 (pbk.)
  1. Perception.   2. Social interaction. I. Title.
  BF311.W26947 2012
  153–dc22   2011009617

ISBN: 978–1–84872–004–6 (hbk)
ISBN: 978–1–84872–005–3 (pbk)

Typeset by Newgen Imaging Systems (P) Ltd, Chennai, India
Cover design by Andy Ward

Printed and bound in the UK by Ashford Colour Press Ltd, Gosport, Hampshire, UK

Contents

About the author   vi
Preface   vii

**1 Introduction to social
   neuroscience**   3
   The emergence of social neuroscience   4
   The social brain?   5
   Is neuroscience an appropriate level of
      explanation for studying social behavior?   7

**2 The methods of social
   neuroscience**   13
   Measuring behavior and cognition:
      Psychological methods   14
   Measuring bodily responses   22
   Electrophysiological methods   25
   Functional imaging:
      Hemodynamic measures   33
   Lesion methods and disruption of
      function using TMS   40

**3 Evolutionary origins of social
   intelligence and culture**   49
   The social intelligence hypothesis   50
   Evolutionary origins of culture   55
   Material symbols: Neuronal
      recycling and extended cognition   62
   Cultural skills: Tools and technology   64

4 Emotion and motivation   71
   Historical perspectives on the emotions   72
   Different categories of emotion
      in the brain   78
   Motivation: Rewards and
      punishment, pleasure and pain   88

5 Reading faces and bodies   101
   Perceiving faces   101
   Perceiving bodies   111
   Joint attention:
      From perception to intention   115

Trait inferences from faces and bodies   120

6 Understanding others   129
   Empathy and simulation theory   130
   Theory of mind and reasoning
      about mental states   139
   Explaining autism   145

7 Interacting with others   157
   Altruism and helping behavior   158
   Game theory and social decision making   166

8 Relationships   179
   All you need is love   180
   Attachment   184
   Separation, rejection, and loneliness   194

9 Groups and identity   201
   Identity and the self-concept   202
   Ingroups, outgroups, and prejudice   213
   Herds, crowds, and religion   220

**10 Morality and antisocial
   behavior**   227
   The neuroscience of morality   229
   Anger and aggression   239
   Control and responsibility:
      'It wasn't me. It was my brain.'   252

**11 Developmental social
   neuroscience**   257
   Social learning during infancy   259
   The social brain in childhood:
      Understanding self, understanding
      others   269
   The adolescent brain   277

References   283
Author index   319
Subject index   333

About the author

Dr Jamie Ward is an Associate Professor at the University of Sussex, UK. He has published over 60 scientific papers and several books including *The Student's Guide to Cognitive Neuroscience* (now in its second edition) and *The Frog who Croaked Blue: Synesthesia and the Mixing of the Senses.* He is the Editor-in-Chief of *Cognitive Neuroscience*, a journal from Psychology Press.

Preface

This textbook came about through a desire to create an accompanying text to *The Student's Guide to Cognitive Neuroscience* specifically in the area of social neuroscience. Cognitive neuroscience may be the parent discipline of social neuroscience, but it was becoming increasingly clear over the last few years that social neuroscience had now grown up and was trying to establish a home of its own. For example, there are now several excellent journals dedicated to it and many universities have introduced social neuroscience onto the undergraduate curriculum as a separate module distinct from cognitive neuroscience. This textbook aims to reflect the new maturity of this discipline. It conveys the excitement of this field to undergraduate and early-stage postgraduate students.

My own interest in the field stemmed from the claims surrounding mirror systems, empathy, and theory of mind. At the start of this project, I imagined that this would form the core of the textbook. However, the more that I delved into the literature, the more I was taken aback by the volume and quality of research in other areas such as prejudice, morality, culture, and neuro-economics. The resulting book is, I hope, a more balanced view of the field than I initially anticipated. As with my previous textbook, it is not an exhaustive summary of the field. It is not my aim to teach students everything about social neuroscience but it is my aim to provide the intellectual foundations to acquire that knowledge, should they wish to become researchers themselves. My ethos is to try to present the key findings in the field, to develop critical thinking skills, and to instill enthusiasm for the subject.

In the absence of previous textbooks on social neuroscience, it was an interesting exercise deciding how to carve the field into chapters, and how to order the chapters. For example, 'Relationships' (Chapter 8) appeared and disappeared several times, at one point being divided amongst 'Interactions' (Chapter 7) and 'Development' (Chapter 11). The first two chapters begin with an overview of the topic (Chapter 1) and a summary of the methods used in social neuroscience (Chapter 2). The 'methods' chapter is a condensed, but updated, version of the more extensive chapters in *The Student's Guide to Cognitive Neuroscience* and uses examples from the social neuroscience literature to illustrate the various methods. The third chapter covers the evolution of social intelligence and culture, and introduces mirror neurons in the context of imitation, social learning, and tool use. The fourth and fifth chapters deal with the 'primitive' building blocks of social processes, namely emotions and motivation (Chapter 4), and recognizing others (Chapter 5). Chapter 6 is concerned with empathy, theory of mind, and autism. The next two chapters consider social interactions (Chapter 7) and relationships (Chapter 8). They deal with issues such as altruism, game theory, attachment, and social exclusion. Chapter 9 is concerned with groups and identity, covering the notion of 'the self', prejudice, and religion. Chapter 10 covers antisocial behavior, aggression, and morality. The final chapter considers social development from infancy through to adolescence.

It will be interesting to see how the field of social neuroscience changes in the coming years. What new chapters will be added to subsequent editions of the book? Which chapters will require revising the most?

Finally, I would like to thank the many reviewers who provided constructive feedback on drafts of the chapters, and also Psychology Press for being so accommodating.

Jamie Ward
Brighton, UK, January 2011

For Katie

CHAPTER 1

CONTENTS

The emergence of social neuroscience   4

The social brain?   5

Is neuroscience an appropriate level of explanation for studying social behavior?   7

Summary and key points of the chapter   10

Example essay questions   10

Recommended further reading   11

Introduction to social neuroscience

1

Imagine two participants lying in different rooms, each with their heads placed in a very large magnetic field. Crucially, the two participants are interacting with each other in order to win money and this interaction requires trust. By trusting money to the other person they stand a greater chance of getting more money returned to them in the future, but they also run the risk of exploitation. As their brains engage in the decision to trust or not to trust, there are subtle changes in blood flow corresponding to these different decisions that can be detected. The fact that different patterns of thought should result in different patterns of brain activity is perhaps not surprising. The fact that we now have methods that can attempt to measure this is certainly noteworthy. What is most interesting about studies such as these is the fact that activity in regions of one person's brain can reliably elicit activity in other regions of another person's brain during this social interaction. For instance, in a trusting relationship, when one person makes a decision the other person's brain 'lights up' their reward pathways, even before any reward is actually obtained (King-Casas et al., 2005). Cognition in an individual brain is characterized by a network of flowing signals between different regions of the brain. However, social interactions between different individuals can be characterized by the same principle: a kind of 'mega-brain' in which different regions in different brains can have mutual influence over each

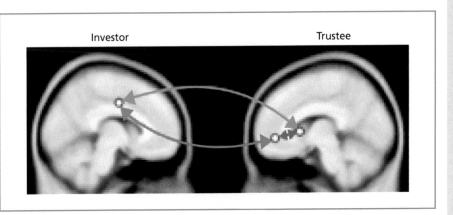

The technique of hyper-fMRI (functional magnetic resonance imaging) records from two different MRI scanners simultaneously: for example, whilst participants in the scanners engage in a social activity. The details of this particular study, involving a game of trust, are not important here (they are covered in Chapter 7). What is of interest is that neural activity in different regions correlates not only within the same brain (due to physical connections; depicted in red) but also across brains (due to mutual understanding; depicted in blue and green). From King-Casas et al. (2005). Copyright © 2005 American Association for the Advancement of Science. Reproduced with permission.

other. This is not caused by a physical flow of activity between brains (as happens between different regions in the same brain) but by our ability to perceive, interpret, and act on the social behavior of others.

This introductory chapter will begin by providing a brief overview of the (brief) history of social neuroscience. It will then go on to consider what kind of mechanisms could constitute the 'social brain' and how they might relate to non-social brain processes. Finally, it will consider how different levels of explanation are needed to derive a complete understanding of social behavior, and it will discuss how neuroscience can be combined with other approaches.

THE EMERGENCE OF SOCIAL NEUROSCIENCE

Allport (1968) defined **social psychology** as 'an attempt to understand and explain how the thoughts, feelings, and behaviors of individuals are influenced by the actual, imagined, or implied presence of others'. By extension, a reasonable working definition of social neuroscience would be:

an attempt to understand and explain, using the methods and theories of neuroscience, how the thoughts, feelings, and behaviors of individuals are influenced by the actual, imagined, or implied presence of others.

Based on this definition, one could regard social neuroscience as being a sub-discipline within social psychology that is distinguished only by its adherence to particular methods and/or theories. Whilst this may be perfectly true, most researchers working within the field of social neuroscience do not have backgrounds within social psychology but tend to be drawn from the fields of **cognitive psychology** and neuroscience. Indeed social neuroscience has also gone by the name 'social cognitive neuroscience' (the term is less commonly used now). Cognitive psychology is the study of mental processes such as thinking, perceiving, speaking, acting, and planning. It tends to dissect these processes into different sub-mechanisms and explain complex behavior in terms of their interaction. Social neuroscience links together all these disciplines: linking cognitive and social psychology, and linking 'mind' (psychology) with brain (biology, neuroscience). Of course, these divisions themselves are arbitrary. They serve as convenient ways of categorizing research programs, and they become embedded in the way they are taught (lecture courses, textbooks, etc.).

The term *social neuroscience* can be traced to an article by Cacioppo and Berntson (1992) entitled 'Social psychological contributions to the decade of the brain: Doctrine of multi-level analysis'. The term appears twice: once in a footnote, and once in a heading and accompanied by a question mark – i.e. 'Social Neuroscience?'. Their particular interest in the topic stemmed from research showing that psychological processes such as perceived social support can affect immune functioning. However, many other areas of study that now fall under the social neuroscience umbrella were already active areas of study prior to 1992. In cognitive psychology, there was a mature literature on face perception. However, this literature was primarily concerned with understanding faces as a type of visual object rather than treating faces as cues to social interactions. There were also detailed accounts of how social behavior might dysfunction as a result of acquired brain damage (Damasio,

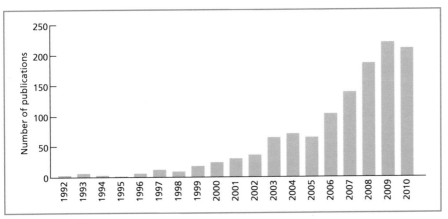

The number of publications using the term 'social neuroscience' has increased dramatically since 2000. The data are based on a search of the *Web of Knowledge* database, searching for the term 'social neuroscience' in the topic field.

Tranel, & Damasio, 1990; Eslinger & Damasio, 1985) or in developmental conditions such as autism (e.g. Frith, 1989). In behavioral neuroscience, there was a longstanding interest in emotional processes such as fear (e.g. Le Doux, Iwata, Cicchetti, & Reis, 1988), aggression (for a review see Siegel, Roeling, Gregg, & Kruk, 1999), and separation distress (e.g. Panksepp, Herman, Vilberg, Bishop, & DeEskinazi, 1980). In social psychology, the field of 'social cognition' applied the approach and methods (e.g. response time) of cognitive psychology to social psychology questions. Finally, the 1990s saw the refinement of the newly established methods of cognitive neuroscience, such as functional magnetic resonance imaging (fMRI) and transcranial magnetic stimulation (TMS), and these methods were directed to social processes as well as to the more traditional areas within cognitive psychology.

By the year 2000, social neuroscience could be recognized as a relatively coherent entity with a core set of research issues and methods and as reflected in prominent reviews of the time (e.g. Adolphs, 1999; Frith & Frith, 1999; Ochsner & Lieberman, 2001). Journals dedicated to the field appeared in mid-2000 and this textbook, published in 2012, represents the first single-authored study guide aimed at undergraduate students.

THE SOCIAL BRAIN?

One overarching issue within social neuroscience is the extent to which the so-called 'social brain' can be considered distinct from all the other functions that the brain carries out – talking, walking, planning, etc. In other words, is the 'social brain' special in any way? This will be a recurring theme in the book, although Chapter 3 considers it in detail from an evolutionary perspective.

One possibility is that there are particular neural substrates in the brain that are involved in social cognition but not in other types of cognitive processing. This relates to the notions of **modularity** and **domain specificity**. A module is the term given to a computational routine that responds to particular inputs and performs a particular computation on them, that is, a routine that is highly specialized in terms of what it does to what (Fodor, 1983). One core property that has been attributed to modules is domain specificity, namely that the module processes only one kind of input (e.g. only faces, only emotions). One contemporary claim is that there is a module that responds to the sight of faces, but not the sight of bodies or the

KEY TERMS

Modularity
The notion that certain cognitive processes (or regions of the brain) are restricted in the type of information they process and the type of processing carried out.

Domain specificity
The idea that a cognitive process (or brain region) is specialized for processing only one particular kind of information.

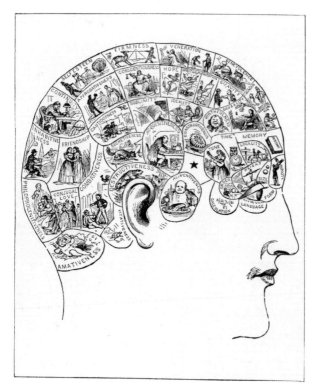

The phrenologist's head was used to represent highly localized functions of the brain in the early nineteenth century. It is an extreme form of a modularity view, albeit based on outdated notions of what the core functions are likely to be (e.g. love of animals, conscientiousness) and how individual differences in these functions represent themselves biologically (larger brain regions giving larger bumps on the skull). To what extent is the 'social brain' a set of specialized modules?

sounds of people's voices or indeed to any non-face stimuli (e.g. Kanwisher, 2000). Another claim is that there is a module for reasoning about mental states (e.g. people's desires, beliefs, knowledge) but not other kinds of reasoning (e.g. Saxe, 2006). Yet another claim is that there is a module for detecting cheating (Cosmides, 1989). In this modular view, the social brain is special by virtue of brain mechanisms that are specifically dedicated to social processes. Moreover, it is claimed that these mechanisms evolved to tackle specific challenges within the social environment (e.g. the need to recognize others, the need to detect when you are being exploited).

The alternative, diametrically opposite, approach is to argue that the 'social brain' is not, in fact, specialized uniquely for social behavior but is also involved in non-social aspects of cognition (e.g. reasoning, visual perception, threat detection). Evolving general neural and cognitive mechanisms that increase intellect, such as having bigger brains, may make us socially smarter too (e.g. Gould, 1991). Of course, it is also possible that the reverse is true – namely that the evolutionary need to be socially smarter leads to general cognitive advances in other domains (e.g. Humphrey, 1976). Under these accounts social cognition and non-social cognition evolved hand-in-hand (albeit with one factor driving the other) but, crucially, they did not necessarily lead to highly specialized routines in the brain for dealing with social problems.

Needless to say, there are other positions that lie in-between these two extremes. Mitchell (2009) notes that there are certain regions of the brain (e.g. the medial prefrontal cortex) that are activated in fMRI studies by a wide range of social phenomena such as evaluating attitudes, interpreting other's behavior, and emotional experience. Rather than arguing for a narrowly defined module in this region, he suggests that social psychology is a 'natural kind' that distinguishes itself from other aspects of cognition because it relates to concepts that are less stable and less definite than those involved in, say, perception and action. In this account, the 'social brain' is special because of the nature of the information that is processed (more fuzzy) rather than because it is social (i.e. inter-personal) per se.

Another possibility is that it is not particular regions of the social brain that are 'special' but rather that there are particular kinds of neural mechanisms especially suited to social processes. For example, Frith (2007) claims: 'I have speculated about the role of various components of the social brain, but in most cases, I believe that these processes are not specifically social. The exception is the brain's mirror system.' Similarly, Ramachandran (2000) predicts that 'mirror neurons will do for psychology what DNA did for biology: they will provide a unifying framework and help explain a host of mental abilities that have hitherto remained mysterious and inaccessible to experiments.' Mirror neurons respond both when an animal sees an action performed by someone else and when they perform the same action themselves (e.g. Rizzolatti & Craighero, 2004). The key insight, with regard to social neuroscience, is that there may be a simple mechanism – implemented at the level of single neurons – that enables a correspondence between self and other. Mirror neurons have been implicated in imitation (see Chapter 3), empathy, and 'mind-reading' (see Chapter 6). Although they were originally discovered for actions, it is possible that mirroring is a general property of many neurons (e.g. those processing pain, emotion, etc.) and they may not be tightly localized to one region (Mukamel, Ekstrom, Kaplan, Iacoboni, & Fried, 2010). Whether they are specifically social (as suggested by Frith, 2007) or derive from general cognitive demands (linking perception and action) is hard to say.

KEY TERM

Reductionism
One type of explanation will become replaced with another, more basic, type of explanation over time.

IS NEUROSCIENCE AN APPROPRIATE LEVEL OF EXPLANATION FOR STUDYING SOCIAL BEHAVIOR?

Perhaps the most general criticism that could be levelled at social neuroscience is that the brain is not the most appropriate level of explanation for understanding social processes. Surely social processes need to be studied and understood at the social level – that is, at the level of interactions between people, groups of people, and societies. There have been some fascinating studies on neural responses to Black faces by White American students, but what could we ever really learn about racism from brain-based measures without situating them in a social, economic, and historical context?

Of course, this presents a distorted view of what social neuroscience is really all about. Most researchers in the field do not take a strongly reductionist approach. **Reductionism** implies that one type of explanation will become replaced with another, more basic, type of explanation over time. In a reductionist framework

Social psychology and neuroscience employ different levels of explanation. Social neuroscience aims to create bridges between these different levels of explanation.

KEY TERMS

Reverse inference
An attempt to infer the nature of cognitive processes from neuroscience (notably neuroimaging) data.

Blank slate
The idea that the brain learns environmental contingencies without imposing any biases, constraints, or pre-existing knowledge on that learning.

the language of social psychology (e.g. attitudes, relationships, conformity) will be replaced by the concepts of neuroscience (e.g. oxytocin, plasticity, medial pre-frontal cortex). However, most researchers in social neuroscience are attempting to create bridges between different levels of explanation rather than replace one kind of explanation with another. For example, social neuroscience studies may combine questionnaire measures (the bread-and-butter of social psychology research) with neuroscience data.

Another common way in which neuroscience data are used to bridge levels of explanation has been termed the **reverse inference** approach (Poldrack, 2006). The reverse inference approach is an attempt to infer the nature of cognitive processes from neuroscience (notably neuroimaging) data. Examples of this abound in the social neuroscience literature. For example, activity within the amygdala may be taken to imply the involvement of a fear-related (or more broadly emotion-related) mechanism in studies of race processing (Phelps et al., 2000). The nature of various moral dilemmas has been inferred on the basis of whether the dilemmas activate regions of the brain implicated in emotion or in higher order reasoning (Greene, Sommerville, Nystrom, Darley, & Cohen, 2001). If the hippocampus is activated, then long-term memory is involved; if the right temporo-parietal junction is activated, then 'theory of mind' is involved. Is reverse inference necessarily good practice? It goes without saying that the reliability of this inference depends on what is known about the functions of given regions. If these regions turned out to have very different functions then the inference would be flawed. Also the function of regions is not resolutely fixed but depends on the context in which they are employed. Poldrack (2006) argues that reverse inference may be improved by examining networks of regions or examining more precise regions (e.g. not just the 'frontal lobes'). Another more general methodological point is the importance of not being over-reliant on neuroimaging data, but to look at other sources of evidence such as TMS in which behavior itself is normally measured (and hence does not suffer from the problems of reverse inference in the same way). Reverse inference is a legitimate approach, but it is not problem free.

Logically, there is one scenario in which brain-based data could have no significant impact on our understanding of social processes – and that is the **blank slate** scenario. In the blank slate scenario, the brain just accepts, stores, and processes whatever information is given to it without any pre-existing biases, limitations, or knowledge. According to the blank slate, the brain is not completely redundant (it still implements social behavior) but the nature of social interactions themselves is entirely attributable to culture, society, and the environment. According to the blank slate, the structure of our social environment is created entirely within the environment itself, reflecting arbitrary but perpetuated historical precedents: culture, society, and the nature of social interactions invent and shape themselves. A more realistic scenario is that the brain, and its underlying processes, creates constraints on social processes. For example, it is claimed that the number of close friends that we have is predicted by the size of the human brain, extrapolating from known group sizes and brain sizes in other primates (Dunbar, 1992). The tendency to form monogamous attachments is causally influenced by brain chemistry (Carter, DeVries, & Getz, 1995). Apparently arbitrary social conventions, such as the rules governing right and wrong (e.g. the law), may not be entirely arbitrary but may reflect a basic tendency to empathize with others and reason about causes and effects (Hauser, 2006). Even in the first few hours of life, infants appear to treat social and non-social stimuli differently. They enter the world with a preference for social stimuli and even appear to have rudimentary knowledge about how faces should be structured (Macchi Cassia, Turati, & Simion, 2004). Social processes are *all* in the brain, but some of them are created by environmental constraints and historical accidents (and learned by the brain) whereas others may be caused by the inherent organization, biases, and limitations of the brain itself.

To give a feel for this debate, consider the topic of aggression. Many current social psychology textbooks (e.g. Hogg & Vaughan, 2011) are dismissive of the role of biological factors in aggression, noting, for instance, the huge variability in levels of aggressive acts such as murder across cultures. However, we can consider this in terms of two questions: What causes aggression and what causes variability in levels of aggression? These questions may generate quite different answers. To give a non-social analogy, the typical number of fingers that we have on our hands (i.e. ten) is almost entirely down to our biology, whereas the *variability* in the number of fingers we have on our hands is almost entirely down to environment, such as industrial accidents (this example is from Ridley, 2003). Moreover, variability in the levels of aggression need not reflect randomness in the situations that tend to trigger aggressive acts (e.g. to gain access to resources, to maintain social order by punishing those perceived to violate it). These triggers are likely to have been shaped by evolution rather than reflecting an arbitrary cultural trend.

The same logic can be applied to other domains, including culture itself. The answer to the question 'what causes culture?' might be something like 'a set of mechanisms that enables people to transfer skills, beliefs, and knowledge from each other and retain these as a relatively stable pattern across individuals' (this being a cognitive mechanistic explanation). A more neuroscientific answer could be 'neural mechanisms that respond to the repeated patterns of behavior in others, whom we affiliate positively with, and increase the likelihood that our own neural mechanisms will generate those behaviors'. This is not intended as a truly accurate answer, but merely conveys what a reductive neuroscience concept of conformity (a central aspect of culture) *might* look like. But note that it would be an entirely circular argument to say that culture creates itself. To take that argument to a logical absurdity, culture cannot create itself in the absence of appropriate biological entities! As to the question of what creates *variability* in culture, the answer could be quite different. It may reflect, for instance, the different environments that people live in and arbitrary historical precedents. However, the number of cultural variants may not be limitless. Hauser (2009) speculates that there could be some cultural forms that will never be created or, if they are, will rapidly die out because they are too difficult to acquire – that is, biology may go as far as to specify which cultural variants are likely, possible, or virtually impossible. This might seem surprising if one thinks of the variety of cultures that exist. For example, slavery is a possible culture (although abhorrent to modern eyes) and some cultures that used slavery, such as the ancient Egyptians, flourished for millennia. However, the existence of a cultural variant such as slavery may require particular kinds of neurocognitive mechanisms: for instance, the switching off of empathic processes towards the slave group and particular kinds of thoughts that drive this switching off (e.g. they are less human). An impossible culture could therefore be a system of slavery associated with high levels of empathy and humane cognitions towards the slave group. The impossibility is created by the nature of brain-based mechanisms, even though it manifests itself in terms of the nature of social processes.

It is worthwhile pointing out that variability is not always attributable to the social or environmental level. For example, different genetic variants and different hormone levels across individuals do contribute to variability in social behavior. In some cases the different levels of explanation cannot be separated from each other. To return to the topic of aggression, testosterone tends to account for more variability in levels of aggression (a neuroscience level of explanation) in people from low socio-economic status backgrounds than those from high socio-economic status backgrounds (a social level of explanation) (Dabbs & Morris, 1990). It is quite possible that these interactions across levels of explanation are likely to be the norm in social neuroscience, but the field is too new to draw that conclusion firmly.

SUMMARY AND KEY POINTS OF THE CHAPTER

- Social neuroscience can be defined as: an attempt to understand and explain, *using the methods and theories of neuroscience*, how the thoughts, feelings, and behaviors of individuals are influenced by the actual, imagined, or implied presence of others.
- There are various ways in which a 'social brain' (i.e. a set of neural routines for dealing with social situations) could be implemented. At one level, there may be domain-specific routines that evolved for serving specific functions. At the other extreme, the same set of routines may be used in both social and non-social situations. Other positions are neural routines that are predominantly used for dealing with social situations but serve a more generic function, or non-modular solutions implemented throughout the brain (e.g. mirror systems).
- Social neuroscience aims to create bridges between different levels of explanation of social behavior. The brain, and its workings, is likely to create causal constraints on the way that social interactions are organized rather than merely soaking up the social world (as a blank slate).

EXAMPLE ESSAY QUESTIONS

- Is the 'social brain' highly modular?
- How can neuroscience and social psychology inform each other?

RECOMMENDED FURTHER READING

- Frith, C. D. (2007). The social brain? *Philosophical Transactions of the Royal Society, 362,* 671–678.

- Hauser, M. D. (2009). The possibility of impossible cultures. *Nature, 460,* 190–196. This article asks whether all variations in culture could be observed or whether some will never be observed due to biological constraints.

- Mitchell, J. P. (2009). Social psychology as a natural kind. *Trends in Cognitive Sciences, 13,* 246–251.

- Willingham, D. T., & Dunn, E. W. (2003). What neuroimaging and brain localization can do, cannot do, and should not do for social psychology. *Journal of Personality and Social Psychology, 85,* 662–671.

CHAPTER 2

CONTENTS

Measuring behavior and cognition: Psychological methods   14

Measuring bodily responses   22

Electrophysiological methods   25

Functional imaging: Hemodynamic measures   33

Lesion methods and disruption of function using TMS   40

Summary and key points of the chapter   46

Example essay questions   47

Recommended further reading   47

The methods of social neuroscience

2

Social neuroscience is too recent a field to have developed a distinct methodology of its own. As such its methods are borrowed from disciplines such as psychology (both cognitive and social psychology) and neuroscience (particularly cognitive neuroscience). The chapter will begin by considering various psychological methods such as performance measures (e.g. response times), observational studies, and questionnaires. It then goes on to consider methods linked to cognitive neuroscience – psychophysiological responses (e.g. skin conductance response) and electrophysiological responses – before turning to functional imaging, effects of brain lesions, and **transcranial magnetic stimulation** (TMS). Most of these methods are covered in more detail in Ward (2010). However, specific examples from the field of social neuroscience are used to illustrate the different methods and to explain the complementary nature of the different methods.

The main methods of cognitive neuroscience can be placed on a number of dimensions:

- The **temporal resolution** refers to the accuracy with which one can measure when an event is occurring. The effects of brain damage are permanent and so this has no temporal resolution as such. Methods such as electroencephalography/event-related potential (EEG/ERP), magnetoencephalography (MEG), TMS, and single-cell recording have millisecond resolution. Positron emission tomography (PET) and functional magnetic resonance imaging (fMRI) have temporal resolutions of minutes and seconds, respectively, that reflect the slower hemodynamic response.
- The **spatial resolution** refers to the accuracy with which one can measure where an event is occurring. Lesion and functional imaging methods have comparable resolution at the millimeter level, whereas single-cell recordings have spatial resolution at the level of the neuron.
- The **invasiveness** of a method refers to whether or not the equipment is located internally or externally. PET is invasive because it requires an injection of a

THE DIFFERENT METHODS USED IN COGNITIVE NEUROSCIENCE

Method	Method type	Invasiveness	Brain property used
EEG/ERP	Recording	Non-invasive	Electrical
Single-cell (and multi-unit) recordings	Recording	Invasive	Electrical
TMS	Stimulation	Non-invasive	Electromagnetic
MEG	Recording	Non-invasive	Magnetic
PET	Recording	Invasive	Hemodynamic
fMRI	Recording	Non-invasive	Hemodynamic

The methods of cognitive neuroscience can be categorized according to their spatial and temporal resolution. Adapted from Churchland and Sejnowski (1988).

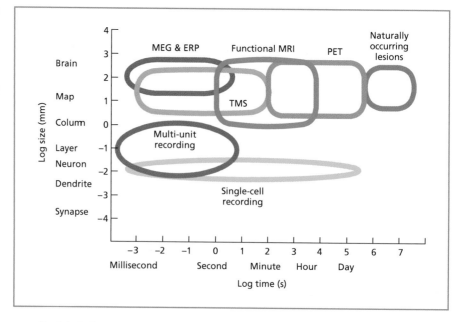

radio-labeled isotope. Single-cell recordings are performed on the brain itself and are normally only carried out in non-human animals. Methods such as TMS are not strictly invasive (because the coil is located entirely outside the body) even though it leads to stimulation of the brain.

MEASURING BEHAVIOR AND COGNITION: PSYCHOLOGICAL METHODS

Almost all experiments in social neuroscience measure behavior in some way, given that it is social behavior that they are trying to explain. In functional imaging experiments, the participant is given a set of instructions on how to respond even if the main dependent measure is brain activity rather than behavior per se. In social neuroscience, it is also common to correlate neurophysiological responses (e.g. during functional imaging) when performing a task with individual differences on a psychological measure such as empathy or personality (assessed outside the scanner using a questionnaire).

In this section, an overview will be provided of three different ways of measuring behavior and cognition: performance-based measures (where the dependent measures are typically response times or error rates); observation-based measures (where the dependent measure is often a frequency count of how often something occurs); and first-person-based measures (where the dependent measure may be scores on a questionnaire).

KEY TERM

Mental chronometry
The study of the time-course of information processing in the human nervous system.

Performance-based measures: Response times and accuracy rates

Mental chronometry can be defined as the study of the time-course of information processing in the human nervous system (Posner, 1978). The basic idea is that changes in the nature or efficiency of information processing will manifest themselves in the time it takes to complete a task. For example, participants are faster at verifying that 4 + 2 = 6 than they are in verifying that 4 + 3 = 7, and this is faster than verifying that 4 + 5 = 9 (Parkman & Groen, 1971). What can be concluded from this? First of all, it suggests that mathematical sums such as these are not just stored as a set of facts. If this were so, then all the reaction times would be expected to be the same because all statements are equally true. It suggests, instead, that the task involves a stage in processing that encodes numerical size together with the further assumption that larger sums place more limits on the efficiency of information processing (manifested as a slower verification time). To give an example more relevant to social neuroscience, it has been found that the response time to identify a face (e.g. by naming it) depends on whether or not the face displays an emotional expression – decisions are faster if the face is smiling (Gallegos & Tranel, 2005). What can we con-

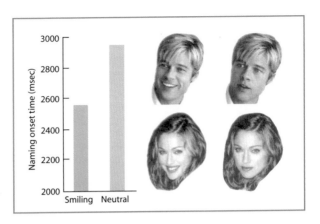

People are faster at identifying faces when they are smiling relative to a neutral pose. How can evidence from response times, such as this, be used to guide theories in social neuroscience? What might such a result mean and how could we explore these hypotheses using other studies? Graph based on data from and images taken directly from Gallegos and Tranel (2005). Copyright © 2005 Elsevier. Reproduced with permission.

clude from this? First of all, many models of face processing assume that recognizing who a person is and recognizing their expression are separate (Bruce & Young, 1986; Haxby, Hoffman, & Gobbini, 2000). These results speak against this, to some extent. However, there are various possibilities. One is that known faces tend to be stored in the brain in an expressive pose (e.g. smiling) rather than a neutral pose as generally assumed. This would make them more efficient to process. Similarly facial identity can be computed, in part, from idiosyncratic facial movements, and smiles (even static smiles) may provide this additional information – that is, it might be motor/movement cues rather than emotion itself that drive the effect. An alternative is that familiar face recognition and expression recognition really are separate but can interact, such that the latter can provide a boost to the former in certain situations. These competing ideas could be explored with further response time studies (e.g. comparing smiling v. angry expressions) or with other methods such as EEG, which can be used to determine whether the effect is early or late in time (i.e. consistent with an interaction at either the perceptual or decision-making stage).

Aside from response times, the other main performance measure is accuracy. This can be measured in terms of error rates, percentage correct, or percentile performance in which individual scores are recalculated relative to the population mean (e.g. IQ scores). Accuracy is obviously crucially related to whether certain knowledge is present/absent rather than to processing efficiency (which is more closely related to response time). However, accuracy and efficiency are related in certain circumstances. For example, if people are forced to respond faster they will tend to be less accurate, a so-called **speed–accuracy trade-off.**

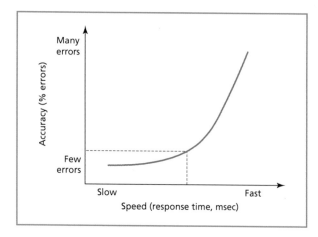

Many cognitive psychology studies instruct participants to be 'as fast and accurate as possible'. In practice, these two factors tend to be in opposition – faster responses tend to be less accurate, and very accurate responses tend to be slower. This is termed speed–accuracy trade-off.

Summary of performance measures

- *Advantages*: They reflect real behavior; they are simple to analyze and interpret.
- *Disadvantages*: They are hard to link directly to neural substrates (unless combined with other measures); there is not always a clear relationship between laboratory tasks and real-world behavior.

Observational measures

If performance-based measures measure 'how well' or 'how fast' something is done, observational measures tend to code 'what' is being done or 'how often' something is done through one person observing the behavior of others. There are certain situations in which observational measures are used in place of the more common performance-based measures:

- Observational measures are the norm in infancy research because the infant cannot be trained or instructed to perform a task.
- Observational methods may often be used for understanding non-human species for similar reasons to those used for human infants. Although training is possible here, there is still a need to know how (untrained) animals behave in the wild. For example, researchers have documented how often different primate species engage in deception in the wild and have correlated this with brain size (Byrne & Corp, 2004).
- Observational measures might be appropriate when the experimenter does not want the participant to know the true nature of a task. For instance, one study in human adults scored their behavior in terms of how often they imitate a given action (e.g. nose rubbing) whilst performing a cooperative task (e.g. Chartrand & Bargh, 1999).

Two specific observational methods in the infant literature are preferential looking and habituation. In **preferential looking** paradigms the infant is presented with a number of stimuli (normally two) and the amount of time that the infant spends looking at each of them is scored. A deviation from chance (e.g. 50/50 for two stimuli) implies that the infant is able to discriminate between the two stimuli (i.e. can tell they are not the same) and has a preference for one (although the reason for the preference is harder to infer). In **habituation** paradigms, the same stimulus (or the same kind of stimulus) is presented repeatedly and the infant's attention towards the stimulus (measured in terms of looking time) diminishes. The critical phase of a habituation experiment occurs when a new stimulus is presented. If the infant's attention is increased it implies that he/she can recognize that it is different, whereas if it is not it implies that he/she treats it as the same. Results using preferential looking and habituation reveal that infants have a preference for social stimuli (e.g. faces) over non-social stimuli (e.g. Johnson, Dziurawiec, Ellis, & Morton, 1991). Coding the imitative behavior of infants (e.g. whether they produce tongue protrusions or lip rounding) is another example of the use of observational methods in this group (Meltzoff & Borton, 1979; Meltzoff & Moore, 1977).

There are several methodological problems that need to be borne in mind when using observational measures, primarily because the scoring system is open to human error. Firstly, there is the issue of **inter-rater (or inter-observer) reliability**, that is, the extent to which two independent observers would generate the same answers. This is typically dealt with by recording the experiment and having two people independently scoring a randomly selected subset of the behaviors. The second issue is whether the observer knows the hypothesis and might be biased to report what they expect to see. In such instances, the observer should perform **blind scoring** of behavior. For example, in preferential looking

"...I can therefore conclude that the primates are indeed social animals."

paradigms the observer typically would not know which stimulus the infant is being presented with on a given trial. To some extent, this problem can be overcome by having computerized scoring methods but this only applies in some domains (e.g. infant eye movements) and not others (e.g. primatology field research).

Summary of observational measures

- *Advantages*: They can be used when it is impossible or inappropriate to give instructions to a participant; they can be used in naturalistic settings.
- *Disadvantages*: There are difficulties associated with scoring behavior and observer biases.

HOW TO MEASURE THE UNCONSCIOUS

Much of what we know is computed prior to us becoming aware of knowing it. Moreover, there is some information that we never become aware of but that can still guide behavior. Here, we will consider how we can measure the effects of stimuli that have been processed unconsciously.

The standard way of presenting a visual stimulus unconsciously is to present it for a brief duration (e.g. less than 50 ms) and follow it with junk visual material (termed **masking**). This prevents an after-image of the briefly presented stimulus from persisting. This method is an example of subliminal perception and research in this area has

www.CartoonStock.com

KEY TERMS

Inter-rater (or inter-observer) reliability
The extent to which two independent observers generate the same answers.

Blind scoring
The observer is unaware of the status of the event that is being scored.

Masking
The presentation of junk visual material after a stimulus (to eliminate persistence of a visual image).

shown that people can detect whether a stimulus was present/absent above chance even when they claim to be guessing (Cheesman & Merikle, 1984) and that subliminally presented stimuli are subsequently judged as more pleasant than non-presented stimuli (Zajonc, 1980). Both studies imply that the stimulus was seen, because it influences behavior, but in the absence of conscious report. An alternative methodology in this literature is to present stimuli for longer durations, but such that they remain outside of the locus of attention (e.g. Simons & Chabris, 1999). This normally requires that the attended task is demanding (e.g. Lavie, 1995). This method can apply to hearing as well as vision.

How can we know whether something was conscious or unconscious? One strategy is to rely on verbal reports: for example, analyzing only those trials in which participants claim to be guessing whether something was seen or not. Another method that has recently been used is wagering in which participants are asked to bet on their performance on a given trial (Persaud, McLeod, & Cowey, 2007). A rather different approach is to use measures of which the participant has no (or very little) volitional control, such as the skin conductance response, electromyography, eyeblink startle responses, and so on. In this case, the participant may (or may not) be conscious of the stimulus but they are unlikely to be able to consciously influence this response.

Survey measures: Questionnaires and interviews

Survey methods involve questioning participants using questions and a set of responses that are fixed in advance (e.g. most questionnaires) or questions and a range of responses that are open-ended (e.g. interviews). These are first-person methods in that the participant is expressing his/her own thoughts that cannot be objectively labeled as right or wrong (in contrast to performance measures described above). For example, contemporary assessments of individual differences in personality (e.g. Costa & McCrae, 1985) or empathy (e.g. Davis, 1980) involve presenting participants with a list of statements (e.g. 'I get easily distressed by the sight of someone else crying') and participants are asked the extent to which they agree or disagree with them.

The **reliability** of questionnaire measures can be assessed by asking participants to repeat the same questionnaire at another time point, and/or by including items in the questionnaire that tap the same knowledge but may require a different response (e.g. 'I like caring for others' / 'I do not like caring for others'). The latter is important because there is a tendency for people to opt for 'agree' more than 'disagree' during surveys. Survey methods can also be used to explore whether lay concepts such as empathy can be fractionated into several underlying variables. For example, if one devises a questionnaire with 40 items on it, one may find 20 questions that are reliably answered in the same way and another 20 questions that also are reliably answered in the same way but differ from the first set. In this example, this would imply an influence of two different underlying variables (statistically, this is assessed using a method called **factor analysis**).

Whereas observational methods measure how people actually behave, survey methods ask people how they think they might behave. As such, one could argue that survey methods have lower **external validity** than observational methods. However, survey methods do have some advantages. Many researchers are interested in what people think and feel, rather than simply how they behave. The fact that our thoughts and behavior may sometimes appear to contradict each other is of interest in its own right, rather than necessarily reflecting a methodological flaw. Pragmatically, questionnaires are easier to carry out, especially using the internet. The external validity may be improved by administering the surveys anonymously and confidentially (the latter being the normal ethical standard). This is because participants may be more inclined to give an honest answer in these situations, rather than presenting themselves in a positive light.

Although survey methods play a central part in social psychology, they have a more supporting role in social neuroscience. In social neuroscience, questionnaire results tend to be correlated with other measures (e.g. from fMRI, EEG) in order to identify the neural correlates of attitudes, feelings, and traits. One particular challenge for this approach stems from the fact that in methods such as fMRI the brain is divided into tens of thousands of regions (termed voxels) and statistical tests may be performed on each and every voxel. As such, the chances of getting a significant, but meaningless, result somewhere in the brain become high – called a **Type I error** (contrast with a **Type II error** in which a null result is obtained even though there is a real effect). Vul, Harris, Winkielman, and Pashler (2009) noted that the reliability of many questionnaire measures is no more than about .8 (i.e. if the same questionnaire is repeated twice, the correlation between answers on the two occasions is about .8). The reliability of fMRI is of a similar magnitude or less, at around .7 (i.e. if the same experiment is done twice then the correlation between activity levels on the different occasions is about .7). However, many studies in social neuroscience report correlations between brain activity and questionnaire measures greatly in excess of what is considered theoretically possible ($\sqrt{.8 \times .7} = .74$). This suggests that some key findings are Type I errors or, at least, an inflation of the true size of the effect. There are various steps that one can take to minimize this when correlating questionnaires with brain imaging data relating, for instance, to whether the correlation is performed on a voxel that has already been selected for its statistical significance (Lieberman & Cunningham, 2009; Vul et al., 2009). However, as a general point it is worth noting that social neuroscience methods – despite technological sophistication – are not invulnerable to flawed designs or analyses.

KEY TERMS

Factor analysis
A statistical method for reducing a data set (e.g. in questionnaires, 20 questions may be grouped into a smaller number of factors).

External validity
The extent to which a measure relates to something useful in 'real life'.

Type I error
Getting a significant result in a statistical test when, in fact, there is no real effect.

Type II error
Getting a non-significant result when in fact there is a real effect.

Summary of survey measures

- *Advantages*: They can be used in situations where an experimental manipulation is not possible or unethical (e.g. exposure to repeated violence); they measure thoughts and attitudes rather than behavior.
- *Disadvantages*: Participants' self-reports may not reflect their true behavior; much social cognition may occur unconsciously.

STRUCTURE AND FUNCTION OF THE NEURON

All **neurons** have basically the same structure. They consist of three components: a cell body (or soma), **dendrites**, and an **axon**. Although neurons have the same basic structure and function, it is important to note that there are some significant differences between different types of neurons in terms of the spatial arrangements of the dendrites and axon. The cell body contains the nucleus and other organelles. The nucleus contains the genetic code, and this is involved in protein synthesis (e.g. of certain neurotransmitters). Neurons receive information from other neurons and they make a 'decision' about this information (by changing their own activity) that can then be passed on to other neurons. From the cell body, a number of branching structures called dendrites enable communication with other neurons. Dendrites receive information from other neurons in close proximity. The number and structure of the dendritic branches can vary significantly depending on the type of neuron (i.e. where it is to be found in the brain). The axon, by contrast, sends information to other neurons. Each neuron consists of many dendrites but only a single axon (although the axon may be divided into several branches called collaterals).

The terminal of an axon flattens out into a disc-shaped structure. It is here that chemical signals enable communication between neurons via a small gap termed a **synapse**. The two neurons forming the synapse are referred to as pre-synaptic (before the synapse) and post-synaptic (after the synapse), reflecting the direction of information flow (from axon to dendrite). When a pre-synaptic neuron is active, an electrical current (termed an **action potential**) is propagated down the length of the axon. When the action potential reaches the axon terminal, chemicals are released into the synaptic cleft. These chemicals are termed **neurotransmitters**. (Note that a small proportion of synapses, such as retinal gap junctions, signal electrically and not chemically.) Neurotransmitters bind to receptors on the dendrites or cell body of the post-synaptic neuron and create a synaptic potential. Depending on the nature of the chemical reaction, the potential can either be excitatory (i.e. promote further firing) or inhibitory (i.e. reduce the likelihood of further firing). The synaptic potential is conducted passively (i.e. without

KEY TERMS

Neurons
A type of cell that makes up the nervous system.

Dendrites
Branching structures that receive information from other neurons.

Axon
A branching structure that carries information away from the cell body towards other neurons and transmits action potentials.

Synapse
The small gap between neurons in which neurotransmitters are released, permitting signaling between neurons.

Action potential
A sudden change in the electrical properties of the neuronal membrane in an axon.

Neurotransmitters
Chemical signals that affect the synaptic functioning of neurons.

creating an action potential) through the dendrites and soma of the post-synaptic neuron. If these passive currents are sufficiently strong when they reach the beginning of the axon in the post-synaptic neuron, then an action potential (an active electrical current) will be triggered in this neuron. It is important to note that each post-synaptic neuron

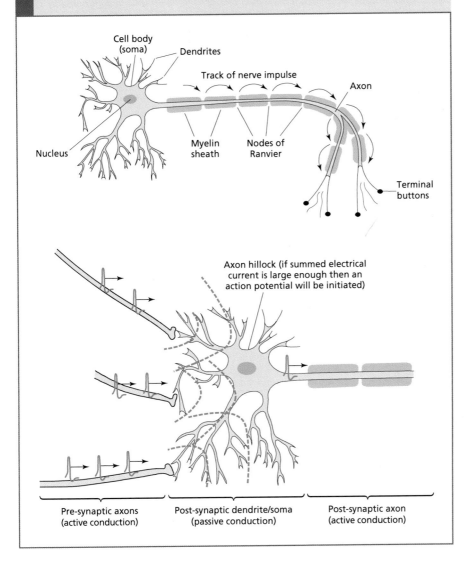

The structure and electrical functioning of neurons (top and bottom, respectively). Neurons consist of three basic features: a cell body, dendrites that receive information and an axon that sends information. In this diagram the axon is myelinated to speed up the conduction time. Electrical currents are actively transmitted through pre-synaptic axons by an action potential. Electrical currents flow passively through dendrites and soma of neurons but will initiate an action potential if their summed potential is strong enough at the start of the axon of the post-synaptic neuron (called the hillock).

sums together many synaptic potentials, which are generated at many different and distant dendritic sites (as opposed to a simple chain reaction between one neuron and the next). Passive conduction tends to be short range because the electrical signal is impeded by the resistance of the surrounding matter. Active conduction enables long-range signaling between neurons by the propagation of action potentials.

The amplitude of an action potential does not vary, but the number of action potentials propagated per second varies along a continuum. This rate of responding (also called the 'spiking rate') relates to the informational 'code' carried by that neuron. For example, some neurons may have a high spiking rate in some situations (e.g. during speech) but not others (e.g. during vision), whereas other neurons would have a complementary profile. Neurons responding to similar types of information tend to be grouped together. This gives rise to the functional specialization of brain regions.

MEASURING BODILY RESPONSES

The central nervous system consists of the brain and the spinal cord. In contrast, the peripheral nervous system consists of nerves sending and receiving signals to other parts of the body. The peripheral nervous system is itself divided into two further systems: the **autonomic nervous system (ANS)** and the **somatic nervous system**. The somatic nervous system coordinates muscle activity whereas the autonomic nervous system controls and monitors bodily functions such as heart rate, digestion, respiration rate, salivation, perspiration, and pupil diameter. The autonomic nervous system is divided into two complementary divisions. The **sympathetic system** increases arousal (e.g. increased heart rate, breathing, pupil size) and decreases functions such as digestion. The **parasympathetic system** has a resting effect (decreased heart rate, breathing, pupil size) and increases functions such as digestion.

Several methods in social neuroscience rely on measurements related to these systems. Electromyography (EMG) is an electrical measure of muscle contraction, implemented by the somatic nerves. Measures of autonomic system functioning include the skin conductance response (SCR), measures of heart rate and breathing (e.g. the traditional lie detector, or polygraph, measures various autonomic functions including these), and also pupilometry (measuring

KEY TERMS

Autonomic nervous system (ANS)
A set of nerves located in the body that controls the activity of the internal organs.

Somatic nervous system
Part of the peripheral nervous system that coordinates muscle activity.

Sympathetic system
A division of the ANS that increases arousal (increased heart rate, breathing, pupil size) but decreases functions such as digestion.

Parasympathetic system
A division of the ANS that has a resting effect (decreased heart rate, breathing, pupil size) but increases functions such as digestion.

changes in pupil dilation). The use of EMG and SCR is considered in more detail here.

The skin conductance response (SCR)

A common way of measuring increased activity of the sympathetic system is to monitor small changes in conductivity as a result of mild sweating (Berry Mendes, 2009). Heightened arousal can lead to more sweat even without overt sweating taking place, and this sweating response (from eccrine glands) is separate from the thermo-regulatory sweating response. The SCR is measured by applying a weak electrical current to the skin. During a sweating response (e.g. elicited by an emotional stimulus) there is decreased conductivity of the skin and the electrical signals flow more easily. This is termed the **skin conductance response (SCR)** or galvanic skin response (GSR). The electrodes are normally placed on two adjacent finger tips with gel in-between the fingers and electrodes to improve contact. A peak SCR occurs between 1 and 5 seconds after stimulus presentation and this is normally recalculated relative to some baseline (e.g. pre-stimulus) activity.

Tranel and Damasio (1995) report how the SCR is affected by a number of brain lesions. Lesions to the ventromedial frontal lobes abolish SCR to psychological stimuli (e.g. risk) but not physical stimuli (e.g. bangs), whereas lesions to the anterior cingulate cortex abolish both. Functional imaging also points to a key role for the anterior cingulate in the production of the SCR (Critchley, Elliott, Mathias, & Dolan, 2000).

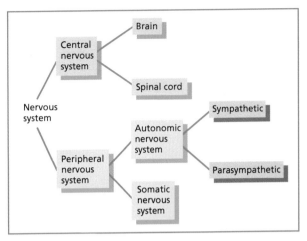

Various methods, including skin conductance response, electomyography (EMG), and pupil size, rely on measures of activity of the peripheral nervous system.

Electromyography (EMG)

Facial **electromyography (EMG)** has been used in social neuroscience research to measure muscle activity associated with emotional expressions in response to seeing expressions in others (e.g. Dimberg, Thunberg, & Elmehed, 2000; Hess & Blairy, 2001) or as a potentially implicit measure of prejudice (Vanman, Paul, Ito, & Miller, 1997). It is also used to measure the **eyeblink startle response**, which is elicited by a startling sound but is further modulated by the participants' present emotional state (Lang, Bradley, & Cuthbert, 1990).

Electromyography (EMG) is a measure of electrical activity associated with muscle contraction (Fridlund & Cacioppo, 1986; Hess, 2009). These changes come about because of an increase in the number of action potentials in muscle fibers during muscle contraction. The greater the force produced by the muscles, the greater the electrical activity (Lawrence & DeLuca, 1983). However, individual action potentials are not measured. Instead the EMG signal is the sum of many such potentials, including those that cancel out because the muscle fibers are not completely aligned. The EMG signal is recorded by placing two small electrodes close to each other and measuring the potential

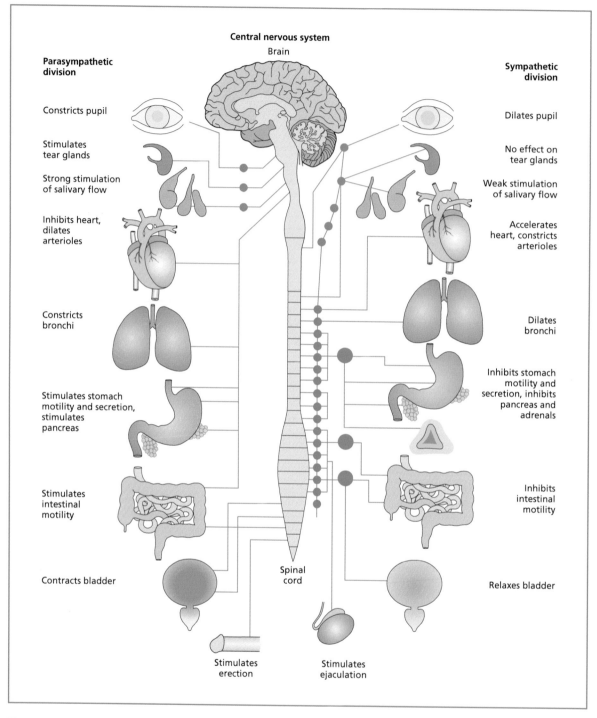

Central nervous system
Brain

Parasympathetic division

Constricts pupil

Stimulates tear glands

Strong stimulation of salivary flow

Inhibits heart, dilates arterioles

Constricts bronchi

Stimulates stomach motility and secretion, stimulates pancreas

Stimulates intestinal motility

Contracts bladder

Spinal cord

Stimulates erection

Sympathetic division

Dilates pupil

No effect on tear glands

Weak stimulation of salivary flow

Accelerates heart, constricts arterioles

Dilates bronchi

Inhibits stomach motility and secretion, inhibits pancreas and adrenals

Inhibits intestinal motility

Relaxes bladder

Stimulates ejaculation

The autonomic nervous system (ANS) is divided into two complementary divisions. The sympathetic system increases arousal and decreases functions such as digestion. The parasympathetic system has a resting effect and increases functions such as digestion.

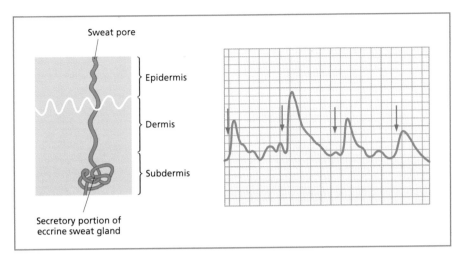

The skin conductance response (SCR) method involves recording changes in electrical conductivity on a person's skin on the hand. A person's SCR can be plotted as a continuous trace throughout the experiment. A peak SCR occurs between 1 and 5 seconds after stimulus presentation.

difference (in microvolts) between them. The frequency and amplitude range of the EMG signal are comparable with those of electrical signals generated by the brain (in EEG) and heart (ECG or electrocardiogram). In order to reduce the influence of the latter two sources, the two measurement electrodes are compared to a third electrode (the ground) placed elsewhere (e.g. the center of the forehead is often used in facial EMG). Fridlund and Cacioppo (1986) offer guidance on the placement of EMG electrodes associated with various facial muscles. In addition to electrical activity from other bodily sources it is also common practice to reference the EMG signal to a baseline measure, such as a rest phase or the period before a stimulus is presented. This is because muscle activity is rarely zero and may fluctuate over time (e.g. due to tension in the participants).

Summary of measures of bodily responses

- *Advantages*: Bodily responses are often present in the absence of awareness of a stimulus and may occur in the absence of a specific task; they are relatively easy to record and analyse.
- *Disadvantages*: It is not straightforward to link bodily responses to brain and cognition.

ELECTROPHYSIOLOGICAL METHODS

By measuring changes in the responsiveness of a neuron to changes in a stimulus or changes in a task,

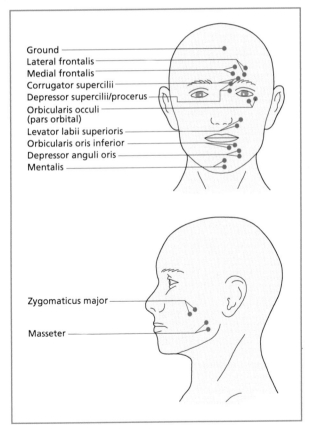

Recommended placement of pairs of electrodes for recording facial EMG. Adapted from Fridlund and Cacioppo (1986).

it is possible to make inferences about the building blocks of cognitive processing. The action potential is directly measured in the method of single-cell recording, whereas EEG is particularly sensitive to post-synaptic dendritic electrical activity.

Single-cell recording

Single-cell recordings can be obtained by implanting a very small electrode either into the axon itself (intracellular recording) or outside the membrane (extracellular recording) and counting the number of times that an action potential is produced (spikes per second) in response to a given stimulus (e.g. a face). This is an invasive method. As such, the procedure is normally conducted on experimental animals only. It is occasionally conducted on humans undergoing brain surgery (Engel, Moll, Fried, & Ojemann, 2005). It is impossible to measure action potentials from a single neuron non-invasively (i.e. from the scalp) because the signal is too weak and the noise from other neurons is too high. Technology has now advanced such that it is possible to simultaneously record from 100 neurons in multi-electrode arrays. This is termed multi-cell recording.

To give one example from the literature, Quiroga, Reddy, Kreiman, Koch, and Fried (2005) recorded the firing rates of neurons in humans undergoing brain surgery. The participants were shown images of famous people and buildings whilst recordings were taken from cells in the medial temporal lobe (a region of the brain implicated in memory). Many neurons showed a high degree of specificity in their response pattern – that is, they tended to respond to some stimuli (measured in spikes per second) more than others. For instance, one neuron responded to images of Jennifer Aniston in a variety of different poses, although not when she appeared next to her ex-husband, Brad Pitt. Another neuron responded to various images of Halle Berry, but not to Jennifer Aniston. It responded to Halle Berry dressed as catwoman, but not to another actress dressed as catwoman, and it even responded to the printed name 'Halle Berry'. Another neuron responded to images of the Sydney Opera House but not the Eiffel Tower or Golden Gate Bridge. Studies such as these show how stimuli in the 'outside' world can be represented in a neural code, using a simple biological parameter (rate of action potentials). However, it need not mean that Jennifer Aniston is represented by a single neuron in our heads (perhaps there are many neurons that respond in this way) and nor does it necessarily mean that these neurons only respond to this particular person (given that the authors only presented a relatively small number of stimuli).

The results of this study can be classified as **rate coding** of information by neurons, in that a given stimulus/event is associated with an increase in the rate of neural firing. An alternative way for neurons to represent information about stimuli/events is in terms of **temporal coding**, in that a given stimulus/event is associated with greater synchronization of firing across different neurons (Engel, Konig, &

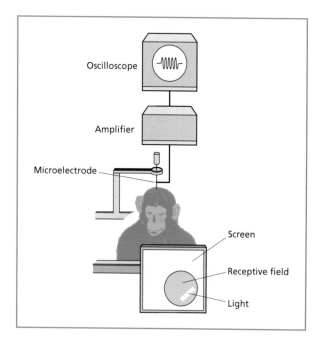

An illustration of a typical experimental set-up for single-cell recording.

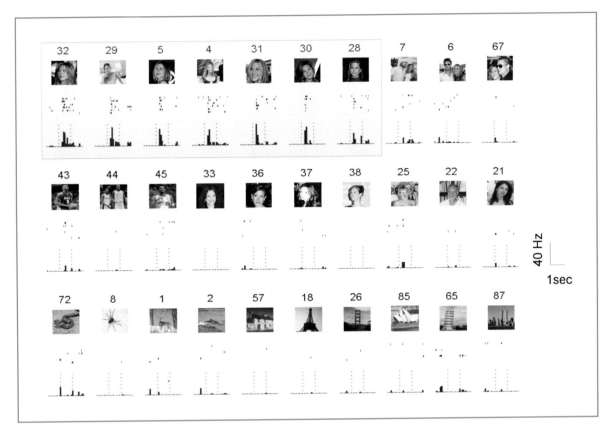

Neurons learn to 'tune in' to (i.e. become specialized in) familiar things in the environment. This neuron, recorded from the human medial temporal lobes, responds to different images of Jennifer Aniston but not to images of other actresses, buildings, or even Jennifer Aniston if she is accompanied by Brad Pitt. The red bars show firing rate (i.e. number of action potentials) across time (on the x-axis). The blue bars above it represent individual trials from which the neuron was presented with that image. From Quiroga et al. (2005). Copyright © 2005 Nature Publishing Group. Reproduced with permission.

Singer, 1991). In temporal coding, the information is contained in the phase of firing (i.e. it depends on whether neurons are firing at the same time).

Summary of single-cell recording

- *Advantages*: This measure is directly related to neural activity (compare fMRI); it has excellent spatial and temporal resolution.
- *Disadvantages*: It is invasive so is very rarely performed on humans; information is limited to the regions probed (not a whole-brain technique).

Electroencephalography (EEG) and event-related potentials (ERPs)

The physiological basis of the EEG signal originates in the post-synaptic dendritic currents rather than the axonal currents associated with the action potential

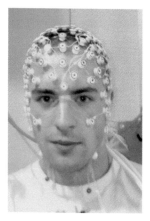

A participant in an EEG experiment.

(Nunez, 1981). **Electroencephalography (EEG)** records electrical signals generated by the brain, through electrodes placed at different points on the scalp. As the procedure is non-invasive and involves recording (not stimulation), it is completely harmless as a method. For an electrical signal to be detectable at the scalp a number of basic requirements need to be met in terms of underlying neural firing. First, a whole population of neurons must be active in synchrony to generate a large enough electrical field. Second, this population of neurons must be aligned in a parallel orientation so that they summate rather than cancel out. Fortunately, neurons are arranged in this way in the cerebral cortex. However, the same cannot necessarily be said about all regions of the brain. For example, the orientation of neurons in the thalamus may render its activity invisible to this recording method.

To gain an EEG measure one needs to compare the voltage between two or more different sites. A reference site is often chosen that is likely to be relatively uninfluenced by the variable under investigation. One common reference point is the mastoid bone behind the ears or a nasal reference; another alternative is to reference to the average of all electrodes. It is important to stress that the activity recorded at each location cannot necessarily be attributed to neural activity near to that region. Electrical activity in one location can be detected at distant locations. In general, EEG is not best equipped for detecting the location of neural activity.

The most common usage of EEG in cognitive neuroscience is in the context of electrophysiological changes elicited by particular stimuli and cognitive tasks. These are referred to as **event-related potentials (ERPs)**. The EEG waveform reflects neural activity from all parts of the brain. Some of this activity may specifically relate to the current task (e.g. reading, listening, calculating) but most of it will relate to spontaneous activity of other neurons that do not directly contribute to the task. As such, the signal-to-noise ratio in a single trial of EEG is very low (the signal being the electrical response to the event and the noise being the background level of electrical activity). The ratio can be increased by averaging the EEG signal over many presentations of the stimulus (e.g. 50–100 trials), relative to the onset of a stimulus. The results are represented graphically by plotting time (milliseconds) on the x-axis and electrode potential (microvolts) on the y-axis. The graph consists of a series of positive and negative peaks, with an asymptote at 0 microvolts. This is done for each electrode, and each will have a slightly different profile. The positive and negative peaks are labeled with 'P' or 'N' and their corresponding number. Thus, P_1, P_2, and P_3 refer to the first, second, and third positive peaks, respectively. Alternatively, they can be labeled with 'P' or 'N' and the approximate timing of the peak. Thus, P300 and N400 refer to a positive peak at 300 ms and a negative peak at 400 ms (not the 300th positive and 400th negative peak!). It is to be noted that the polarity of the peaks (i.e. whether positive or negative) is of no real significance either cognitively or neurophysiologically (e.g. Otten & Rugg, 2005). It depends, for instance, on the baseline electrical activity and position of the reference electrode. What is of interest in ERP data is the timing and also the amplitude of the peaks.

How can ERP recordings be used to inform theory? Consider one example from the face processing literature.

There is evidence for an ERP component that is relatively selective for the processing of faces compared with other classes of visual objects. This has been termed the **N170** (a negative peak at 170 ms) and is strongest over right posterior temporal electrode sites (e.g. Bentin, Allison, Puce, Perez, & McCarthy, 1996). This, in itself, is interesting for several reasons. First, it suggests that there is a mechanism in the brain that is relatively specialized for faces more than objects. In this respect it is on a par with other methods such as fMRI or patient-based neuropsychology. However, in other respects it reveals something new: it suggests *when*, in time, faces become treated as special. This cannot be revealed by fMRI or neuropsychology. Nor can it be revealed by response time methods. Unlike response time measures, ERP measures can tell *when* something is happening before any response to it has been made. They are not, however, the best method for revealing *where* in the brain something is happening because the electrical activity conducts itself through the brain to distant sites.

One can then ask additional questions about the N170 to understand its nature and – by inference – to understand the nature of early mechanisms of face processing. For example, is it just found for human faces? No it is not (Rousselet, Mace, & Thorpe, 2004). Is it found for schematic 'smiley' faces? Yes, it is (Sagiv & Bentin, 2001). Does it depend on whether a face is famous or not? No, it does not (Bentin & Deouell, 2000). The details of these studies are not of relevance here. The main point is that ERPs do not just tell us *when* things are happening in the brain; they can be taken a step further in order to understand *how* things happen in the brain.

Summary of ERPs

- *Advantages*: Excellent temporal resolution; direct measure of neural activity.
- *Disadvantages*: Poor spatial resolution; some subcortical brain regions are impossible to investigate.

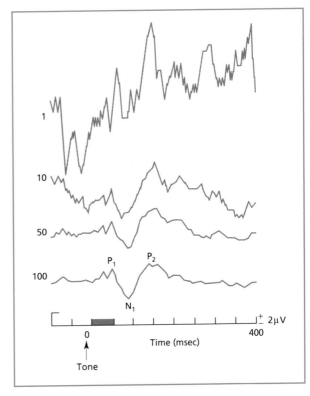

When different EEG waves are averaged relative to presentation of a stimulus (e.g. a tone) then the signal-to-noise ratio is enhanced and an ERP is observed. From Kolb B., & Whishaw I. Q. (2002). *Fundamentals of human neuropsychology* (5th edition). Copyright © 2002 by Worth Publishers. Reproduced with permission.

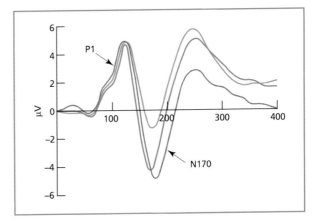

The N170 is observed for both human faces (purple) and animal faces (blue) but not other objects (green). From Rousselet et al. (2004). Reproduced with permission of ARVO.

THE ORGANIZATION AND STRUCTURE OF THE BRAIN

Neurons are organized within the brain to form **white matter** and **gray matter**. Gray matter consists of neuronal cell bodies. White matter consists of axons and support cells (**glia**). The brain consists of a highly convoluted folded sheet of gray matter (the cerebral cortex), beneath which lies the white matter. In the center of the brain, beneath the bulk of the white matter fibers, lies another collection of gray matter structures (the subcortex), which includes the basal ganglia, the limbic system, and the diencephalon. White matter tracts project between different cortical regions within the same hemisphere, between different cortical regions in different hemispheres (the most important being the corpus callosum), and between cortical and subcortical structures.

Anterior and posterior refer to directions towards the front and the back of the brain, respectively. These are also called rostral and caudal, respectively, particularly in other species that have a tail (caudal refers to the tail end). Directions towards the top and the bottom are referred to as superior and inferior, respectively; they are also known as dorsal and ventral, respectively. The terms anterior, posterior, superior, and inferior (or rostral, caudal, dorsal, and ventral) enable navigation in two dimensions: front–back and top–bottom. Needless to say, the brain is

Terms of reference in the brain, and the four lobes of the lateral surface. Note also the terms lateral (referring to the outer surface of the brain) and medial (referring to the central regions).

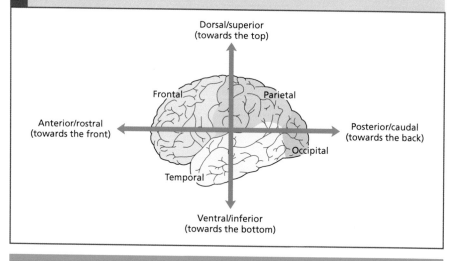

KEY TERMS

White matter
Tissue of the nervous system consisting primarily of axons and support cells.

Gray matter
Tissue of the nervous system consisting primarily of neuronal cell bodies.

Glia
Support cells of the nervous system involved in tissue repair and in the formation of myelin (amongst other functions).

three-dimensional and so a further dimension is required. The terms lateral and medial are used to refer to directions towards the outer surface and the center of the brain, respectively.

The cerebral cortex consists of two folded sheets of gray matter organized into two hemispheres (left and right). The raised surfaces of the cortex are termed **gyri** (or gyrus in the singular). The dips or folds are called **sulci** (or sulcus in the singular). The lateral surface of the cortex of each hemisphere is divided into four lobes: the frontal, parietal, temporal, and occipital lobes. Other regions of the cortex are observable only in

KEY TERMS

Gyri
The raised folds of the cortex.

Sulci
The buried grooves of the cortex.

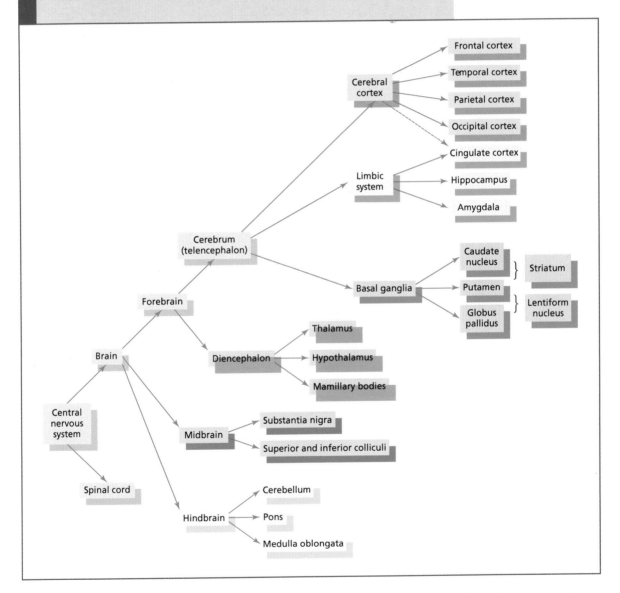

The central nervous system (CNS) is organized hierarchically. The upper levels of the hierarchy, corresponding to the upper branches of this diagram, are the newest structures from an evolutionary perspective.

a medial section, for example the cingulate cortex. Finally, an island of cortex lies buried underneath the temporal lobe; this is called the insula (which literally means 'island' in Latin).

Beneath the cortical surface and the intervening white matter lies another collection of gray matter nuclei termed the subcortex. The subcortex is typically divided into a number of different systems with different evolutionary and functional histories. These include the basal ganglia, the limbic system, and the diencephalon:

* The **basal ganglia** are large rounded masses that lie in each hemisphere. The main structures comprising the basal ganglia are the caudate nucleus (an elongated tail-like structure), the putamen (lying

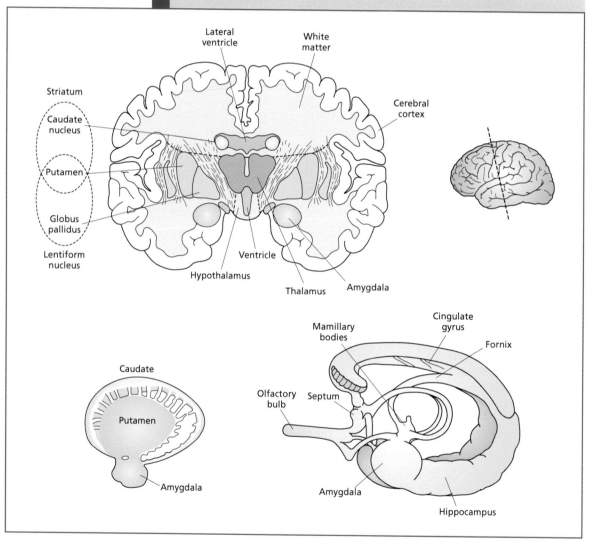

A coronal cross-section of the brain showing various subcortical regions (top). The bottom panels show the structure of the basal ganglia from a single hemisphere in a lateral view (left) and the structure of the limbic system across both hemispheres in a lateral view (right).

more laterally), and the globus pallidus (lying more medially). Different circuits passing through different regions in the basal ganglia connect to the thalamus and prefrontal cortex and midbrain structures. These serve somewhat different functions. For example, one circuit is involved in regulating motor activity and the learning of actions (e.g. skills and habits) and another is involved in processing rewards.

- The **limbic system** is important for relating the organism to its environment based on current needs and the present situation, and based on previous experience. It is involved in the detection and expression of emotional responses. For example, the amygdala has been implicated in the detection of fearful or threatening stimuli and parts of the cingulate gyrus have been implicated in the detection of emotional and cognitive conflicts. The hippocampus is particularly important for learning and memory.

- The two main structures that make up the **diencephalon** are the **thalamus** and the **hypothalamus**. The thalamus consists of two interconnected egg-shaped masses that lie in the center of the brain and appear prominent in a medial section. The thalamus is the main sensory relay for all senses (except smell) between the sense organs (eyes, ears, etc.) and the cortex. It also contains projections to almost all parts of the cortex and the basal ganglia. At the posterior end of the thalamus lie the lateral geniculate body and the medial geniculate body. These are the main sensory relays to the primary visual and primary auditory cortices, respectively. The hypothalamus lies beneath the thalamus and consists of a variety of nuclei that are specialized for different functions primarily concerned with the body. These include body temperature, hunger and thirst, sexual activity, and regulation of endocrine functions.

FUNCTIONAL IMAGING: HEMODYNAMIC MEASURES

Functional imaging methods such as PET and fMRI measure the dynamic physiological changes in the brain that are associated with different patterns of thought and behavior. This can be contrasted with structural imaging, which measures the stable properties of the brain (e.g. the distribution of white and gray matter) and includes CT scanning and conventional MRI. It is important to emphasize at the outset that PET and fMRI are not measuring the activity of neurons directly but, rather, are measuring a downstream consequence of neural activity: namely, changes in blood flow/blood oxygen to meet metabolic needs of neurons. It is for this reason that they are termed **hemodynamic methods**, in contrast to methods such as EEG that measure the electrical fields generated by the activity of neurons themselves.

In fMRI experiments, a strong magnetic field is applied constantly during the scanning process. This is harmless, although it is important to follow safety procedures (e.g. removing metal from pockets).

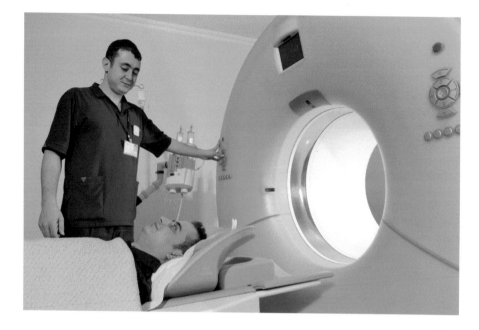

The main advantage of these methods is their spatial resolution. The whole brain is divided into tens of thousands of regions, termed **voxels**, each of the same size (e.g. 3 × 3 × 3 mm) and the 'activity' of each of these voxels in various tasks can be assessed in order to draw inferences about the functioning of different brain regions. In fMRI the larger the magnet, the smaller the voxel size that can be obtained. The strength of the magnetic field is measured in units called tesla (T). Typical scanners have field strengths between 1.5 T and 3 T; the Earth's magnetic field is of the order of 0.0001 T.

Basic physiology underpinning functional imaging

The brain consumes 20% of the body's oxygen uptake; it does not store oxygen and it stores little glucose. Most of the brain's oxygen and energy needs are supplied from the local blood supply. When the metabolic activity of neurons increases, the blood supply to that region increases to meet the demand (Attwell & Iadecola, 2002; Raichle, 1987). Techniques such as PET measure the change in blood flow to a region directly, whereas fMRI is sensitive to the concentration of oxygen in the blood.

The brain is always physiologically active. Neurons would die if they were starved of oxygen for more than a few minutes. This has important consequences for using physiological markers as the basis of neural 'activity' in functional imaging experiments. It would be meaningless to place someone in a scanner, with a view to understanding cognition, and simply observe which regions were receiving blood and using oxygen, because this is a basic requirement of all neurons, all of the time. As such, when functional imaging researchers refer to a region being 'active', what they mean is that the physiological response in one condition is greater relative to another. There is a basic requirement in all functional imaging studies that the physiological response must be compared to one or more control responses. A good understanding of the hypothesized mechanisms underlying the behavior is needed to ensure that the baseline task is appropriately matched to the experimental task, otherwise the results will be very hard to interpret.

KEY TERM

Voxel
A three-dimensional version of a pixel; the brain is divided into tens of thousands of these during functional imaging analysis.

What is the 'signal' measured in functional imaging experiments? In PET studies, the participant is injected with a radioactive tracer and the signal is the amount of radioactivity in each voxel of the brain. The major breakthrough in fMRI came from the realization that no tracer needed to be introduced for this method, but rather one could measure a magnetic resonance signal that is affected by the amount of deoxyhemoglobin in the blood in different regions of the brain (Ogawa, Lee, Kay, & Tank, 1990). This signal is termed the **BOLD response**, standing for Blood Oxygen-Level Dependent. When neurons consume oxygen they convert oxyhemoglobin to deoxyhemoglobin. Deoxyhemoglobin has strong paramagnetic properties and this introduces distortions in the local magnetic field. This distortion can itself be measured to give an indication of the concentration of deoxyhemoglobin present in the blood. The way that the BOLD signal evolves over time in response to an increase in neural activity is called the **hemodynamic response function (HRF)**. The HRF has three phases, as plotted and discussed below (see also Hoge & Pike, 2001):

1. *Initial dip*. As neurons consume oxygen there is a small rise in the amount of deoxyhemoglobin, which results in a reduction of the BOLD signal (this is not always observed in the standard 1.5 T magnets).
2. *Overcompensation*. In response to the increased consumption of oxygen, the blood flow to the region increases. The increase in blood flow is initially greater than the increased consumption, which means that the BOLD signal increases substantially. This is the component that is normally measured in fMRI.
3. *Undershoot*. Finally, the blood flow and oxygen consumption dip before returning to their original levels. This may reflect a relaxation of the venous system, causing a temporary increase in deoxyhemoglobin again.

The hemodynamic signal changes are small – approximately 1–3% of the total signal with moderately sized magnets (1.5 T). The HRF is relatively stable across sessions with the same participant in the same region, but is more variable across different regions within the same individual and more variable between individuals (Aguirre, Zarahn, & D'Esposito, 1998). However, the shape of the HRF in different regions and different people can be estimated during the fMRI analysis and this can be entered into the analysis.

The temporal resolution of fMRI is several seconds and related to the rather sluggish hemodynamic response. The temporal resolution of PET is related to the

KEY TERMS

BOLD response
The change in blood oxygenation that accompanies neural activity (BOLD = Blood Oxygen-Level Dependent).

Hemodynamic response function (HRF)
The change in the BOLD response over time.

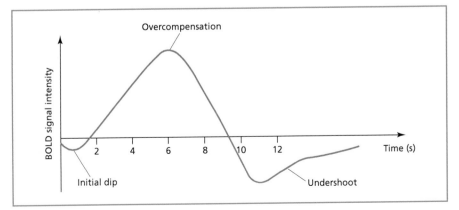

The hemodynamic response function (HRF) is the change in the fMRI BOLD signal over time as a result of an increase in neural activity in that region. It has a number of distinct phases that reflect the changing levels of deoxyhemoglobin in response to oxygen consumption and increased blood supply to that region.

time it takes for radioactivity levels in the blood stream to peak (around 30 s for a 'heavy water' isotope). Although the temporal resolution is better in fMRI than PET it is still slow compared to the speed at which cognitive processes take place. However, at this temporal resolution fMRI offers far more freedom in the choice of experimental designs (enabling event-related designs, described below). The spatial resolution of MRI is also better than PET, at around 1 mm depending on the size of the voxel.

Over the last 10 years, fMRI has largely taken over from the use of PET in functional imaging experiments. The key advantages of fMRI over PET are the better temporal and spatial resolution, the fact that event-related designs are possible, and the fact that it does not use radioactivity. PET, however, can be used to trace the pathways of certain chemicals in vivo by, for example, administering radio-labeled pharmacological agents.

Constraints on experimental design

In fMRI the sluggishness of the hemodynamic response to peak and then return to baseline does place some constraints on the way that stimuli are presented in the scanning environment that differ from equivalent tasks done outside the scanner. However, it is not the case that one has to wait for the BOLD response to return to baseline before presenting another trial, as different HRFs can be superimposed on each other. In general during fMRI, there may be fewer trials that are more spaced out in time, and it is common to have 'null events' (e.g. a blank screen) in the experiment to allow the HRF to dip towards baseline. In standard cognitive psychology experiments (e.g. using response time measures) the amount of data is effectively the same as the number of trials and responses. In the equivalent fMRI experiment, the amount of data is related to the number of *brain volumes* acquired rather than the number of trials or responses.

The way that different kinds of trials are grouped together can be broadly classified into either **event-related designs** or **block designs**. In a block design, trials that belong together are grouped together during stimulus presentation. In an event-related

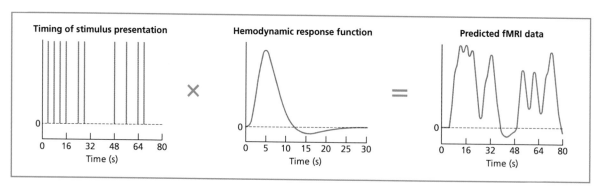

Unless the stimuli are presented far apart in time (e.g. every 16 s) the predicted change in BOLD response will not resemble a single HRF but will resemble many superimposed HRFs. Statistically, we are trying to find out which voxels in the brain show the predicted changes in the BOLD response over time, given the known design of the experiment and the estimated shape of the HRF. To achieve this there has to be sufficient variability in the predicted BOLD response (big peaks and troughs). Figure from http://imaging.mrc-cbu.cam.ac.uk/imaging/DesignEfficiency. Reproduced with permission from the author.

design, all trials are randomly (or semi-randomly) interspersed during stimulus presentation but are then treated separately at the analysis stage. Block designs are the only option in PET studies, but fMRI studies may use either.

To give an example, imagine that one wanted to investigate the neural basis of gender judgments on names and faces (i.e. deciding whether a name or face was male or female). A block design could involve the presentation of a block of 20 faces, followed by a separate block of 20 names. An event-related design might involve presenting all 40 stimuli (20 faces, 20 names) randomly. There is no objectively 'right' or 'wrong' design to choose from but there are important differences. In this example, the event-related design would mean that the participant could not predict whether he/she would see a face or name, whereas in the block design it would be predictable (and this predictability, or lack of, will have its own neural substrate). Block designs have more statistical power (i.e. are better able to detect small but significant effects) than truly random designs (e.g. Josephs & Henson, 1999). However, in many situations event-related designs are favored or are the only option. Sometimes there is no way of knowing in advance how trials should be blocked. For example, one could present a participant with a list of 40 words to remember during scanning. Maybe 10 words will be subsequently forgotten and 30 remembered, and one could then go back to the scanning data and see if there were any differences in brain activity at the initial stage that are predictive of later remembering versus forgetting (e.g. Wagner et al., 1998).

Finally, perhaps the most important aspect of experimental design in functional imaging research is in the selection of control conditions. For example, the simple experiment concerning gender categorization is inadequate in this regard. If one were to compare activity when performing gender judgments to faces relative to gender judgments to names (i.e. names is the control condition) then it would only reveal activity in regions implicated in face perception (and perception of images) and not in regions that may be involved in gender judgments. To test for the neural substrates of gender judgments to faces, better control conditions could be either presenting faces and requiring no judgment at all or presenting faces and

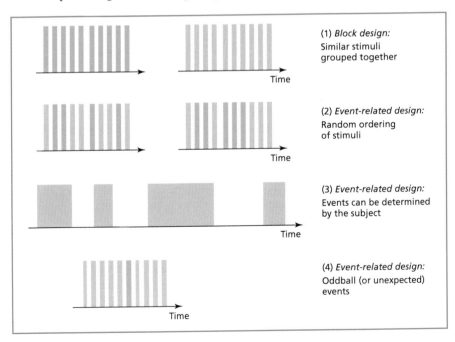

(1) *Block design:*
Similar stimuli grouped together

(2) *Event-related design:*
Random ordering of stimuli

(3) *Event-related design:*
Events can be determined by the subject

(4) *Event-related design:*
Oddball (or unexpected) events

Different conditions (either stimuli or tasks) can be presented in two ways in fMRI experiments: either in a block design or an event-related design.

requiring some other judgment (e.g. age). The 'no judgment at all' condition could be problematic if it is compared with a condition in which an actual response is made (e.g. a button press) because the comparison would reveal lots of regions involved in motor production, which probably has little to do with gender judgment. It is also unclear what instructing a participant to do nothing, other than passively watch, would mean (do they really do nothing?). The claims that can be made depend crucially on the control condition. In one case we would learn something about gender categorization relative to face perception, and in the other case we would learn something about gender categorization relative to age categorization. But it is a moot point as to whether we would learn anything about gender categorization itself in a context-free way. Of course, as more studies are done employing different designs and contrasts, such a picture may indeed emerge.

Analysis of fMRI data

The analysis of fMRI data involves determining whether there is a statistically significant relationship between the changes in BOLD signal over time based on what is expected from the study design (i.e. given the known timings of stimulus presentations in the various conditions). This is done at each and every voxel. Each voxel can then be 'colored in' according to its level of significance and in this way one builds up a picture of how the brain is activated by the various experimental conditions. This is an important point because the images shown in functional imaging papers are not literal pictures of the workings of the brain: they depict levels of statistical significance that are superimposed onto a (structural) image of the brain for depictive purposes.

To get from the raw data (which consists of a set of two-dimensional images from one volume of the brain) to being able to perfom a statistical analysis of the data involves a number of stages that are termed **pre-processing**. The following stages of pre-processing normally occur:

1. One needs to link the brain images (in each brain volume) to the timing of the stimulus presentation.
2. Correction for head movement. If a person moves their head, even a milimeter or so, then the regions of activity will also shift around, making them harder to detect. Fortunately, one can use the raw images themselves in order to infer how the person has moved over time and then use these so-called movement parameters to partial out these effects from the data.
3. **Stereotactic normalization** maps the voxels on an individual's brain onto the equivalent regions in a standard brain. This is needed in any analysis in which several different brains are entered, and is needed because of anatomical variability in brain size and shape. For example, the position of the sulci can vary by a centimeter or more in different people (Thompson, Schwartz, Lin, Khan, & Toga, 1996). The standard brain that is often used is based on an average of 305 brains provided by the Montreal Neurological Institute – the **MNI template** (Collins, Neelin, Peters, & Evans, 1994). Thus, a voxel in one person's brain can be compared to an equivalent voxel in another person's brain if they share the same coordinates in this standard space. Another template that is used, particularly for reporting data in journals, is based on the atlas of Talairach and Tournoux (1988). These are referred to as **Talairach coordinates** and are based on detailed data from a single postmortem brain.

4. **Smoothing** involves increasing the spatial extent of activity in voxels. Specifically, a normal distribution is superimposed on each voxel such that most activity remains in the voxel itself but some spreads to neighbors. One reason why this is done is to cope with individual differences in the site of brain activity. Common regions of activity between participants are easier to find when the activity itself is more spatially diffuse. A second reason why smoothing is done is to improve the signal-to-noise ratio. If there are collections of voxels nearby with high activity then that activity will be enhanced by smoothing, whereas if there is a single active voxel with no active neighbors then its activity will be diminished. Smoothing is almost always done when comparing several brains in an analysis, but it need not be done if the analysis is done on one individual's brain. For example, there are various techniques that explore how the *pattern* of activity in a region relates to cognition (rather than the *amount* of activity in a region) in which smoothing is not used (e.g. Haynes & Rees, 2006).

> **KEY TERMS**
>
> **Smoothing**
> Increasing the spatial extent of activity in voxels.
>
> **Familywise error (FWE)**
> A statistical threshold used in functional imaging based on assumptions of spatial smoothness.
>
> **False discovery rate (FDR)**
> A statistical threshold used in functional imaging based on random permutations of the data.

After pre-processing each person, it is then possible to conduct statistical tests. The fact that the brain is divided into tens of thousands of voxels creates a statistical challenge in itself. If one were to use the standard threshold of statistical significance used in psychology ($P < .05$, meaning that there is a 5% probability of getting a significant result by chance) then there could be thousands of brain regions active by chance. However, the activities in different voxels are not strictly independent from each other (voxels tend to have similar activity to their neighbors). This has led to the development of mathematical models that choose a level of significance based on assumptions of spatial smoothness using so-called random field theory. This generates a statistical threshold called the **familywise error (FWE)**. This threshold is related to the number of statistical tests that are run. An alternative method is based on considering the actual number of positive results obtained, and is termed the **false discovery rate (FDR)**. A comparison that produces lots of positive results (i.e. lots of activity) would have a different proportion of expected false positives than a study with only a few positive results. The FDR method takes this into account, whereas the FWE method only takes into account the number of tests performed.

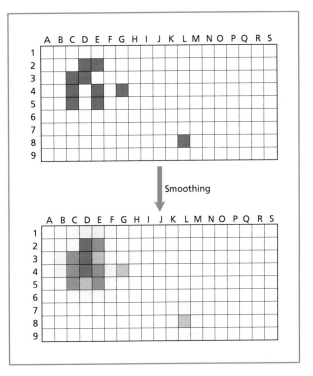

Smoothing spreads the activity across voxels – some voxels (e.g. D4) may be enhanced whereas others (e.g. L8) may be reduced.

Summary of fMRI

- *Advantages*: Very good spatial resolution; generally safe and non-invasive.
- *Disadvantages*: Poor temporal resolution; not a direct measure of neural activity.

SAFETY ISSUES IN fMRI RESEARCH

Before entering the scanner, all participants should be given a checklist that asks them about their current and past health. People with metal body parts, cochlear implants, embedded shrapnel, or pacemakers will not be allowed to take part in fMRI experiments. In larger magnets, eye make-up should not be worn (it can heat up, causing symptoms similar to sunburn) and women wearing contraceptive coils are not normally tested. Before going into the scanner both the researcher and participant should put to one side all metal objects such as keys, jewellery, and coins, as well as credit cards, which would be wiped by the magnet. Zips and metal buttons are generally okay, but metal spectacle frames should be avoided. It is important to check that participants do not suffer from claustrophobia because they will be in a confined space for some time. Participants wear ear protectors, given that the scanner noise is very loud. Larger magnets (> 3 T) can be associated with dizziness and nausea and participants need to enter the field gradually to prevent this. Participants have a rubber ball that can be squeezed to signal an alarm to the experimenter, who can terminate the experiment if necessary. The standard safety reference is by Shellock (2004) and updates can be found at www. magneticresonancesafetytesting.com.

LESION METHODS AND DISRUPTION OF FUNCTION USING TMS

The basic premise behind the approach is that, by studying the abnormal, it is possible to gain insights into normal function. This is a form of 'reverse engineering', in which one attempts to infer the function of a region by observing what the rest of the cognitive system can and cannot do when that region is removed. The three methods that will be considered in this section are effects of naturally occurring brain damage in humans, experimental lesions in animals, and TMS. Although superficially these are very different approaches (e.g. TMS is considered harmless) they are conceptually related methods bound together by the 'reverse engineering' principle.

Neuropsychology: The effects of naturally occurring lesions in humans

The effects of brain damage on cognition are explored in the field of cognitive neuroscience known as **neuropsychology**. The experimenter obviously has no experimental control over the nature and location of brain damage, but can exert control in terms of the selection criteria for participation in studies. There are various causes of brain damage, including:

1. **Strokes or CVAs (cerebrovascular accidents)**. These are disruptions to the blood supply to the brain due to blockages or ruptures, leading to neuronal death. The probability of suffering a stroke is age related (increasing sharply in the elderly).

2. Traumatic head injuries may be of either an 'open' nature (the skull is fractured) or 'closed' nature (as commonly found in road traffic accidents). They tend to be most prevalent in younger males. Brain damage typically results from compression and bruising.
3. Tumors are caused when new cells are formed in a poorly regulated manner and in the brain they are formed from the supporting cells, such as the meninges and glia (termed meningioma and glioma, respectively). The extra material puts pressure on neurons, resulting in possible cell death.
4. Neurosurgery tends not to be favored over pharmacological treatments but is sometimes used (e.g. to remove the focus of an epileptic seizure) and results in localized brain damage.
5. Viral infections may target specific cells in the brain, possibly leading to cell death.
6. Neurodegenerative disorders such as dementia of Alzheimer's type (DAT) are becoming increasingly more prevalent as the population ages.

The logic of patient-based neuropsychology rests, to some degree, on the notion of functional specialization (i.e. the notion that different regions of the brain perform different kinds of computation or process different kinds of information). In this respect, it is similar to other methods such as functional imaging. In other respects it differs in important ways from the logic of functional imaging. The data in neuropsychology are behavioral (typically error rates, or degree of impairment) and the independent variable (i.e. conditions manipulated) is the lesion itself. In functional imaging, the data are localized differences in brain activity and the independent variable is behavior itself (i.e. the instructions given to the participant). These differences are important because it suggests that one method cannot simply substitute for the other in a simple way. It has also been argued that neuropsychology, but not functional imaging, is able to make claims that a region is *necessary* for a task to be performed successfully (Kosslyn, 1999). This is because there is an assumed causal connection between the lesion and the impaired behavior. In functional imaging, activity in a particular brain region can depend on many factors, such as how well the control condition is matched (e.g. for difficulty) and the particular strategy that a person uses (the strategy itself would have its own neural activity but may be unrelated to how well the task is performed).

The notion of functional specialization leads to the prediction that localized brain damage will impair some but not all cognitive functions (unless, of course, the region is crucial to all aspects of cognition). These are termed **dissociations**. For example, patients have been reported who – after sustaining damage to the amygdala – have difficulties in recognising facial expressions of fear (Adolphs, Tranel, Damasio, & Damasio, 1994; Calder et al., 1996). This could be taken to

Functional brain imaging and patient-based neuropsychology (or TMS) are logically different types of methodology. It is unlikely that one will supplant the other.

	Functional imaging	Neuropsychology
Dependent measure (i.e. your data)	Brain regions	Behavior (task performance)
Independent variable (i.e. conditions manipulated)	Behavior (task performance)	Brain regions

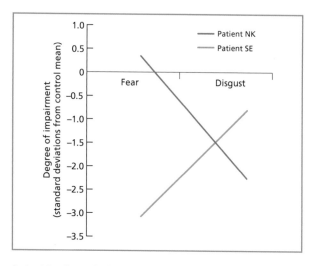

A double dissociation: Patient NK is impaired at recognizing disgust (but not fear) whereas patient SE is impaired at recognizing fear (but not disgust). The y-axis shows the degree of impairment relative to controls measured in terms of standard deviations from the control mean (0 = no impairment, negative = impaired). Data from Calder et al. (1996, 2000).

imply that there is a specialized region for processing fear. However, it is not the only conclusion that could be reached. It could be the case that facial expressions of fear are more difficult to recognize than other facial expressions. Evidence against this comes from **double dissociation,** in which recognition of fear expressions is preserved but other emotions (such as disgust) are hard to recognize (Calder, Keane, Manes, Antoun, & Young, 2000). However, even with a double dissociation one cannot necessarily conclude that there is a specialized region for recognizing fear expressions. It could be, for example, that recognizing fear in faces requires attention to particular regions of a face and that disrupting this face-attention mechanism will disrupt fear recognition (Adolphs et al., 2005). Nor can one conclude that this brain region is not involved in any other cognitive function (one would need to test in other domains to establish this). It would be wrong to take away from this argument that nothing can be concluded from double dissociations from patients. We can draw some conclusions using these methods (e.g. that fear recognition and disgust recognition use partially distinct neural resources). However, the conclusions we draw do depend on the theories we are aiming to test (i.e. what those neural resources are believed to be computing).

One interesting aspect of patient-based neuropsychology is that it can still be possible to draw inferences in the absence of knowing where the lesion is. From the double dissociation mentioned previously we can conclude that fear recognition and disgust recognition are partially separable without knowing *where* in the brain they may be. This does not mean that neuropsychology cannot be used to address the 'where' question but it is not the only question that can be asked. However, to find out where the crucial region (or regions) is typically involves either patients with very focal brain lesions or large groups of patients (with non-focal lesions) who can be assessed on a particular task and their group performance linked (e.g. on a voxel-by-voxel basis) to their known lesions (e.g. Rorden & Karnath, 2004). In this way, patient-based neuropsychology benefits from the advances in imaging science in the same way as fMRI.

Finally, the logic of patient-based neuropsychology may not hold for brain damage sustained in childhood relative to adulthood. Lesions in the same region may have different consequences in childhood because other regions may be better able to take on lost functions (e.g. Ballantyne, Spilkin, Hesselink, & Trauner, 2008). However, it remains unclear whether moving a function to a different location in the brain (e.g. the intact hemisphere) necessarily results in a change in *how* the function is performed. Is language qualitatively the same if it shifts into the right hemisphere as a result of a childhood left lesion?

Summary of human neuropsychology

- *Advantages*: Lesion methods enable inferences of brain–behavior causality.
- *Disadvantages*: There is no experimental control over lesions.

Experimentally induced lesions in animal models

Although lesion methods in humans rely on naturally occurring lesions, it is possible – surgically – to carry out far more selective lesions on other animals. Unlike human lesions, each animal can serve as its own control by comparing performance before and after the lesion. It is also common to have control groups of animals that have undergone surgery but received no lesion, or a control group with a lesion in an unrelated area. There are various methods for producing experimental lesions in animals (Murray & Baxter, 2006):

1. *Aspiration.* The earliest methods of lesioning involved aspirating brain regions using a suction device and applying a strong current at the end of an electrode tip to seal the wound. These methods could potentially damage both gray matter and the underlying white matter that carries information to distant regions.
2. *Transection.* This involves the cutting of discrete white matter bundles, such as the corpus callosum (separating the hemispheres) or the fornix (carrying information from the hippocampus).
3. *Neurochemical lesions.* Certain toxins are taken up by selective neurotransmitter systems (e.g. for dopamine or serotonin) and once inside the cell they create chemical reactions that kill the cell. A more recent approach involves toxins that bind to receptors on the surface of cells, allowing for even more specific targeting of particular neurons.
4. *Reversible 'lesions'.* Pharmacological manipulations can sometimes produce reversible functional 'lesions'. For example, scopolamine produces a temporary amnesia during the time in which the drug is active. Cooling of parts of the brain also temporarily suppresses neural activity.

Whilst the vast majority of behavioral neuroscience research is conducted on rodents, some research is still conducted on non-human primates. In many countries, including in the EU, neuropsychological studies of great apes (e.g. chimpanzees) are not permitted. More distant human relatives used in research include three species

A family of macaque monkeys.

of macaque monkeys (rhesus monkey, cynomolgus monkey, and Japanese macaque) and one species of New World primate (the common marmoset).

Summary of animal neuropsychology

- *Advantages*: Lesion methods enable inferences of brain–behavior causality; precision is possible through anatomically or chemically selected lesions.
- *Disadvantages*: The range of behaviors studied is limited relative to humans; there are concerns over animal welfare.

Transcranial magnetic stimulation (TMS)

Attempts to stimulate the brain electrically and magnetically have a long history. Electrical currents are strongly reduced by the scalp and skull and are therefore more suitable as an invasive technique on people undergoing surgery. In contrast, magnetic fields do not show this attenuation by the skull. However, the limiting factor in developing this method has been the technical challenge of producing large magnetic fields, associated with rapidly changing currents, using a reasonably small stimulator (for a historical overview, see Walsh & Cowey, 1998). It was not until 1985 that adequate technology was developed to magnetically stimulate focal regions of the brain (Barker, Jalinous, & Freeston, 1985). Since then, the number of publications using this methodology has increased rapidly.

Typically, the effects of TMS are small, such that they alter reaction time profiles rather than elicit an overt behavior. But there are instances of the latter. If the coil is placed over the right visual cortex, then the subject may report visual sensations or 'phosphenes' on the left side (given that the right visual cortex represents the left side of space). Even more specific examples have been documented. Stewart, Battelli, Walsh, and Cowey (1999) stimulated a part of the visual cortex dedicated to

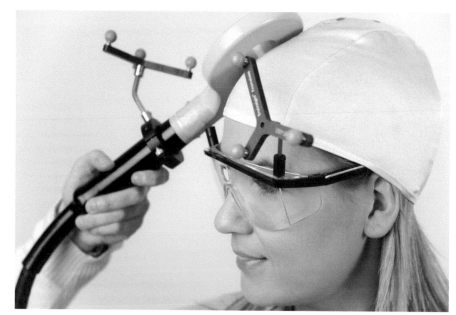

The TMS coil is held against the participant's head, and a localized magnetic field is generated during performance of the task. University of Durham/ Simon Fraser/Science Photo Library.

motion perception (area V5) and reported that these particular phosphenes tended to move. Stimulation in other parts of the visual cortex produces static phosphenes.

TMS works by virtue of the principle of electromagneic induction that was first discovered by Michael Faraday. A change in electrical current in a wire (the stimulating coil) generates a magnetic field. The greater the rate of change in electrical current, the greater the magnetic field. The magnetic field can then induce a secondary electrical current to flow in another wire placed nearby. In the case of TMS the secondary electrical current is induced, not in a metal wire, in the neurons below the stimulation site. The induced electrical current in the neurons is caused by making them 'fire' (i.e. generate action potentials) in the same way as they would when responding to stimuli in the environment. The use of the term 'magnetic' is something of a misnomer because the magnetic field acts as a bridge between an electrical current in the stimulating coil and the current induced in the brain. Pascual-Leone, Bartres-Faz, and Keenan (1999) suggest that 'electrodeless, non-invasive electric stimulation' may be more accurate.

TMS causes neurons underneath the stimulation site to be activated. If these neurons are involved in performing a critical cognitive function, then stimulating them artificially will disrupt that function. Although the TMS pulse itself is very brief (less than 1 ms), the effects on the cortex may last for several tens of milliseconds. As such, the effects of a single TMS pulse are quickly reversed. Although this process is described as a 'virtual lesion' or a 'reversible lesion', a more accurate description would be in terms of interference. In the cognitive psychology literature, dual-task interference paradigms are used to determine whether two tasks share cognitive resources. For example, it is hard to pat your head whilst rubbing your tummy (although it is easy to do each in isolation). This suggests that they share some cognitive/neural mechanisms. In contrast, it is easy to pat your head and read aloud, which suggests little sharing of cognitive/neural mechanisms. TMS uses a comparable logic to infer whether a given brain region is critical. If a region is critical for a task, then there is likely to be interference because of the dual use of the region in terms of the computational demands of the task together with the activity ensuing from the applied stimulation.

TMS has a number of advantages over traditional lesion methods (Pascual-Leone et al., 1999). The first advantage is that real brain damage may result in a reorganization of the cognitive system, whereas the effects of TMS are brief and reversible. This also means that within-subject designs (i.e. with and without lesion) are possible in TMS that are very rarely found with organic lesions (neurosurgical interventions are an interesting exception, but in this instance the brains are not strictly 'normal' given that surgery is warranted). In TMS, the location of the stimulated site can be removed or moved at will. In organic lesions, the brain injury may be larger than the area under investigation and may affect several cognitive processes.

Summary of TMS

- *Advantages*: TMS can be used to investigate the timing of cognition as well as the location of cognition; a 'virtual lesion' can be moved within the same participant.
- *Disadvantages*: TMS can only stimulate certain regions; it is hard to predict how TMS affects the functioning of distant sites.

SAFETY ISSUES IN TMS RESEARCH

Whereas single-pulse TMS is generally considered to be safe, repetitive-pulse TMS carries a very small risk of inducing a seizure (Wassermann, Cohen, Flitman, Chen, & Hallett, 1996). Given this risk, participants with epilepsy or a familial history of epilepsy are normally excluded. Participants with pacemakers and medical implants should also be excluded. Credit cards, computer discs, and computers should be kept at least 1 meter away from the coil. The number of pulses that can be delivered to a participant in a given testing session has been established, by consensus (Pascual-Leone et al., 1993; Wassermann, 1996). Up-to-date information about safety is available at http://pni.unibe.ch/MailList.htm (note that the web address is case sensitive). The intensity of the pulses that can be delivered is normally specified with respect to the 'motor threshold' – the intensity of the pulse, delivered over the motor cortex, that produces a just noticeable motor response (for a discussion of problems with this, see Robertson, Theoret, & Pascual-Leone, 2003).

During the experiment, some participants might experience minor discomfort due to the sound of the pulses and facial twitches. Although each TMS pulse is loud (~100 dB), the duration of each pulse is brief (1 ms). Nonetheless, the ears should be protected with earplugs or headphones. When the coil is in certain positions, the facial nerves (as well as the brain) may be stimulated, resulting in involuntary twitches (e.g. blinking, jaw clamping). Participants should be warned of this and told they can exercise their right to withdraw from the study if it causes too much discomfort.

SUMMARY AND KEY POINTS OF THE CHAPTER

- The neuroscientific methods used by social neuroscience (e.g. fMRI) do not preclude the use of more traditional psychological methods (e.g. response time, questionnaires) and a combination of the two approaches is commonplace (e.g. linking questionnaire data to neural function).
- Methods relying on activity of the peripheral nervous system include the skin conductance response (SCR), electromyography (EMG), and pupillometry. These measures are often used as 'unconscious' measures of behavior or affective processes.
- Electrophysiological methods offer excellent temporal resolution. Single-cell recordings additionally offer excellent spatial resolution and are important for understanding how neurons code information, but they have the disadvantage of being usually limited to non-human animals. EEG methods measure electrical activity at the scalp and have an

advantage over response time measures in that changes in neural activity can be detected without waiting for (or requiring) a response.

- The functional imaging methods of PET and fMRI are not direct measures of neural activity but are hemodynamic methods, dependent on changes in blood flow and blood oxygenation resulting from neural activity. They offer good spatial resolution but poor temporal resolution.
- Neuropsychological methods in humans rely on naturally occurring lesions to the brain that may selectively impair certain aspects of cognition; in non-human animals, more selective lesions are possible. They enable researchers to establish a causal relationship between brain structure and function (rather than functional imaging, which is correlational).
- TMS creates localized neural interference by creating a brief magnetic field over the skull that temporarily disrupts performance (measured by errors or response time). Unlike neuropsychology, it enables within-subject designs (the 'virtual lesion' can be moved) and can explore the time-course of cognition.

EXAMPLE ESSAY QUESTIONS

- Do questionnaires and other methods relying on subjective report have any place in social neuroscience?
- Describe how three different types of methodology in social neuroscience can provide complementary insights into our understanding of face processing.
- What can functional imaging reveal about the social brain that other methods cannot?
- Do the methods of social neuroscience make it impossible to study naturalistic social behavior?

RECOMMENDED FURTHER READING

- Harmon-Jones, E., & Beer, J. S. (2009). *Methods in social neuroscience*. New York: Guilford Press. A good collection of chapters that include some methods not covered here (e.g. salivary hormones, genomic imaging, neuroendocrine manipulations).

- Senior, C., Russell, T., & Gazzaniga, M. S. (2009). *Methods in mind*. Cambridge, MA: MIT Press. A good collection of chapters that include some methods not covered here (e.g. magnetoencephalography, MEG).

- Ward, J. (2010). *The student's guide to cognitive neuroscience* (2nd edition). New York: Psychology Press. The methods chapters in this book have been condensed to form the basis of this chapter.

CHAPTER 3

CONTENTS

The social intelligence hypothesis   50

Evolutionary origins of culture   55

Social learning versus imitation   57

Material symbols: Neuronal recycling and
extended cognition   62

Cultural skills: Tools and technology   64

Summary and key points of the chapter   68

Example essay questions   69

Recommended further reading   69

Evolutionary origins of social intelligence and culture

3

Modern humans, Homo sapiens, emerged as a distinct species only 200,000 years ago. Over time, this new species developed a variety of tools, produced elaborate art, and began to bury their dead in ornate rituals. In the last few hundred years, they invented computers, visited the moon, and discovered the basic physical laws that govern the universe. We are separated from our nearest living ancestor, the chimpanzee, by only 1.6% of DNA (King & Wilson, 1975) and we shared a common ancestor with the chimpanzee around 6 or 7 million years ago. What is it in this 1.6% of DNA that has enabled humans to achieve such a level of technological and cultural complexity? According to one idea, the main evolutionary pressure for human

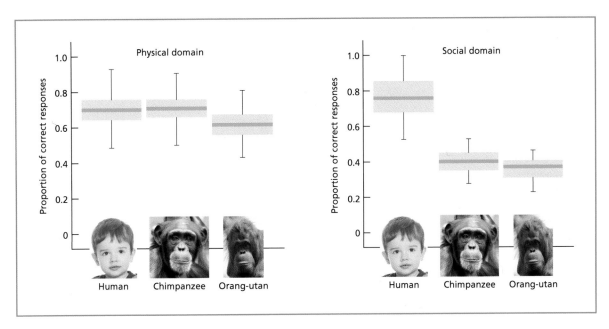

Hermann, Call, Hernàndez-Lloreda, Hare, and Tomasello (2007) compared the physical and social intelligence of 2–3-year-old human children with that of somewhat older chimpanzees and orang-utans. They found that humans excelled in the social domain on tasks such as social learning (solving a problem after a solution is demonstrated), communication (pointing to receive a hidden reward), and gaze following. They did not, however, excel in the physical domain on tasks such as spatial memory (to locate a reward), quantity discrimination, and tool use (using a stick to get a hidden reward). This was taken to support the view that the evolution of human intelligence has primarily occurred in the social domain. Figure adapted from Herrmann et al. (2007). Copyright © 2007 American Association for the Advancement of Science. Reproduced with permission.

Social intelligence
The ability to understand and predict complex social interactions and to outwit our peers.

Culture
A shared set of values, skills, artifacts, and beliefs amongst a group of individuals.

Social learning
The transmission of skills and knowledge from person to person.

Social intelligence hypothesis
Evolutionary pressures to be socially smarter lead to more general changes (e.g. increased brain size) resulting in increased intellect in non-social domains.

intellectual development is not the ability to be smarter per se but rather the ability to understand and predict complex social interactions and to outwit our peers – so-called **social intelligence** (Humphrey, 1976). According to this view, evolutionary pressures to be socially smarter would lead to more general changes (e.g. larger brain size) that would lead to increased intellect in other, non-social, domains. The complex culture that we have today, and that sets us apart from all other species, would then be viewed as a by-product of these earlier, more general adaptations.

Culture can be defined as a shared set of values, skills, artifacts, and beliefs amongst a group of individuals. There is an inherent social dimension to culture. Culture is *shared* amongst members of a group via a process of **social learning** from person to person, both within and across generations. Differences in culture also form one way of distinguishing between social groups. It is to be noted that culture in this context does not mean 'high culture' (opera, art, etc.), although this would be a component of it. Culture encompasses far more than this, including skills (e.g. literacy), technology (e.g. tool-making from spears to computers), and beliefs (including but not limited to religious beliefs). One tends to think of culture as an environmental rather than a genetic effect. We are not born predisposed to speak a particular language or believe a particular religion. However, whilst the *differences* between cultures are entirely attributable to environmental factors (our time and place of birth), the *similarities* between cultures (including the fact that we are all immersed in one) are almost certainly down to biology and evolution. Our brains have developed in such a way that allows us to both create and absorb shared knowledge, skills, and beliefs.

This chapter begins by examining the hypothesis that primate intelligence evolved to deal with increasing social complexity. It will then go on to consider how this could lead to non-genetic social transmission of knowledge via culture, and it will consider evidence for culture in other species. Finally, the chapter will discuss how certain aspects of culture are represented in the brain through a consideration of material symbols (written words and numbers, symbolic art) and tool use. Other aspects of culture, such as culturally different perceptions of self, are dealt with in Chapter 9.

THE SOCIAL INTELLIGENCE HYPOTHESIS

As already noted, the **social intelligence hypothesis** argues that evolutionary pressures to be socially smarter lead to more general changes (e.g. larger brain size) resulting in increased intellect in non-social domains (Humphrey, 1976). This has also been termed the social brain hypothesis (Dunbar, 1998) and the Machiavellian intelligence hypothesis (Whiten & Byrne, 1988). (Machiavelli was a Renaissance politician renowned for cunning and deceit.) However, when one comes to consider this hypothesis in detail it can be shown to have several different meanings. Whiten and van Schaik (2007) considered three different interpretations of this basic idea:

1. 'Intelligence is manifested in social life.' This is the weakest interpretation of the hypothesis. Historically, animal intelligence was studied by looking at the ability of individual animals in a laboratory to solve problems. This weaker interpretation merely states that intelligence should be more broadly construed to include problem solving in one's social life. This idea is now widely accepted (Whiten & van Schaik, 2007).

2. 'Complex society selects for enhanced intelligence.' This stronger interpretation argues that there is something particularly demanding about problem solving in the social realm that leads to a need for greater intelligence. The more complex the society, the greater the need. In this interpretation, 'intelligence' is regarded as a more general capacity rather than a specialized set of functions that deal with social life.
3. 'Complex society selects the specific characteristics of intelligence.' This interpretation is a stronger form of the one above. It suggests that social pressures select not only for the *amount* of intelligence but also the *type* of intelligence. For example, one might imagine relatively specialized mechanisms in the brain for dealing with social problems (e.g. imitation, theory of mind) that are not necessarily reducible to general intellect.

To some extent, the debate between the second and third interpretations will be a recurring theme of this book. This chapter will consider the issue with particular reference to evidence from non-human species.

Social intelligence and brain size

Several studies have tested the social intelligence hypothesis by examining the relationship between social intelligence and brain size. The prediction is that the more complex the species' social world, the larger the brain will need to be to cope with such complexity. As Seyfarth and Cheney (2002) put it, primates 'live in large groups where an individual's survival and reproductive success depends on its ability

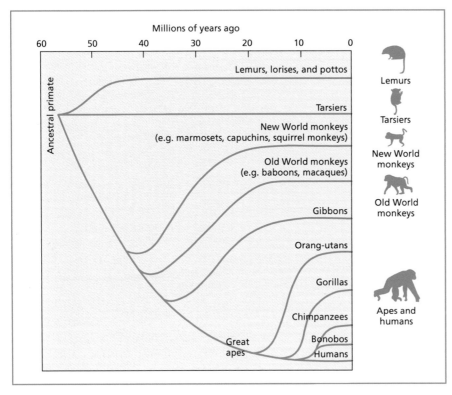

A history of primate evolution. The extinction of the dinosaurs occurred around 65 million years ago and, at this time, the primates' ancestral relative was like a tree-dwelling shrew. Changes in the genetic code between species act as a 'molecular clock' that enables more accurate timing of branching.

to manipulate others within a complex web of kinship and dominance relations' (p. 4141).

There are several inherent problems with examining a link between brain size and social intelligence. Firstly, there are different ways of defining 'social intelligence' as noted above. Secondly, it is not easy to measure it in the natural social settings of, say, an orang-utan or capuchin monkey. Finally, this approach assumes that brain size is a useful index of general intellect. Typically, the size of the whole brain is not measured but a ratio is taken between the amount of neocortex (most of the gray matter surface of the brain) and the rest of the brain. The reliance on these measures of brain size has been criticized by some (e.g. Healy & Rowe, 2007) and others have noted that brain size may not be very important at all (Chittka & Niven, 2009). Chittka and Niven (2009) note that many insects live in complex large-scale social networks but have tiny brains. Despite these differences and difficulties, several studies using somewhat different measures have converged on a common answer – that there is a strong correlation between social intelligence and brain size.

In one of the earliest attempts to explore this issue, Dunbar (1992) used social group size of various primates as an approximate measure of social complexity and found a significant correlation with neocortex ratio. The implication is that the larger the brain, the greater the number of social relationships that can be sustained. Extrapolating from other species to humans, Dunbar (1992) estimated that humans are adapted to an optimal group size of about 150 people. Although most of us know many more people than this, the claim is that our brains can only support *active* relationships (based on regular exchanges) with around 150 others. There is some evidence that supports this number, from clan sizes in traditional societies (see Dunbar, 1998) to exchange of Christmas cards in Western societies (Hill & Dunbar, 2003).

Other research has used a more direct measure of social intelligence than group size. Byrne and Corp (2004) found a correlation between frequency of tactical deception in different primate species and neocortex ratio. Deception is a complex social skill involving an appreciation of another's knowledge and the ability to manipulate it. In another study, Reader and Laland (2002) measured the number of times that researchers had documented, in natural habitats, examples of social learning, innovation (coming up with novel solutions), and tool use in 106 species of primates. These figures, scaled according to number of opportunities for observation (i.e. some species are studied more than others), were then correlated with a measure of proportional brain size. All three of these variables correlated strongly with brain size. This suggests the co-evolution of an aspect of social intelligence (social learning) and non-social intelligence (innovation). These results, therefore, do not support the view that social factors were more important than other factors in leading to increased intellect/brain size. It suggests, instead, that both were crucial. Being able to come up with innovative ideas will have limited impact on cultural development if the ideas die out with the inventor (i.e. social learning is required). Similarly, being able to learn from each other is only important if there is something worth learning (i.e. innovation is required). Tool use could, perhaps, be considered a product of both of these processes.

The studies above could potentially be criticized on the grounds that they give an over-emphasis to one particular measure – namely brain size. A finer grained analysis (e.g. based on individual regions) plus using other types of measure (e.g. brain connectivity, cortical thickness) may yet reveal a quite different picture.

One important factor that goes hand-in-hand with evolutionary increases in brain size is the length of immaturity (Joffe, 1997). Humans take an unusually long time to reach adulthood and this comes at a significant cost in terms of provision of food and protection of offspring. One likely benefit to this is that it provides an extended window for learning and adapting to one's environment and culture. Greater dependency in early life provides rich opportunities for learning from one another. The emphasis on social learning puts a new twist on the nature–nurture debate. If intelligence is related (at least in part) to our ability to learn from each other, then it is equally a product of our nature (a genetic disposition to learn from each other) and nurture (our accumulating knowledge of the world) – without one, it is not possible to have the other.

Did language evolve from increased social and cognitive intellect?

There are a number of interesting parallels to be made between various theories of language evolution and the social intelligence hypothesis. For instance, there is a debate concerning whether language arose from non-specific evolutionary changes such as increased brain size. This view has been most famously championed by the evolutionary biologist Stephen Jay Gould (e.g. Gould, 1991) and the linguist Noam Chomsky (e.g. Chomsky, 1980). According to this view, language arose out of general selection pressures to be smarter – either socially, cognitively, or probably both. This pressure then led to general changes (e.g. in brain size) from which language emerged. One specific idea along these lines is Dunbar's (2004) proposal that language evolved to facilitate the bonding of large social groups. He argues that language evolved due to social pressures to live in large groups and that language enables greater cohesion of groups. Dunbar draws an explicit comparison between grooming behavior in primates and social use of language in humans. Human language, according to Dunbar, evolved to enable humans to keep 'in touch' without literally being in touch! The contrary view to Chomsky, Gould, and Dunbar, espoused for example by Pinker and Bloom (1990), is that language did arise from selection pressures relating specifically to communicative needs rather than as a by-product of other, more general changes such as brain size or social group size.

Recent papers take the more realistic position that language should be considered not as a single entity but as being multi-faceted, consisting of speech production,

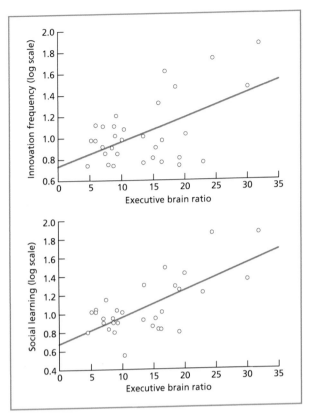

The frequency of social learning and frequency of innovation (both on logarithmic scales) correlate with executive brain ratio (defined here as ratio of neocortex and striatum volume relative to brainstem volume). Each data point represents a different primate species and the frequency counts are corrected according to 'research effort' (i.e. taking into account the fact that some species are studied more than others giving more opportunity for positive observations). From Reader and Laland (2002). Copyright © 2002 National Academy of Sciences, U.S.A. Reproduced with permission.

Grooming serves a social function in primate societies. It is related to social complexity (larger groups engage in more grooming) and alliance formation (grooming releases endorphins, the brain's natural opiates). Why do humans not engage in this activity? According to Dunbar (2004), human language serves the social function of keeping in touch (without literally 'keeping in touch') and sharing information about who is doing what to whom.

syntax, semantic concepts, and so on (Fitch, Hauser, & Chomsky, 2005; Pinker & Jackendoff, 2005). Each of these different aspects of language can then be considered from either a generalist or a language-specific perspective on the merits of its own case. To give a flavor of the debate, consider two key aspects of language: the descent of the human **larynx**; and **syntax**. Humans have evolved a descended larynx that is not present in other primates and comes at a significant survival cost. Unlike most other animals, adult humans cannot breathe and swallow at the same time and thus have elevated risk of choking. However, a descended larynx is also crucial for human speech. This is acknowledged by both camps. Where the disagreement lies is in whether larynx descent occurred because of the need to increase the repertoire of speech (i.e. it evolved for the function of communication) or whether it occurred for some other reason not related to communication. For example, a descended larynx makes an animal sound bigger, thus making it more attractive to a mate and better able to achieve social dominance (e.g. Reby et al., 2005). This would set up an 'arms race' in which the larynx descends further and further until it ceases to have benefit. It is found in at least one other species that lacks language – red deer (Fitch & Reby, 2001). Thus, whilst a descended larynx enables human speech, it is impossible to say for certain whether it evolved for that specific reason. Similarly for syntax (i.e. the grammatical rules that specify how words are combined), one suggestion is that the hierarchical nature of syntactic representation (in which words are clustered into phrases, and phrases into sentences, etc.) may have been driven by the need to mentally represent complex social groups and hierarchies rather than for communicative purposes (Hauser, Chomsky, & Fitch, 2002).

Evaluation

There is good evidence that evolutionary increases in relative brain size have been accompanied by increased complexity in the social domain – larger social groups, more deception, and more social learning. This provides support for the social

KEY TERMS

Larynx
An organ in the neck of mammals involved in sound production.

Syntax
The rules by which words are combined to make meaningful sentences.

intelligence hypothesis. What remains unclear, and keenly debated, is the extent to which specific processes (such as language or theory of mind) arose out of these more general changes or were specifically shaped during the course of evolution.

SOCIAL INTELLIGENCE IN BOTTLENOSE DOLPHINS

Most of the evidence in this chapter comes from primates, both human and non-human. This may reflect, at least in part, an anthropocentric agenda. There is an enduring interest in the question 'what makes humans special?', which naturally leads to a comparison with our nearest living ancestors. Methods of studying humans are also often easier to adapt to primates than to other species. Of course humans are not necessarily special, and a consideration of non-primate animals reveals some surprising evidence of social intelligence in other species.

Dolphins, like the great apes, have unusually large brains (particularly the evolutionary newer neocortex) relative to their size. Whilst brain enlargement in great apes and dolphins reflects separate evolutionary events it is possible that it arose through similar selection pressures, such as the need to deal with social complexity (for a review see Connor, 2007). Connor (2007) states that dolphins have around 60–70 associates, which is comparable to the largest non-human primate groups, but the dolphin associates are not sharply demarcated into stable groups. Dolphins may affiliate to each other via gentle rubbing with their pectoral fins (Connor, Wells, Mann, & Read, 2000) and through synchronous displays such as joint surfacing (Herman, 2002). Such behavior may be analogous to primate grooming and imitation, respectively. Dolphins, like apes but unlike monkeys, are able to recognize themselves in mirrors to investigate marked parts of their body (Reiss & Marino, 2001). This has been linked to self-awareness (Schilhab, 2004).

Is it a coincidence that dolphins have unusually large brains (for their size) and also live in large and complex social groups?

EVOLUTIONARY ORIGINS OF CULTURE

It is strange, but almost certainly true, that if you were to take (with the aid of a time machine) a newborn baby born tens of thousands of years ago and raise it in the modern age it would have no difficulties in learning our language, performing algebra, driving a car, or surfing the net. What has primarily evolved in this period of time is not our biology but our culture. However, we should not fall into the trap of thinking that some human societies have 'more' culture than others, rather they have different systems of culture. Hunter–gatherers are cultured – they have certain skills, tools, and beliefs that enable them to survive in *their* environment.

KEY TERM

Meme
Units of culture that
are transmitted from
person to person
according to their own
perceived fitness (e.g.
usefulness).

A Westerner transplanted into their environment may quickly die of poisoning, predation, or starvation. However, one also cannot gloss over the fact that some cultures have developed far more advanced forms of technology than others. It has been claimed (Blackmore, 1999; Diamond, 1997) that certain cultural trends will tend to dominate over others when they come into contact with each other in a kind of cultural 'survival of the fittest' – think guns versus spears, or the ease with which one can perform sums with modern number notation (e.g. 54 x 10 = 540) compared to Roman numerals (e.g. LIV × X = DXL). Blackmore (1999; following Dawkins, 1976) uses the term **meme** as a deliberate analogy to the term gene, to denote cultural ideas that pass themselves, socially, from person to person according to their 'fitness' level (i.e. the benefits they convey or are believed to convey). Certain skills or ideas may be more valued by particular members of a group than others and these skills or ideas may be more likely to be passed on until surpassed by something more appropriate.

Culture in non-human species

Other species do not have the complexity of culture found in humans, but important similarities can nonetheless be found. To give one concrete example, Whiten, Horner, and de Waal (2005) taught two individual chimpanzees from two different groups one of two ways of obtaining food from the same apparatus – either by poking with a stick to remove a block, or using the stick to lift the block. When these individuals were released back into the group, almost all group members adopted the particular skill of their group rather than the other group. Even when subsequently a new way of getting the food was taught, individuals tended to conform to the social norm of their group.

Whiten and van Schaik (2007) described the evolution of culture in terms of a 'culture pyramid' consisting of four tiers. The use of a pyramid shape conveys the fact that those aspects of culture on the broader, lower tiers are more pervasive in the natural world than those higher up. At the lowest tier there is 'social information transfer' in which animals may, for example, learn from each other by

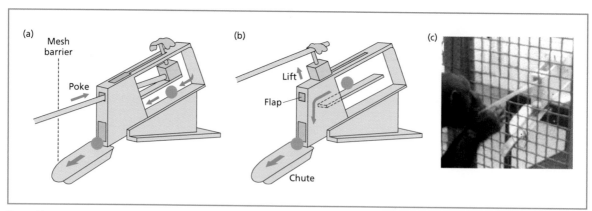

Two chimpanzees were taught how to obtain food from the same device either using a poke action (a, c) or a lift action (b). When introduced to two groups, other individuals learned using the conventional method and were resistant to change even when shown the alternative. This study demonstrates social learning of traditions and conformity to social norms once learned. From Whiten et al. (2005). Copyright © 2005 Nature Publishing Group. Reproduced with permission.

watching where they hide food or forage for food. This information tends to be used temporarily and then discarded. The next level up is termed **traditions**. A tradition is considered to be a distinctive pattern of behavior shared by two or more individuals in a social group. For example, the same species of bird in one location may have a different song structure from the same birds living in another location. The use of two ways of achieving the same reward, perpetuated within a group, is another example of a tradition as in the study of chimpanzees described above. The tier above traditions in the 'culture pyramid' consists of culture itself, which is construed here as a collection of traditions. Cumulative cultures are those in which traditions are gradually enhanced or modified over time, such as moving from stone-based to metal-based tools or from Roman to Arabic numbers.

The most basic level of the pyramid, social information transfer, is found in many mammals, birds, fish, and even invertebrates such as bees (Whiten & van Schaik, 2007). Evidence for traditions in a wide variety of species has not been collected. Whilst there are many examples of social learning and traditions in birds (e.g. Lefebvre & Bouchard, 2003), there is little evidence of multiple patterns of behavior being transmitted together. According to the criteria above, birds would have traditions but not culture. One potential exception is the New Caledonian crow (*Corvus moneduloides*), which exhibit two types of tool use involving both twigs and strips cut from leaves (e.g. Hunt & Gray, 2003). However, others have suggested that this may be an innate skill rather than socially learned (Kenward, Weir, Rutz, & Kacelnik, 2005). The most convincing examples of multiple traditions come from apes but there is evidence from some species of monkeys too. Perry et al. (2003) studied social traditions in capuchin monkeys (*Cebus capucinus*) involving a variety of social games, such as the 'toy game' (putting non-food objects in each other's mouths and removing them, taking-turns), the 'hand sniff game' (place another's hand or foot over own face and, with eyes closed, inhale deeply and repeatedly for more than 1 minute), and so on. These multiple traditions had unique distributions amongst different capuchin communities. This suggests that the games are culturally learned rather than part of their innate repertoire of behaviors. Evidence for cumulative culture, for example as evidenced by stepwise changes in tool development, is lacking in species other than humans (e.g. Tomasello, 1999).

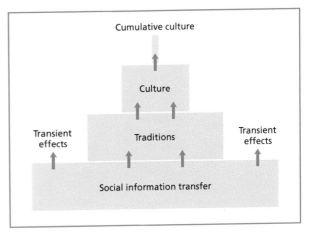

Whiten and van Schaik (2007) categorize socially learned behavior amongst different species into a 'culture pyramid'. Lower levels are more prevalent in nature, and higher levels develop from the lower levels. From Whiten and van Schaik (2007). The evolution of animal 'cultures' and social intelligence. *Philosophical Transactions of the Royal Society B, 362,* 603–620. Copyright © 2007 The Royal Society. Reproduced with permission.

Social learning versus imitation

Just because birds, dolphins, monkeys, apes, and humans (to name a few) are all capable of social learning, this does not mean that all have exactly the same mechanism of social learning in place. Humans certainly have one unique option available to them in that they can acquire traditions via language. Language aside, it has been suggested that the mechanisms of social learning in humans

KEY TERM

Biological anthropology
Study of the behavioral and anatomical evolution of the human species.

PREHISTORIC ORIGINS OF HUMAN CULTURE

Although comparisons between humans and our closest living relatives offer a window into social and cognitive evolution, another crucial line of evidence comes from comparisons with our now *extinct* ancestral relatives, in the field of **biological anthropology** (Leakey, 1994, offers an accessible review and the information here comes from this source unless otherwise stated). The evolutionary branch that separates the different species of humans that once existed from other great apes is characterized by bipedalism (walking upright). It was initially speculated by Darwin (1871) and others that bipedalism may have arisen due to a selection pressure for using the hands for tools – that is, those early apes that were better at using their hands for tools would have been more likely to survive and so this trait would be gradually enhanced over time. This view is no longer accepted. The first bipedal hominids emerged around 6–7 million years ago, whereas the first evidence of stone tools emerged only 2–3 million years ago. The emergence of stone tool use was associated with an evolutionary expansion in brain size, not bipedalism. The earliest bipedal hominids had small brains and probably had similar cognitive and social intellect to modern-day great apes.

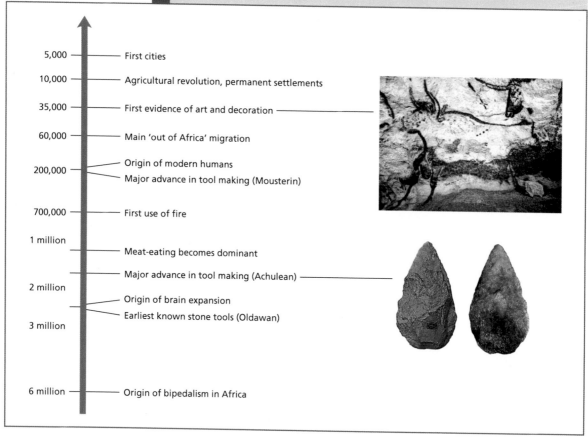

Time	Milestone
5,000	First cities
10,000	Agricultural revolution, permanent settlements
35,000	First evidence of art and decoration
60,000	Main 'out of Africa' migration
200,000	Origin of modern humans
	Major advance in tool making (Mousterin)
700,000	First use of fire
1 million	Meat-eating becomes dominant
	Major advance in tool making (Achulean)
2 million	Origin of brain expansion
	Earliest known stone tools (Oldawan)
3 million	
6 million	Origin of bipedalism in Africa

Some significant milestones in human evolution and human culture.

With the arrival of the big-brained Homo erectus around 2 million years ago, there was a major advance in stone tool manufacture and evidence of dietary change towards meat-eating (from the tooth fossil record). Both the increase in brain size and the need to cooperate when hunting for meat are likely to have been associated with a shift in social complexity. There is evidence of right-handedness in tool manufacture in this period, suggesting laterality changes in brain organization and possibly proto-language (Toth, 1985). The other great cultural skill in this period, aside from stone tools, was the use of fire around 700,000 years ago. This would have enabled cooking of meat and boiling of water, as well as offering warmth and protection. It is not unreasonable to imagine that those who were better at cooperating during hunting and those better at learning new skills (tools, fire) would have increased their survival chances, leading to further enhancement of these abilities in future generations.

Modern humans (Homo sapiens) did not emerge until around 200,000 years ago in Africa and this was accompanied by another major advance in tool manufacture. The most significant radiation out of Africa, from which all other present-day humans descend, was as recent as 60,000 years ago (Forster, 2004). For many prominent anthropologists, the most significant human cultural change cannot be linked to a genetic modification but rather to a change in lifestyle that occurred independently in several places in the world from 10,000 years ago – namely, the shift from hunting–gathering to permanent settlements (Mithen, 2007; Renfrew, 2007). This would have led to a more diverse division of labor, requiring greater organization and institutional control. It would have created opportunities for trade and also a greater reliance on material culture than that found in portable communities in which the material world had to be transported or created anew in each location.

are fundamentally different to other species. According to some, only humans are capable of social learning based upon an understanding of the goals, intentions, and mental states of other individuals (Penn & Povinelli, 2007; Tomasello, 1999). This type of social learning, termed **imitation**, is not straightforward to spot. The challenge lies in finding ways to observe, via behavior, the unobservable (i.e. mental states).

Imitation involves the understanding and reproduction of the actions of others and could be regarded as a more sophisticated form of social learning. It involves reproducing the goals of the other person and this is likely to entail an understanding of his/her intentions. As such, sociocognitive mechanisms as well as sensorimotor mechanisms are implicated in imitation. In this view, imitation is regarded as one of several forms of social learning (see Heyes & Galef, 1996). Non-imitative social learning, by contrast, could arise from a number of different mechanisms, including:

- Copying the action without understanding the goal of the action (also called mimicking). An example would be 'talking' parrots.

KEY TERM

Imitation
Social learning based on an understanding of the goals, intentions, and mental states of other individuals.

- **Stimulus enhancement** or **local enhancement**. Having another individual draw attention to an object or location may increase the likelihood that the observer will engage with that object/location, and that he/she will then learn to perform the action. In this view, each individual engages in self-discovery and the social element of learning is limited to drawing attention to certain important features in the environment.
- **Contagion.** This refers to the repetition of behaviors that are innate rather than learned, such as yawning and laughing. If you see someone else yawning you are also likely to yawn; and canned laughter on comedy shows can lead to more smiling and laughter (Provine, 1996).

In humans there is evidence for 'true' imitation. For instance, if asked to reproduce a complex action sequence, participants often reproduce the end-state but not the means to the end (Wohlschlager, Gattis, & Bekkering, 2003). Even human infants show evidence of goal-based imitation (Gergely, Bekkering, & Kiraly, 2002). In this study, the infants watched an adult press a button on a table by using their forehead. In one condition, the adult's hands and arms are bound up under a blanket and in the other condition the adult's hands are free. When the adult's hands are free, the infants copy the action directly – they use their foreheads too. But when the adult's hands are not free the infants imitate the goal but not the action (i.e. the infants use their hands rather than their head). The implication is that the infants understand that the goal of the action is to press the button and they assume that the adult would have used his/her hands if they had been free. This is often called 'taking the **intentional stance**', in that it involves attributing intentions to another person to account for their actions (Dennett, 1983).

What about other primates? According to Tomasello (1999) cultural traditions such as washing sand off potatoes in a nearby stream by Japanese macaque monkeys (Kawai, 1965; Kawamura, 1959) may arise via self-discovery facilitated by stimulus enhancement and location enhancement. When using tools, Tomasello (1999) argued that chimpanzees use trial-and-error learning to achieve a goal, rather than imitation. For example, when watching another chimp being rewarded with food (e.g. after poking a stick in a hole), he/she will attempt to get a reward too, but without necessarily inferring that the other chimp *knew* there was food in the hole or that the other

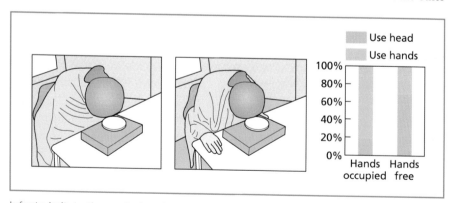

Infants imitate the goal of actions, rather than the motor aspects of actions. If the experimenter presses a button with his/her head because their arms are occupied, the infants 'copy' the action by using their hands rather than heads – they appear to infer that the experimenter would have used his/her hands to achieve the goal if they had been free. Drawing based on Gergely et al. (2002).

chimp *intended* to get it. One problem that arises from linking imitation closely with inferring of mental states (or 'theory of mind') is that whereas imitation emerges in humans in the first year of life, accurate performance on most tests of reasoning about mental states emerges between 3 to 4 years. The problem could be resolved by arguing, as Tomasello (1999) did, for a distinction between conscious attribution of mental states (e.g. as assessed in many theory-of-mind tasks) and unconscious goal attribution required for imitation. This idea is discussed more extensively in Chapter 11.

Other evidence suggests that chimpanzees and other apes are capable of 'true' imitation and some former skeptics have now changed their position (Call & Tomasello, 2008). Chimpanzees (Custance, Whiten, & Bard, 1995) but not macaque monkeys (Mitchell & Anderson, 1993) are capable of learning a 'do-as-I-do' game to produce complex actions (e.g. grab thumb of other hand). This does require considerable training (humans do not need to be trained to imitate) but this research represents a 'proof of principle' that arbitrary acts can be imitated by non-human apes. Buttelmann, Carpenter, Call, and Tomasello (2007) adapted the human infant study of Gergely et al. (2002) in which actions were performed with an unusual body part. When the hands are occupied by the demonstrator then the monkeys did use their hands, implying that not only did they understand the goal but they also understood why the other person did not use their hands. Similarly, Horner and Whiten (2005) studied imitation in young chimpanzees. They observed a familiar person ram a stick several times into a hole in the top of a box, and then insert the stick into a front hole in order to extract a food reward. In one condition, the top of the box was transparent and it could be seen that the first stage was meaningless (i.e. the top hole was not connected to the reward). In another condition, the top of the box was covered except for the hole. Young chimpanzees in the transparent condition omitted the first step and went straight for the reward by putting the stick in the front. Young chimpanzees in the covered-box condition performed both steps. It suggests that the chimpanzees in this task are imitating, based on an understanding of goals and perhaps intentions.

If human imitation and ape imitation are cognitively equivalent, as many researchers now agree, one then needs to ask why human imitation (and culture) is far more prolific than that found in apes? One possibility is human ability in other domains. If humans are more creative and innovative (and the evidence suggests so) then there could simply be more things that are worth imitating. Another possibility is that there are different rewards to imitation in humans versus apes. Apes may imitate in order to obtain a material reward, such as food. Humans may imitate each other because imitating, and being imitated, is a reward in itself. As such, imitation may serve to bind human social groups together in ways that are less apparent in other species (Dijksterhuis, 2005). As the proverb goes, 'imitation is the sincerest form of flattery' and there may be truth in that.

Evaluation

Natural selection, brought about by variations in the gene pool, enables species to adapt slowly to their environments. However, humans and other species are also able to adapt to their environments via a much faster mechanism – social learning. When coupled with innovation and other cognitive skills, it enables complex systems of culture to evolve a 'life of their own' insofar as they are modified over time and come and go according to how useful they are. These cultural traditions expand our cognitive capacities (as described in the next section) and physical capacities (through tools and technology), and provide a means for establishing group and individual identities.

MATERIAL SYMBOLS: NEURONAL RECYCLING AND EXTENDED COGNITION

Although language may have been selected for by evolutionary pressures, humans have created a wide range of material symbols that are products of culture – including writing, number systems, and art. For example, writing was first invented about 5000–6000 years ago by the Babylonians and was a skill possessed by a minority of humans until quite recently. One interesting question is how the brain is able to adapt to incorporate such information. For example, to what extent do different cultural manifestations of these symbols (e.g. in different writing systems) lead to different brain-based solutions? One might imagine that a purely cultural invention, such as literacy, may end up using different brain circuits in different people if, for example, we imagine that our brains are highly plastic, such that any new information can be slotted into any under-used region. However, this does not appear to be the case and the neural circuits for writing and calculation appear to be quite conserved across individuals and across cultures (Dehaene & Cohen, 2007). Dehaene and Cohen (2007) refer to this as 'neuronal recycling'. Their assumption is that neural resources, set aside for other functions in our evolutionary past, may be recruited by cultural knowledge.

Reading involves a number of cognitive capacities: for recognizing written words, for translating these words into speech, and for understanding the meaning of the words. In their review, Dehaene and Cohen (2007) concentrate particularly on the system for recognizing written words and on a region in the left ventral visual stream termed the **visual word form area (VWFA)**. This region responds to visual presentation of letter strings more than other objects, including made-up letters (so-called false fonts) and letters from unfamiliar writing systems (Cohen et al., 2002). A number of cross-cultural studies now show that the same region is activated by Roman script, Chinese characters, and Japanese Kana and Kanji (Bolger, Perfetti, & Schneider, 2005). Dehaene and Cohen (2007) speculate that this region may have evolved for certain types of object recognition. They suggest, for example, that the common developmental confusion between b/d stems from the fact that object recognition systems tend to treat mirror-images as the same (e.g. a cup is a cup irrespective of where the handle is) whereas this is not true for letters.

Similarly for numerical cognition, there is a region in the parietal lobes (intraparietal sulcus) that responds during arithmetic tasks, and when viewing different types of numerical symbols (digits, dot patterns, number words) both within (e.g. Piazza, Izard, Pinel, Le Bihan, & Dehaene, 2004) and across cultures (Tang et al., 2006). This may represent a core semantic representation of number (i.e. an approximate code for 'how many?'). This basic system may not only be cross-cultural but may also exist in other species. For example, monkeys contain neurons in this region that respond to different numbers of objects such as dots in an array (Nieder, 2005). However, humans can augment this basic ability through the additional use of numerical symbols (represented in other regions of the brain) such as written digits, number names, tallies, and so on, which extends their numerical abilities beyond other species (Dehaene, Dehaene-Lambertz, & Cohen, 1998) – that is, cognition itself is transformed by the availability of certain culturally learned symbols. For example, cultures that lack number words for numbers above four (using a term corresponding to 'many' for all quantities greater than four) appear to have some difficulties in understanding large exact quantities but can understand large approximate quantities (Pica, Lemer, Izard, & Dehaene, 2004). For example, if asked to add together 5 stones with 7 stones they may choose an answer that is approximately 12 (e.g. 11, 12, or 13) but are less likely to choose a more distant number (e.g. 8 or 20). Similarly, even in highly numerate cultures, certain forms of higher maths (e.g. algebra, multi-digit calculations) can be performed with ease using pen and paper (or calculator and computer) because these systems effectively function as externalized working memories, enabling humans to escape the confounds of our own limited capacity and error-prone memory systems. When viewed in this way, symbols and tools are quite literally 'mind expanding' (Clark, 2008). By offloading certain cognitive capacities (e.g. for remembering, calculating, reaching) onto external technology, it is claimed that we create an **extended cognition** that bridges the brain-based and material-based worlds (Clark, 2008).

Although systems of writing and number representation can be considered social in the narrow sense of having been invented and passed on by the collective action of many minds, they have had more direct influences on the nature of social interactions. The most obvious example is money. Indeed most of the earliest written records were for trade transactions rather than, say, poetry or stories. According to some contemporary thinkers (e.g. Lea & Webley, 2006), the function of money is essentially social. It may serve two broad social functions: as a means of social exchange (related to the notion of reciprocal altruism: 'if you scratch my back, I'll scratch yours') and also as a way of displaying or achieving a higher social standing through conspicuous consumption or benevolence. This is an interesting example of how our culture mirrors our biology.

Although certain cognitive abilities are shaped by biological evolution (e.g. the ability to judge the approximate number of items in an array), the cultural 'evolution' of ideas and symbols can extend this capacity. In this example, the numerical abilities of an Amazonian tribe (the Mundurukú) with no names for large numbers (but a generic name meaning 'many') is assessed on tasks involving putting different numbers of counters into a tin (for addition) and/or taking them out (for subtraction). It is still possible to perform approximate arithmetic without any words for large numbers, but not exact arithmetic with large numbers. Adapted from Pica et al. (2004).

CULTURAL SKILLS: TOOLS AND TECHNOLOGY

If culturally based symbols enable us to escape the constraints of our own minds (e.g. escaping the limits of our working memory), then cultural tools could be said to free us from the constraints of our own bodies. They enable us to fly (e.g. in an airplane or spacecraft) and they enable us to perform extraordinary feats of strength (e.g. chopping trees, killing larger animals). The human body is not adapted for flying or chopping trees but our brains are adapted to create useful objects (**tools**) and transmit this information, socially, from person to person. As noted above, some have argued that cultural use of symbols and tools enables new kinds of thought. For instance, Clark (2003) dismisses the notion, popular in evolutionary psychology, that modern-day humans are stuck with the Stone Age minds that were selected for in our earliest ancestors. For Clark (2003) we are 'natural-born cyborgs' capable of soaking up and creating complex technologies. The technology and the ideas behind them are themselves passed from person to person (and modified over time), not in the genes but by social and cultural transmission. But, crucially for his argument, in taking on such technology our minds and brains are themselves transformed.

Modifying the brain by using tools and technology

There are certain neurons in the brain that respond both when a particular body part is touched and when a visual stimulus is moved near to the same body part. These neurons are found in both frontal (Graziano, 1999) and parietal (Graziano, Cooke, & Taylor, 2000) regions and they can be said to be multi-sensory insofar as they receive input from more than one sensory system. They are normally studied in monkeys via the method of single-cell electrophysiology, which records how often a neuron 'fires' (i.e. produces an action potential) in response to a particular stimulus. If the neuron produces a large response (relative to some baseline, such as spontaneous activity), then it is concluded that the neuron codes information related to that stimulus. Some of

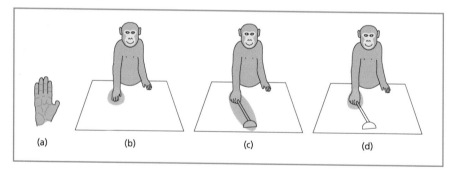

(a)   (b)   (c)   (d)

Some neurons respond to touch to the hand (a) and also to the sight of a visual stimulus near the hand, shown in green (b). After the monkey has been trained to use a tool the neuron may also respond to visual stimuli along the length of the tool (c). It is as if the tool is an extension of the body. Merely holding a tool does not have this effect (d). Note that (c) and (d) are physically equivalent but have very different neural responses. From Iriki, A., & Sakura, O. (2008). The neuroscience of primate intellectual evolution: Natural selection and passive and intentional niche construction. *Philosophical Transactions of the Royal Society B, 363*, 2229–2241. Reproduced with permission.

these multi-sensory neurons might fire when the monkey's hand is touched, even if the hand cannot be seen, and even when the hand is moved around in space. These same neurons also fire when a visual stimulus is placed on or near the hand, again irrespective of where the hand is. The region of space that elicits a neuronal response is termed the neuron's **receptive field**, and in this example we can say that the receptive field is centered on the hand rather than being at some fixed coordinate relative to the eyes.

Iriki, Tanaka, and Iwamura (1996) noted that the visual receptive fields of these neurons changed as a result of the monkey using a tool (a rake for getting peanuts out of reach). As a result of using the tool, the receptive field was no longer centered on the arm but was elongated down the length of the tool itself. It was as if the monkey's neural representation of its body had been stretched to incorporate the tool. There was also an important control condition in which the monkey passively held the tool but did not use it. In this condition, the receptive field was not extended. This control condition physically resembles the tool-use condition but is cognitively equivalent to the no-tool-use condition.

In humans there is evidence that multi-sensory processing of space is extended by tool use. When sighted people are trained to use a blind-person's cane there is evidence that multi-sensory space becomes expanded along the length of the cane (Serino, Bassolino, Farne, & Ladavas, 2007). A sound emanating near the end of the cane (1.25 m away) can facilitate detection of a weak tactile stimulus applied to the hand. However, when the sighted person passively holds the cane (without using it as a tool) this does not occur. Thus, their brains temporarily adapt to cane use. Blind people who have extensive experience of cane use show evidence of extended body space even from passive holding, suggesting that their brains have undergone more permanent adaptation. In the visual, rather than auditory, domain flashes of light both near the hand and at the end of a tool can facilitate detection of a tactile stimulus on the hand after tool use (Holmes, Calvert, & Spence, 2007). Prior to tool use, only a flash of light on or near the hand (but not the end of the tool) has this effect.

Having considered one example of how an individual's brain may be modified via tool use, I shall go on to consider how this process may spread, at the neural level, via imitation.

Mirror neurons and imitation

One of the most fascinating discoveries in cognitive neuroscience over the last decade has been of the **mirror neuron** system. Rizzolatti and colleagues found a group

KEY TERMS

Receptive field
The region of space that elicits a neuronal response.

Mirror neurons
Neurons that respond to both self-initiated actions and the observed actions of others.

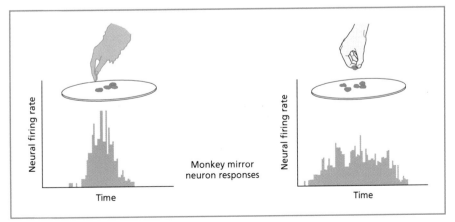

Monkey mirror neuron responses

Mirror neurons respond both when the animal performs an action (left) and when the animal sees someone else perform the action (other). This self–other similarity, operating at the neural level, has been very influential in social neuroscience, including for theories of social learning and imitation. From Rizzolatti et al. (2006). Reproduced with permission from Lucy Reading-Ikkanda for Scientific American Magazine.

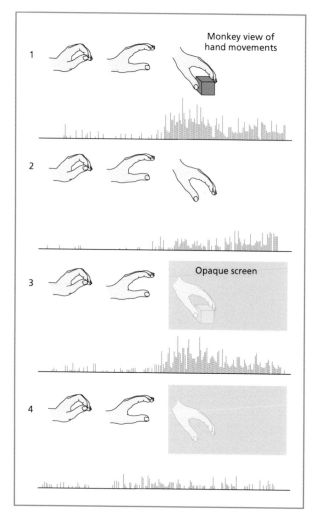

Monkey view of hand movements

Opaque screen

Mirror neurons respond to *inferred* goal-directed actions as well as those observed. In this example: (1) the monkey *sees* a goal-directed action to an object; (2) the monkey *sees* the same action but without an object; (3) the monkey *knows* that the object is there, because it has previously seen it, even though it cannot see it now; and (4) the monkey knows the object is not there. Note that conditions (3) and (4) are visually identical but only in condition (3) does the mirror neuron respond. The hand and object in (3) and (4) cannot be seen, but are drawn here for illustrative purposes. From Rizzolatti et al. (2006). Reproduced with permission from Lucy Reading-Ikkanda for Scientific American Magazine.

of neurons in the monkey premotor cortex (area F5) that respond both during the performance *and* during observation of the same action (e.g. di Pellegrino, Fadiga, Fogassi, Gallese, & Rizzolatti, 1992; Rizzolatti, Fadiga, Fogassi, & Gallese, 1996). Thus, the response properties of mirror neurons disregard the distinction between self and other, and this may provide a crucial basis for imitation. They respond to actions performed by the experimenter or another monkey as well as to actions performed themselves. The response properties of these neurons are quite specific. They are often tuned to precise actions (e.g. tearing, twisting, grasping) that are goal-directed. They do *not* respond to mimicked action in the absence of an object, or if the object moves robotically without an external agent. This suggests that it is the purposeful nature of the action rather than the visual/motoric elements that is critical. As such, mirror neurons have been likened to 'intention detectors' (Iacoboni et al., 2005). Mirror neurons have also been reported in the parietal lobes (Fogassi et al., 2005). By contrast, other regions such as the superior temporal sulcus also respond to specific movements of body parts but have a purely visual component (Perrett et al., 1989).

Moreover, mirror neurons respond if an appropriate action is implied as well as directly observed. Umilta and colleagues compared viewing of a whole action versus viewing of the same action in which a critical part (the hand–object interaction) was obscured by a screen (Umilta et al., 2001). These findings suggest that the premotor cortex contains abstract representations of action intentions that are used both for planning one's own actions and interpreting the actions of others.

The evidence above is derived from non-human primates. What is the evidence that humans possess such a system? The human analogue of area F5 is believed to be in Broca's area (specifically in Brodmann's area 44) extending into the premotor area (Rizzolatti, Fogassi, & Gallese, 2002). This region is activated by the observation of hand movements, particularly when imitation is required (Iacoboni et al., 1999), and also the observation of lip movements within the human repertoire (e.g. biting and speaking but not barking) (Buccino et al., 2004).

Why monkeys do not use tools

There is one potentially fatal flaw in the story of mirror neurons, imitation, and tool use – namely, that mirror neurons are assumed to be present in monkeys, chimpanzees, and humans but evidence for tool use in the wild is virtually non-existent for monkeys, common in chimpanzees, and extensive in humans. Macaque monkeys are able to perform some kinds of imitation-like behavior. For instance, neonate monkeys reproduce basic facial gestures such as tongue protrusion and mouth opening (Ferrari et al., 2006). Within the taxonomy presented above, this could perhaps reflect contagion (of innate motor programs) rather than imitation based on goals. Chimpanzees are capable of more complex forms of imitation (see above), but it perhaps does not serve the same social functions as imitation in humans. So what is missing between a macaque with mirror neurons but minimal imitation, a chimp with mirror neurons and some evidence of imitation, and humans with mirror neurons and boundless imitation?

In a number of recent papers, Iriki and colleagues have suggested a potential solution to this problem (Iriki, 2006; Iriki & Sakura, 2008). Under their account, mirror neurons may be a necessary precursor to imitative tool use but are not suf-ficient (see also Rizzolatti, 2005). First of all, macaque monkeys *can* use tools in the laboratory but only after extensive training. It takes a minimum of 10–14 days of intensive training using a specially developed training regime (e.g. Hihara et al., 2006). This training regime involves systems of reward and gradual modification of behavior rather than imitation or social learning of tool use. This long length of time plus the nature of the training may be sufficient to prevent all but very minimal tool use in the wild by these monkeys (e.g. bending a branch to get a fruit at the end) but, as a proof of principle, they can use tools. Moreover, when they have achieved it their performance is swift and effortless and shows some degree of flexibility. Without hesitation, the monkeys can use a short rake to pull a long rake to get a more distant food (i.e. a chaining process of successive tools) (Hihara, Obayashi, Tanaka, & Iriki, 2003). The key question is what are the differences in the brains of macaque mon-keys who have acquired tool use versus those that have not? The answer, according to Iriki and Sakura (2008), lies in the way that two particular regions are connected. In monkeys who are proficient tool-users there are extra connections between the intraparietal sulcus and the temporo-parietal junction that are absent in monkeys who cannot use tools (Hihara et al., 2006). The intraparietal region contains neurons whose visual receptive fields are extended via tool use and also mirror neurons. The temporo-parietal junction, in humans, has been implicated both in theory of mind (Frith & Frith, 2003) and in feelings of embodiment, for instance when contrasting physical perspectives between self and other (Blanke et al., 2005). Changes in gene expression in the intraparietal region accompany learning of tool use (Ishibashi et al., 2002) and presumably trigger the connectivity changes. The implication of this finding, in evolutionary terms, is that the human brain may have evolved (via genetic modification) stable connections between these two regions that are normally absent in many other primates. This may enable humans to link neural mechanisms related to tool use (e.g. multi-sensory visuo-tactile neurons) with mechanisms related more closely to social cognition (perspective taking, theory of mind). The question of whether this adaptation is specific to tool use or a response to more general evolu-tionary pressures is unknown.

Evaluation

Tool use is a particularly interesting example of culture for a number of reasons. Firstly, there are obvious parallels between human tool use and that found in other

This monkey has been trained to use a single tool to reach a food reward. However, when the tool is too short (purple) but can be used to reach a longer tool (green) the monkey – without further training – is able to use the short tool to get the long tool to get the food reward. From Iriki, A., & Sakura, O. (2008). The neuroscience of primate intellectual evolution: Natural selection and passive and intentional niche construction. *Philosophical Transactions of the Royal Society B, 363,* 2229–2241. Reproduced with permission.

species. However, there are important differences too. Secondly, tool use is now beginning to be understood in terms of basic neuroscience, offering the possibility of linking together different levels of explanation from genes through to the behavior of individuals and groups. Mirror neurons may be an important starting point for imitation and the spread of tools and technology, but recent evidence suggests that other aspects of brain function are crucial too.

SUMMARY AND KEY POINTS OF THE CHAPTER

- The social intelligence hypothesis argues that evolutionary pressures to be socially smarter lead to more general changes (e.g. larger brain size) resulting in increased intellect in non-social domains. Evidence for this comes from the correlation between relative brain size in different primates and factors such as the size of social groups, the degree of deception, and amount of social learning.
- It remains keenly debated whether the need to deal with social complexity led to general changes linked to intellect (e.g. bigger brains) or shaped intelligence in a more precise way (e.g. by creating specialized neural circuits to represent the social world).
- Culture can be defined as a shared set of values, skills, artifacts, and beliefs amongst a group of individuals. Many species have elements of culture, including social learning (e.g. of food caches) and traditions (e.g. different tribes of chimpanzee use stick tools in different ways).
- There are likely to be different mechanisms for social learning. Humans, and possibly some other primates, may learn from each other by inferring intentions (i.e. imitation).
- Mirror neurons respond both when an animal performs an action and when it sees someone else performing the action. Their response

depends on the goal rather than the movement per se and it has therefore been suggested that it provides the neural basis of imitation.

- Monkeys, who are known to have mirror neurons, do not necessarily use tools in the wild and they show limited evidence for spontaneous imitation. This suggests that mirror neurons may be necessary but not sufficient for imitative tool use by animals.

EXAMPLE ESSAY QUESTIONS

- What is the 'social intelligence hypothesis' and what is the evidence for and against it?
- Do non-human animals have culture?
- Modern human culture is too recent to have been shaped by genetic evolution, so what kind of primitive mental capacities have made it possible?
- What kind of neural and cognitive mechanisms enable imitation by humans and other animals?

RECOMMENDED FURTHER READING

- Emery, N., Clayton, N., & Frith, C. (2008). *Social intelligence: From brain to culture.* Oxford: Oxford University Press. This extensive collection of papers was originally published in *Philosophical Transactions of the Royal Society B* (2007, vol. 362, pp. 485–754).

- Hurley, S., & Chater, N. (2005). *Perspectives on Imitation: From Neuroscience to Social Science.* Cambridge, MA: MIT Press. A collection of essays in two volumes dealing with both humans and other animals.

CHAPTER 4

CONTENTS

Historical perspectives on the emotions   72

Different categories of emotion in the brain   78

Motivation: Rewards and punishment, pleasure and pain   88

Summary and key points of the chapter   98

Example essay questions   99

Recommended further reading   99

Emotion and motivation

The classic science fiction depiction of androids such as C3PO in *Star Wars* and Data in *Star Trek: Next Generation* is of super-human intelligent beings able to speak many languages and store vast amounts of information. Nevertheless, such intelligence does not enable them to fully understand the eccentric behaviors of their human colleagues who constantly place themselves in danger, fall in love, and tell jokes. These androids lack **emotions**. Reading between the lines of these popular depictions we might conclude that: emotions are what 'makes us human'; that we could probably do without emotions if we were to be redesigned from scratch; and that emotions force us to make decisions that are, in some sense, illogical. Needless to say, these conclusions are incorrect. Emotional processes have a long evolutionary history and are by no means unique to humans. What may 'make us human' is our ability to consciously reflect on our emotions and share them socially via our language and culture, but not our emotions per se. If a new organism were redesigned from scratch it would still be helpful to have early warning routines for danger and fast-acting mechanisms that prepare it to fight or flee. It would still be helpful to devote greater attention to stimuli that are necessary for survival. These are all considered functions of emotions. Finally, emotions do sometimes lead to decisions that may not have occurred via more deliberative reasoning but this, by itself, does not make them illogical. For instance, in cooperative games with another person we often make decisions based on social values of fairness rather than maximizing individual financial gain (e.g. Sanfey, Rilling, Aaronson, Nystron, & Cohen, 2003).

SOME CHARACTERISTICS OF EMOTIONS

- An emotion is a state associated with stimuli that are **rewarding** (i.e. that one works to obtain) or **punishing** (i.e. that one works to avoid). These stimuli often have inherent survival value.
- Emotions are transient in nature (unlike a **mood**, which is where an emotional state becomes extended over time), although the emotional status of stimuli is stored in long-term memory.
- An emotional stimulus directs attention to itself, to enable more detailed evaluation or to prompt a response.
- Emotions have a **hedonic value**, that is, they are subjectively liked or disliked.
- Emotions have a particular 'feeling state' in terms of an *internal* bodily response (e.g. sweating, heart rate, hormone secretion).
- Emotions elicit particular *external* motor outcomes in the face and body, which include emotional **expressions**. These may prepare the organism (e.g. for fighting) and send signals to others (e.g. that one intends to fight).

KEY TERMS

Emotions
States associated with stimuli that are rewarding or punishing.

Reward
An outcome that one is willing to work to obtain.

Punishment
An outcome that one is willing to work to avoid.

Mood
An emotional state that is extended over time (e.g. anxiety is a mood and fear is an emotion).

Hedonic value
The subjective liking or disliking of a stimulus/event.

Expressions
External motor outcomes in the face and body associated with emotional states.

This is not necessarily illogical – there may be good survival reasons, honed by evolution, that promote such cooperation. For humans, many social stimuli and situations are rewarding (e.g. imitation, cooperation) or punishing (e.g. social exclusion). As such, both social stimuli and non-social stimuli are likely to have been selected as having survival value in our evolutionary past.

An emotion can be regarded as a state that can have various facets – conscious and unconscious; internal and external; automatic or controlled. The precise nature of the state may vary according to the stimulus, learned history, and current context. An emotional stimulus may also affect processing in other more basic cognitive mechanisms – for example, by making a memory more memorable (e.g. Cahill, Prins, Weber, & McGaugh, 1994) and by directing attention to certain objects or locations (e.g. Vuilleumier, 2005). As such, a theory of the 'neuroscience of emotions' is likely to entail a range of different interacting brain processes, in the same way as contemporary theories of vision divide processing amongst routines specialized for shape, color and motion, visually guided action, and so on.

Why are some stimuli associated with emotions and others are not? The standard answer to this question is that some stimuli are more important than others (e.g. because they enhance or threaten survival chances). Emotions are one way of tagging these stimuli to ensure that they receive priority treatment and are responded to appropriately. Broadly speaking, they can be tagged in one of two ways: either as something that is to be sought (i.e. a rewarding stimulus) or avoided (i.e. a punishing stimulus). As such, many theories closely tie emotions with the concept of **motivation**. For example, Rolls (2005) defines emotions as states elicited by rewards and punishers, whereas motivation is defined as states in which rewards are sought and punishers are avoided. Importantly, emotions are not just tied to stimuli but also to predicted stimuli. Thus, the omission of an expected reward can lead to emotions (e.g. anger), as can omissions of expected punishment (e.g. relief). For humans, we can make the further claim that we *like* rewards and we *dislike* punishers, and that we are motivated to seek the things we like and avoid the things we dislike. (For animals, we tend to avoid the terms like/dislike and adopt more neutral terminology such as seek/avoid because we cannot know their subjective feelings.) Although we may be born with a core set of basic likes and dislikes (e.g. we like sweet things and dislike pain), it is possible to arbitrarily learn new emotional associations by pairing neutral stimuli with emotive responses. We may come to be afraid of flying in airplanes, or we may come to like certain painful stimuli (e.g. eating chillies, fetishes). As such, emotional learning is a highly flexible system that is not limited to stimuli in our evolutionary past and extends beyond stimuli with obvious survival value.

This chapter will first consider different historical accounts of emotion. It will then consider whether or not there are discrete 'basic emotions' in the brain. Finally, it will go on to consider the role of emotions in motivation and goal-directed behavior. Throughout, I will present examples of emotional processes in social and non-social contexts and consider how they may be related.

HISTORICAL PERSPECTIVES ON THE EMOTIONS

Darwin's evolutionary theory of emotion

In 1872, Charles Darwin published 'The Expression of the Emotions in Man and Animals' (Darwin, 1872/1965). For much of this work Darwin was concerned with

EMOTION AND THE 'RIGHT BRAIN'

This notion of the right hemisphere being more emotional than the left has had an enduring influence on popular scientific views of the brain. Following Broca's discovery that language is a predominantly left-hemisphere faculty, nineteenth-century neurologists speculated on possible complementary specializations for the right hemisphere – emotion being one of them. However, this 'right brain hypothesis' of emotions is essentially incorrect.

Contemporary theories concerning the neural basis of emotions assume that both hemispheres are crucially involved in emotional processing. However, there is evidence of subtle laterality differences between the cerebral hemispheres. When *producing* emotional expressions, there is evidence from muscle recordings that the left side of the face (controlled by the right hemisphere) is more expressive than the right side of the face (Dimberg & Petterson, 2000). This was found for both a positive expression (smiling) and a negative one (anger). When *recognizing* emotional expressions, there is evidence that the valence of the emotion is important. The left side of a face is judged to be sadder and the right side happier (Nicholls, Ellis, Clement, & Yoshino, 2004). This implies a right-hemisphere bias for sadness recognition and a left-hemisphere bias for happiness recognition. Note that in these examples, both hemispheres are implicated in producing and recognizing emotions but one hemisphere may have a small relative advantage over the other.

documenting the outward manifestations of emotions – expressions – in which animals produce facial and bodily gestures that characterize a particular emotion such as fear, anger, or happiness. Darwin noted how many expressions are conserved across species: anger involves a direct gaze with mouth opened and teeth visible, and so on. He claimed that such expressions are innate 'that is, have not been learnt by the individual'. Moreover, such expressions enable one animal to interpret the emotional state of another animal – for example, whether an animal is likely to attack, or is likely to welcome a sexual advance. For Darwin, an emotional expression was

Darwin argued that many emotional expressions have been conserved by evolution.

a true reflection of an inner state: 'They reveal the thoughts and intentions of others more clearly than do words, which may be falsified.'

Darwin's contribution was to provide preliminary evidence as to how emotions may be conserved across species. His reliance on expressions resonates with some contemporary approaches, such as Ekman's attempts to define 'basic' emotions from cross-cultural comparisons of facial expressions (e.g. Ekman, Friesen, & Ellsworth, 1972). This is covered in detail later in the chapter.

Freud and unconscious emotional motivations

For Freud, our minds could be divided into three different kinds of mechanisms: the id, the ego, and the super-ego. The **id** was concerned with representing our 'primitive' urges that connect us to non-human ancestry. It includes motivations to meet our basic emotional needs for sex, food, warmth, and so on. The id was concerned with unconscious motivations but these ideas would sometimes be consciously accessible via the **ego**, which operates according to reason rather than passion. The **super-ego**, by contrast, represents the ideal self, such as our cultural norms and our aspirations. For Freud (and many of his clients), there was a perceived conflict between the super-ego, for which sexual behavior was tightly regulated by cultural norms, and the id, for which more unbridled sexual impulses were considered as desirable.

The specific details of Freud's theory no longer have contemporary currency. Many of his ideas (e.g. relating to childhood sexual fantasies) were derived from anecdotes and speculation rather than scientific testing. However, the basic idea that emotions are an unconscious bias in our behavior is very much relevant. For example, simple emotional reactions can be elicited from stimuli that are presented too briefly to be consciously seen (Tamietto & De Gelder, 2010). Most cognitive models of emotions assume that the majority of emotional processing occurs unconsciously. Freud's other enduring influence is the notion that many psychiatric disorders can be understood as emotional disturbances. Freud was particularly interested in neuroses, or what would now be called anxiety disorders, and today many of these are understood as emotional disturbances (e.g. Le Doux, 1996).

The James–Lange theory

One of the founding fathers of psychology, William James, proposed a theory of emotion that placed the somatic (i.e. bodily) response of the perceiver at its center (James, 1884). This theory later became known as the **James–Lange theory** of emotion. According to this theory, it is the self-perception of bodily changes that produces emotional experience. Thus, changes in bodily state precede the emotional experience rather than the other way around. We feel sad because we cry, rather than we cry because we feel sad. This perspective seems somewhat radical compared to the contemporary point of view elaborated thus far. For instance, it raises the question of what type of processing leads to the change in bodily states and whether or not this early process could itself be construed as a part of the emotion. Changes in the body are mediated by the autonomic nervous system (ANS), a set of nerves located in the body that controls activity of the internal organs.

There is good empirical evidence to suggest that changes in somatic state, in themselves, are not sufficient to produce an emotion. Schacter and Singer (1962) injected participants with epinephrine (also termed adrenaline), a drug

KEY TERM

Id
Unconscious motivations that represent 'primitive' urges from our non-human ancestry (in Freudian theory).

Ego
The conscious self operating according to reason rather than passion (in Freudian theory).

Super-ego
The ideal self such as our cultural norms and our aspirations (in Freudian theory).

James–Lange theory
The perception of our own bodily changes produces emotional experience.

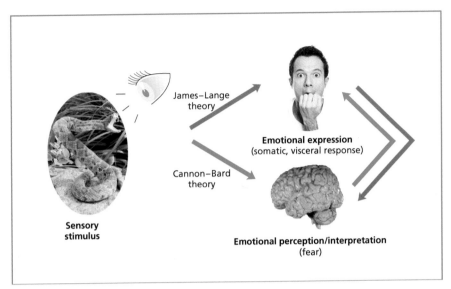

According to the James–Lange theory, bodily reactions occur first and emotional processing occurs after (as the perception/interpretation of those reactions). According to the Cannon–Bard theory, the emotional perception/ interpretation occurs first and the bodily reaction occurs after.

that induces autonomic and visceral changes. They found that the presence of the drug by itself did not lead to self-reported experiences of emotion, contrary to the James–Lange theory. However, in the presence of an appropriate cognitive setting (e.g. an angry or happy man enters the room), the participants did self-report an emotion. A cognitive setting, but without epinephrine, produced less intense emotional ratings. This study suggests that bodily experiences do not create emotions (contrary to the James–Lange theory) but they can enhance conscious emotional experiences.

There are several contemporary theories that bear similarity to the James–Lange theory, most notably Damasio's (1994) suggestion that bodily responses linked to emotions guide decision making. Although the James–Lange theory states that these bodily responses must be consciously perceived, Damasio (1994) takes the different view that they are unconscious modifiers of behavior.

The Cannon–Bard theory

The **Cannon–Bard theory** of emotions that emerged in the 1920s argued that bodily feedback could not account for the differences between the emotions (Cannon, 1927). According to this view, the emotions could be accounted for solely within the brain and bodily responses occur after the emotion itself. The Cannon–Bard theory was inspired by neurobiology. Earlier research had noted that animals still exhibit emotional expressions (e.g. of rage) after removal of the cortex. This was considered surprising given that it was known that cortical motor regions are needed to initiate most other movements (Fritsch & Hitzig, 1870). In a series of lesion studies, Cannon and Bard concluded that the hypothalamus is the centerpiece of emotions. They believed that the hypothalamus received and evaluated sensory inputs in terms of emotional content, and then sent signals to the autonomic system (to induce the bodily feelings discussed by James) and to the cortex (giving rise to conscious experiences of emotion).

Although it has not stood the test of time (e.g. the hypothalamus is not a central nexus of emotions, although it does regulate bodily homeostasis), the theory was important historically in providing an alternative to the James–Lange theory and also for the development of another important theory: namely the Papez circuit and the limbic brain hypothesis.

The Papez circuit and the limbic brain

Papez (1937) drew upon the work of Cannon and Bard in arguing that the hypothalamus was a key part of emotional processing, but extended this into a circuit of other regions that included the regions of the cingulate cortex, hippocampus, hypothalamus, and anterior nucleus of the thalamus. Papez argued that the feeling of emotions originated in the subcortical **Papez circuit**, which was hypothesized to be involved in visceral regulation. A second circuit, involving the cortex, was assumed to involve a deliberative analysis that retrieved memory associations about the stimulus. The work of MacLean (1949) extended this idea to incorporate regions such as the amygdala and orbitofrontal cortex, which he termed the 'limbic brain'. The different regions were hypothesized to work together to produce an integrated 'emotional brain'.

There are a number of reasons why these earlier neurobiological views are no longer endorsed by contemporary cognitive neuroscience. First, some of the key regions of the Papez circuit can no longer be considered to carry out functions that relate primarily to the emotions. For example, the role of the hippocampus in memory was not appreciated until the 1950s (e.g. Scoville & Milner, 1957). Second, contemporary research places greater emphasis on different types of emotion (e.g. fear versus disgust). Each basic emotion may form part of its own circuit, and different parts of the circuit may make different cognitive contributions.

A QUICK TOUR OF THE EMOTIONAL BRAIN

In this chapter, five regions of the brain are considered in detail and their basic architecture and functions are summarized below for reference.

The amygdala

- A collection of nuclei buried bilaterally in the anterior temporal poles.
- It receives connections mainly from the overlying temporal lobes. These correspond to higher stages of sensory processing and 'semantics', although there may be some inputs from early sensory areas (e.g. auditory inputs via medial geniculate nucleus; Le Doux, 1996). It has some subcortical inputs, including parts of the hippocampus, hypothalamus, and olfactory structures. Its outputs include the hypothalamus, ventral striatum, and temporal, orbitofrontal and insula cortex (for a review see Amaral, Price, Pitkanen, & Carmichael, 1992).
- Involved in learning the emotional value of stimuli (e.g. via classical conditioning, in which a neutral stimulus is paired with an emotion-evoking stimulus).

- In humans there is some evidence of a preferential involvement in fear perception, but recent evidence from functional imaging contradicts this view.

The insula

- A region of cortex lying beneath the temporal lobes.
- Connects anteriorly to the orbitofrontal cortex, limbic structures, and basal ganglia; the posterior region receives connections from sensory thalamus and parietal and temporal association cortex.
- Anterior portion of the insula is considered to be involved in interoceptive awareness (e.g. detection of heartbeat) and bodily feelings in general.
- Some evidence of a preferential involvement in disgust perception.

The anterior cingulate cortex

- Located around the anterior corpus callosum on the medial surface of the brain and often divided into dorsal regions and ventral regions.
- Connections from medial thalamic areas (concerned with pain perception), orbitofrontal cortex, amygdala, and insula. It has output connections to the periaqueductal gray area (linked to pain), dorsal motor nucleus of the vagus (elicits autonomic effects), and ventral striatum (for a review see Van Hoesen, Morecraft, & Vogt, 1993).
- Involved in the production of certain bodily responses elicited by an emotional stimulus, such as skin conductance response (Tranel & Damasio, 1995) and changes in heart rate and blood pressure (Critchley et al., 2003).
- Dorsal region involved in monitoring of responses, for instance in terms of whether a response is incorrect and possibly in terms of whether responses are rewarded or punished.
- Ventral region of cingulate is adjacent to medial prefrontal cortex region implicated in 'mentalizing' but the specific role of this cingulate region is not agreed upon.

The orbitofrontal cortex

- Located on the ventral surface of the frontal lobes, above the eye sockets (orbits).
- Receives connections from cortical sensory areas, and has reciprocal connections with areas such as the amygdala, hippocampus, insula, and cingulate cortex (Cavada, Company, Tejedor, Cruz-Rizzolo, & Reinoso-Suarez, 2000).
- Computes the motivational value of rewards (e.g. whether I would like chocolate now rather than whether I like chocolate per se), and changes the value of rewards according to context (e.g. reversal learning).

KEY TERM

Basic emotions
Different categories
of emotions assumed
to be independent of
culture and with their
own biological basis
(in terms of evolution
and neural substrate).

The ventral striatum

- Part of the basal ganglia and includes the nucleus accumbens.
- Involved in a 'limbic circuit' connecting the orbitofrontal cortex, basal ganglia, and thalamus.
- Important for operant conditioning, for example learning to press a lever when a certain tone is heard in order to obtain a reward.
- Responds to rewards and the anticipation of rewards. The latter has been used to argue that it computes a reward prediction error (i.e. the discrepancy between actual reward and expected reward).

DIFFERENT CATEGORIES OF EMOTION IN THE BRAIN

One of the first challenges faced by the empirical study of emotions is how to go about categorizing them or, indeed, whether it makes more sense to treat all emotions as a single entity. Are some types of emotion (e.g. happy, sad) more basic or primary than other types (e.g. love, jealousy)? Are emotional categories independent of language and culture?

Are some emotions more 'basic' than others?

One of the most influential ethnographic studies of the emotions concluded that there are six **basic emotions** that are independent of culture (Ekman & Friesen, 1976;

Ekman has argued
that there are six basic
emotions that manifest
themselves as universal
(i.e. cross-cultural) facial
expressions: happy, sad,
fear, anger, disgust, and
surprise. Can you match
the expression with the
emotion? Copyright ©
Paul Ekman. Reproduced
with permission.

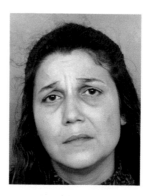

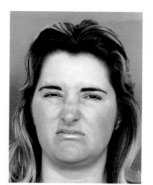
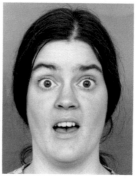
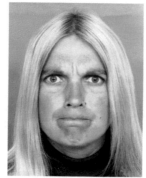

Ekman et al., 1972): happy, sad, disgust, anger, fear, and surprise. This study was based on comparisons of the way that facial expressions are categorized and posed across diverse cultures. Ekman (1992) considers other characteristics for classifying an emotion as 'basic' aside from universal facial expressions, such as: each emotion having its own specific neural basis; each emotion having evolved to deal with different survival problems; and occurring automatically. The list of emotions is not considered closed. For instance, Ekman (1992) considers adding embarrassment, awe, and excitement as basic emotions and dropping surprise. Johnson-Laird and Oatley (1992) examined the words that we have for emotions and came up with a list of five basic emotions that overlap closely with Ekman's but does not contain surprise.

What of other candidate emotions? One possibility is that different candidate emotions are different shades of the same basic emotion. For example, happiness might include amusement, relief, pride, satisfaction, and excitement (Ekman, 1992). Another possibility is to consider some emotions as being comprised of two or more basic emotions. Plutchik (1980) offers a detailed account along these lines. He proposes eight basic emotions (surprise, sadness, disgust, anger, anticipation, joy, acceptance, fear) that may be combined in various ways: for example, joy + fear = guilt and fear + surprise = alarm. A third possibility is that some emotions should be construed in terms of a basic emotion(s) plus a non-emotional cognitive appraisal. These cognitive + emotional blends might be needed to account for complex emotions such as jealousy, pride, embarrassment, and guilt. Such emotions might involve attribution of mental states that imply awareness of another person's attitude to oneself, or awareness of oneself in relation to other people. As such, they have been referred to as **moral emotions** (e.g. Haidt, 2003). Along these lines, Smith and Lazarus (1990) argue that pride, shame, and gratitude might be uniquely human emotions. Darwin (1872/1965) also believed that blushing (linked to shame or embarrassment) might be a uniquely human expression.

Not all models of emotion assume that some emotions are more basic. Two different accounts along these lines are considered here: that of Ortony and colleagues (Ortony, Clore, & Collins, 1988; Ortony & Turner, 1990) and that of Rolls (2005). Ortony and colleagues argue that *all* emotions are appraisals based on a valenced reaction (i.e. positive vs. negative) to a given stimulus and event. The range of emotions is limited by the range of appraisals that one can deploy, rather than consisting of some pre-determined number. These appraisals can occur unconsciously as well as consciously. For example, an emotion such as 'shame' would be an outcome of various appraisals such as: it has a negative valence; it refers to the action of people; and it is self-focused.

Rolls (2005) offers an account of different emotions that arise out of different aspects of reward and punishment but he does not assume a core set of basic emotions in the same way as many, indeed most, other theories do. Instead, he argues that different types of emotion emerge from a consideration of a small set of principles, including:

- Whether a reward or punishment is applied (e.g. pleasure vs. fear); whether a reward is taken away (e.g. anger) or a punishment is taken away (e.g. relief).
- The intensity of the above (e.g. rage, anger, sadness, or frustration) could be different emotional outcomes arising out of having a rewarding stimulus removed or unexpectedly not appearing.
- Different combinations of the above (e.g. guilt) may be a combination of reward and punishment learning.

• The context in which an emotional stimulus appears. For example, whether the stimulus is social or not (i.e. related to other people) may determine whether the emotion feels like love, anger, jealousy (emotions implying another agent) versus enjoyment, frustration, or sadness (emotions that need not imply another agent). Indeed the eliciting stimulus is considered part of the emotional state, so love for one person may be different to love for another person just because the individual is different.

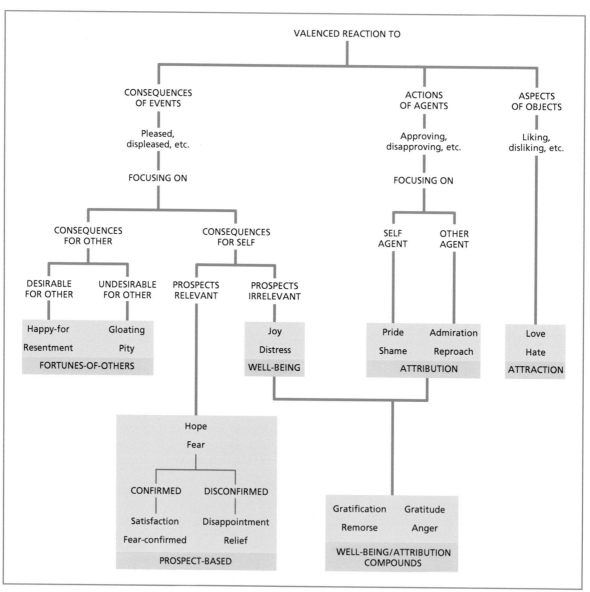

Researchers such as Ortony categorize emotions in terms of qualitatively different kinds of cognitive appraisal that may occur either consciously or unconsciously. From Clore and Ortony (2000). Copyright © 2000 Oxford University Press. Reproduced with permission.

The following sections will present evidence for and against the 'basic emotions' hypothesis, by considering whether different emotional categories have their own neural substrate (favoring the 'basic emotions' position) or not (favoring more distributed models of the emotions).

The amygdala and fear

The **amygdala** (from the Latin word for almond) is a small mass of gray matter that lies buried in the tip of the left and right temporal lobes. It lies to the front of the hippocampus and, like the hippocampus, is believed to be important for memory – particularly for the emotional content of memories (Richardson, Strange, & Dolan, 2004) and for learning whether a particular stimulus/response is rewarded or punished (Gaffan, 1992). In monkeys, bilateral lesions of the amygdala have been observed to produce a complex array of behaviors that have been termed the **Kluver–Bucy syndrome** (Kluver & Bucy, 1939; Weiskrantz, 1956). These behaviors include an unusual tameness and emotional blunting, a tendency to examine objects with the mouth, and dietary changes. This is explained in terms of objects losing their learned emotional value. The monkeys typically also lose their social standing. In humans, the effects of amygdala lesions are not as profound. This may reflect either a greater cortical influence on emotional and social behavior or the fact that the earlier monkey studies are likely to have produced lesions extending beyond the amygdala.

Evidence for the role of the amygdala in fear

The role of the amygdala in fear conditioning is well established (Le Doux, 1996; Phelps, 2006). If a stimulus that does not normally elicit a fear response, such as

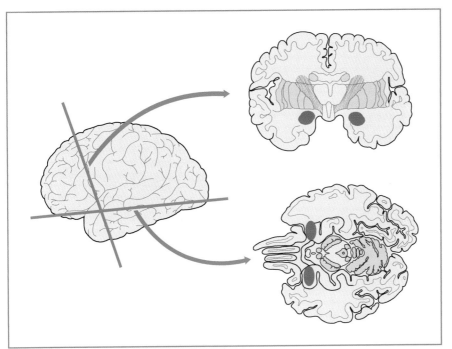

The amygdala lies buried in the tip of the temporal lobes, bilaterally.

Dave, Zeke, and Riva all had bilateral removal of the amygdala region (each animal was operated on in that order, 2 months apart) resulting in changes in the social hierarchy. The amygdala is involved in fear processing but also modulates the fight-or-flight response linked to aggressive behavior. From Rosvold, Mirsky, and Pribram (1954). Copyright © 1954 American Psychological Association. Reproduced with permission.

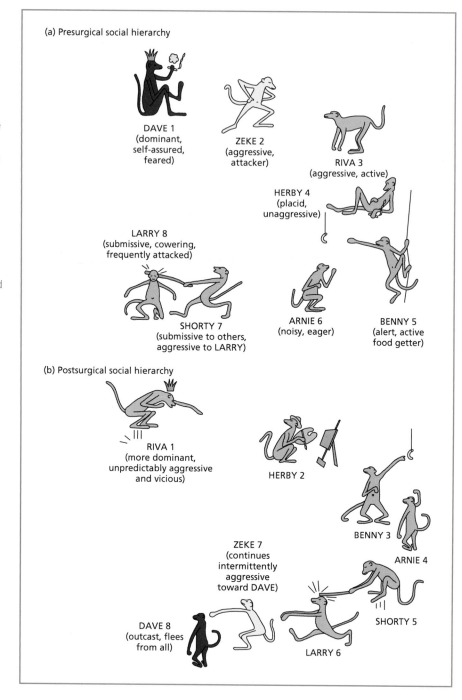

(a) Presurgical social hierarchy

DAVE 1
(dominant, self-assured, feared)

ZEKE 2
(aggressive, attacker)

RIVA 3
(aggressive, active)

HERBY 4
(placid, unaggressive)

LARRY 8
(submissive, cowering, frequently attacked)

SHORTY 7
(submissive to others, aggressive to LARRY)

ARNIE 6
(noisy, eager)

BENNY 5
(alert, active food getter)

(b) Postsurgical social hierarchy

RIVA 1
(more dominant, unpredictably aggressive and vicious)

HERBY 2

BENNY 3

ARNIE 4

ZEKE 7
(continues intermittently aggressive toward DAVE)

SHORTY 5

DAVE 8
(outcast, flees from all)

LARRY 6

an auditory tone (unconditioned stimulus, CS–), is paired with a stimulus that does normally evoke a fear response (termed conditioned response), such as an electric shock, then the tone will come to elicit a fear response by itself (it becomes a conditioned stimulus, CS+). If the amygdala is lesioned in mice (specifically the basolateral nucleus of the amygdala) then the animal does not show this learning, and if the lesion is performed after the animal has been trained then this learned association

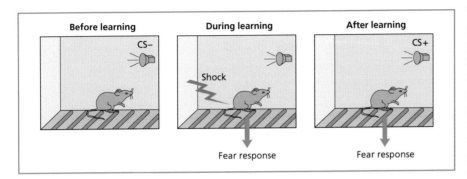

Before learning	During learning	After learning
CS−	Shock	CS+
	Fear response	Fear response

The basic procedure in fear conditioning involves presenting an initially neutral stimulus (the CS−, e.g. a tone) with a shock. After sufficient pairings, the stimulus will elicit a fear response without an accompanying shock (it has become a CS+).

is lost (e.g. Phillips & Le Doux, 1992) – that is, the amygdala is important for both learning and storing the conditioned fear response (although for a different view see Cahill, Weinberger, Roozendaal, & McGaugh, 1999). Single-cell recordings suggest that different cells within the amygdala could be involved in learning versus storage of the association (Repa et al., 2001). Animals with lesions to the amygdala still show a fear response to normal fear-evoking stimuli (such as shocks), which suggests that its role is in learning and storing the emotional status of stimuli that are initially emotionally neutral.

In humans, a comparison of learned fear responses to a shock (CS+) with neutral stimuli (CS−) reveals amygdala activation during fMRI that correlated with the degree of conditioned response, in this instance a skin conductance response (LaBar, Gatenby, Gore, Le Doux, & Phelps, 1998). Bechara et al. (1995) report that humans with amygdala damage fail to show this conditioned response, but nevertheless are able to verbally learn the association ('when I saw the blue square I got a shock'), whereas amnesic patients with hippocampal damage show a normal conditioned response but cannot recall the association. This suggests that the association is stored in several places: in the amygdala (giving rise to the conditioned fear response) and also in the hippocampus (giving rise to declarative memories of the association). fMRI studies also show that the amygdala may also be important for fear-related conditioning in social settings in which participants learn fear associations by watching someone else receive a shock (Olsson & Phelps, 2004).

In humans, amygdala lesions can selectively impair the ability to recognize facial expressions of fear but not necessarily the other Ekman categories of emotion (e.g. Adolphs et al., 1994; Calder et al., 1996). For example, patient DR suffered bilateral amygdala damage and subsequently displayed a particular difficulty with recognizing fear (Calder et al., 1996). She was also impaired to a lesser degree in recognizing facial anger and disgust. She could imagine the facial features of famous people, but not of emotional expressions. She could recognize famous faces and match different views of unfamiliar people, but could not match pictures of the same person when the expression differed (Young, Hellawell, Van de Wal, & Johnson, 1996). DR also shows comparable deficits in recognizing vocal emotional expressions, suggesting that the deficit is related to emotion processing rather than modality-specific perceptual processes (Scott et al., 1997).

Functional imaging studies generally support, and extend, these conclusions. Morris et al. (1996) presented participants with morphed faces on a happy–neutral–fearful continuum. Participants were required to make male–female classifications (i.e. the processing of emotion was incidental). Left amygdala activation was found only in the fear condition; the happy condition activated a different neural circuit.

KEY TERM

Phobia
Long-term fear and
avoidance of particular
stimuli or situations.

Winston, O'Doherty, and Dolan (2003) report that amygdala activation was independent of whether or not subjects engaged in incidental viewing or explicit emotion judgments. However, other regions, including the ventromedial/orbitofrontal cortex, were activated only when making explicit judgments about the emotion. This was interpreted as reinstatement of the 'feeling' of the emotion.

Some researchers have argued that the ability to detect threat is so important, evolutionarily, that it may occur rapidly and without conscious awareness (Le Doux, 1996). Ohman, Flykt, and Esteves (2001) report that people are faster at detecting fear-related stimuli such as snakes and spiders amongst flowers and mushrooms than the other way around. When spiders or snakes are presented subliminally to people with spider or snake **phobias**, then participants do not report seeing the stimulus but show a skin conductance response indicative of emotional processing (Ohman & Soares, 1994). In these experiments, arachnophobics show the response to spiders, not snakes; and ophidiophobics show a response to snakes but not spiders. In terms of neural pathways, it is generally believed that there is a fast subcortical route from the thalamus to the amygdala and a slow route to the amygdala via the visual cortical pathways (Adolphs, 2002; Morris, Ohmann, & Dolan, 1999). Functional imaging studies suggest that the amygdala is indeed activated by unconscious fearful expressions, presented too briefly to be consciously seen (Morris et al., 1999). This is consistent with a subcortical/fast route to the amygdala, although it is to be noted that the temporal resolution of fMRI does not enable any direct conclusions to be drawn about processing speed.

Le Doux has argued that the amygdala has a fast response to the presence of threatening stimuli such as snakes.

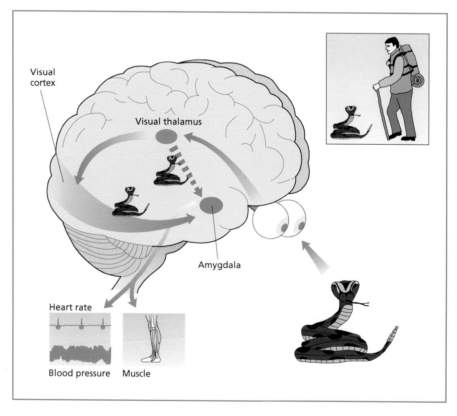

Activation of a fear response by the amygdala may trigger changes elsewhere in the brain that enable the threat to be evaluated and responded to, if necessary. There are connections from the amygdala to the autonomic system (Le Doux et al., 1988). These may help prepare the body for fight and flight by increasing the heart and breathing rates. The anterior cingulate is believed to be involved in this process (Critchley et al., 2003), and it too is selectively activated by fear relative to happiness (Morris et al., 1996). In addition, there is a strong relationship between the level of fear and increases in activity in regions of the visual cortex (Morris et al., 1998). Thus, the detection of potential threat by the amygdala may trigger more detailed perceptual processing of the threatening stimulus, enabling further evaluation. Other, more frontal, regions may also be important for deciding whether to act on this information. In conclusion, although the amygdala may be essential for the evaluation of potential danger, its role should be construed in terms of its influence upon a wider circuit of emotional processing.

Evidence for the role of the amygdala in other emotions

The most convincing evidence for a specialized role of the amygdala in fear comes from functional imaging studies that compare fear expressions with other emotional expressions, and studies of human patients with damage to the amygdala who show relatively selective deficits in recognizing fear. However, the conclusion that the amygdala is specialized for fear may still be premature. Firstly, the interpretation of these findings hinges on the assumption that the stimuli were appropriately matched. If fear-related stimuli are simply more difficult or more arousing (e.g. because a fearful face has more survival value than happy or sad faces) then the data could be explained without assuming specialization for fear. However, evidence that damage to other regions does *not* selectively affect fear (e.g. insula lesions and disgust) speaks against this more general account. Secondly, the amygdala might be specialized for some other process that just happens to be more relevant for fear. For instance, it has been suggested that selective impairments in fear may arise because of a failure to attend closely to the eyes (Adolphs et al., 2005). However, evidence that the amygdala is involved in fear in other domains (music, speech) speaks against this account (Gosselin, Peretz, Johnsen, & Adolphs, 2007; Scott et al., 1997).

With regard to learning of stimulus–emotion associations there is evidence that the amygdala is involved in learning positive associations, based on food rewards, as well as fear conditioning (Baxter & Murray, 2002). However, the amygdala system for positive associations operates somewhat differently to fear conditioning (and, hence, could be argued to be independent of the fear-based system). For example, selective lesions of the amygdala in animals do *not* affect learning of classically conditioned light–food associations (i.e. the animal learns to approach the food cup when the light comes on) (Hatfield, Han, Conley, Gallagher, & Holland, 1996), although such lesions are known to affect learning that a light predicts a shock. However, amygdala lesions do affect other aspects of reward-based learning, such as second-order conditioning in which a light + tone is subsequently paired with the absence of food after learning that a light alone predicts food or that the food is devalued (Hatfield et al., 1996). Different nuclei within the amygdala also have rather different roles in fear learning relative to reward learning (Baxter & Murray, 2002).

Recent functional imaging studies that compare stimuli with learned positive and negative associations relative to emotionally neutral ones but do not rely on facial expressions have revealed amygdala activation to negative and positive

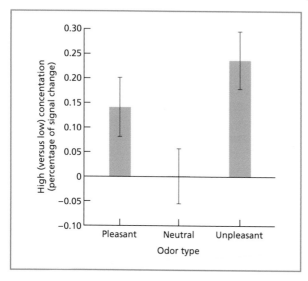

There is evidence that the amygdala responds to pleasant and unpleasant smells (but not neutral smells). This suggests a wider role of the amygdala in emotion processing, in contrast to the commonly held assumption that it is specific to fear. From Dolan, R. J. (2007). The human amygdala and orbitofrontal cortex in behavioural regulation. *Philosophical Transactions of the Royal Society of London Series B, 362,* 787–799. Reproduced with permission.

affective stimuli: for instance, comparing positive, negative, and neutral tastes (Small et al., 2003), smells (Winston, Gottfried, Kilner, & Dolan, 2005), pictures, and sounds (Anders, Eippert, Weiskopf, & Veit, 2008), and comparing the emotional intensity of personally held attitudes to concepts such as 'welfare' and 'abortion' (Cunningham, Raye, & Johnson, 2004). One interesting line of research that may explain some of the discrepancies between studies is to consider individual differences in, say, gender or personality (Hamann & Canli, 2004). For example, Canli et al. (2001) found amygdala activation in response to happy faces but only in individuals with an extravert personality type. This may be consistent with a wider role of the amygdala in emotional intensity (both positive and negative), but modulated by individual differences.

Finally, although most of the studies on the role of the amygdala in (non-conditioned) episodic memory have not compared fear-related memories to other categories of emotion, they show that the amygdala has a broader role here. Patients with amygdala lesions show impaired memory for the emotional details of scenes but not other peripheral details of scenes (Adolphs, Tranel, & Buchanan, 2005). Activation of the amygdala and hippocampus is correlated during learning of emotional scenes that are subsequently better remembered (Dolcos, LaBar, & Cabeza, 2004).

Summary

There is good evidence that the amygdala is crucial for the perception of fear. This includes facial expressions of fear but is not limited to faces. In addition, it appears to have a more general role in learning and storing the emotional value of stimuli. This includes both positive associations to stimuli (e.g. pleasant smells and tastes) as well as negative associations (e.g. stimuli paired with pain). These emotional associations extend to the emotional content of episodic memories.

The insula and disgust

The **insula** is a small region of cortex buried beneath the temporal lobes (it literally means 'island'). It is involved in various aspects of bodily perception, including important roles in pain perception and taste perception. The word **disgust** literally means 'bad taste' and this category of emotion may be evolutionarily related to contamination and disease through ingestion.

KEY TERMS

Insula
A region of cortex lying beneath the temporal lobes.

Disgust
A category of emotion that may be evolutionarily related to contamination and disease through ingestion.

Patients with Huntington's disease can show selective impairments in recognizing facial expressions of disgust (Sprengelmeyer et al., 1997) and relative impairments in vocal expressions of disgust (Sprengelmeyer et al., 1996). Huntington's disease is a genetic disorder with symptoms arising in mid-adulthood and including excessive movements, cognitive decline, and structural atrophy in the brain, particularly in regions such as the basal ganglia. However, the degree of the disgust-related impairments in this group correlates with the amount of damage in the insula (Kipps, Duggins, McCusker, & Calder, 2007). Selective lesions resulting from brain injury to the insula can affect disgust perception more than recognition of other facial expressions (Calder et al., 2000). In healthy participants undergoing fMRI, facial expressions of disgust activate this region but not the amygdala (Phillips et al., 1997). Feeling disgust oneself and seeing someone else disgusted activates the same region of the insula (Wicker et al., 2003).

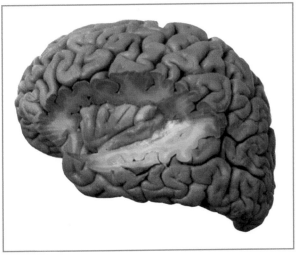

The insula is an island of cortex lying, bilaterally, underneath the temporal lobes. It is implicated in the creation of bodily feelings associated with emotions, and in the perception of disgust in particular. From Singer, Critchley, and Preuschoff (2009). Copyright © 2009 Elsevier. Reproduced with permission.

According to a 'basic emotion' viewpoint, a separate neural substrate for disgust may have evolved to deal with one particular situation – contamination. This may also explain why disgust has its particular anatomical location, close to the primary gustatory cortex involved in early cortical processing of taste. However, we use the word 'disgust' in at least one other context, namely to refer to social behavior that violates moral conventions. Disgusting behavior is said, metaphorically, to 'leave a bad taste in the mouth'. But is there more to this than metaphor? Some have argued that moral disgust has evolved out of non-social, contamination-related disgust (e.g. Tybur, Lieberman, & Griskevicius, 2009). Moral disgust also results in activity in the insula (Moll, de Oliveira-Souza et al., 2005) and moral disgust is associated with subtle oral facial expressions characteristic of disgust more generally (Chapman, Kim, Susskind, & Anderson, 2009).

The insula is generally considered to have a wider role in emotional processing, in addition to a more specific involvement in disgust. Specifically, it is regarded as monitoring (probably both consciously and unconsciously) for bodily reactions that are characteristic of emotional states (Craig, 2009; Singer et al., 2009). When the bodily reactions are consciously perceived they may constitute the 'feeling' of an emotion. Damasio et al. (2000) report insula activity in response to recalling emotional memories from various categories (sadness, happiness, anger, fear) relative to emotionally neutral memories (note that disgust was not studied). This monitoring of bodily states does not occur in isolation but rather attempts to link actual bodily states with those predicted from the current context and sensory inputs (Critchley, Wiens, Rotshtein, Ohman, & Dolan, 2004). For example, it shows greater activity in risky decisions in which outcomes are less certain (Paulus, Rogalsky, Simmons, Feinstein, & Stein, 2003).

Anger

A selective deficit in recognizing anger has been reported following damage to the ventral striatal region of the basal ganglia (Calder, Keane, Lawrence, & Manes, 2004). The dopamine system in this region has been linked to the production of aggressive displays in rats (van Erp & Miczek, 2000), as well as in reward-based motivation more generally. In this latter context anger/aggression could be construed as a motivated behavior to obtain or defend rewards. Anger and aggression are considered in detail in Chapter 10.

Evaluation

The concept of a 'basic emotion' rests, at least in part, on the assumption that there are separate neural foundations for different emotions. The clearest examples concern the role of the amygdala in fear, and the role of the insula in disgust. In both instances, the evidence suggests that the particular brain region is critically involved in that emotion. However, other evidence suggests that both the amygdala and insula are involved in the processing of other emotions too. How can these seemingly contradictory findings be reconciled? One possibility is that the processing of some emotions is more distributed across the brain than others. Thus, damage to one part of the circuit for a more distributed emotion could be partly compensated for elsewhere. Another (related) possibility is that the same brain region is involved in processing of different emotions but performs different computations for each emotion. For example, there is some evidence that the amygdala performs rather different roles in fear conditioning relative to reward conditioning, even though it is relevant to them both.

The idea of basic emotions is hard to definitively prove correct or incorrect. In general, there are a number of key difficulties with the 'basic emotion' approach. These have been discussed by Feldman Barrett (2006) and Panksepp (2007). Some of the salient points are listed here:

- Emotions can be 'basic' in one respect but not another (e.g. love does not have a facial expression but may have evolved to meet specific needs).
- Some 'basic' emotions appear to be more basic than others – for example, fear has more specialized neural substrates than happiness even though both are considered equally basic.
- The tendency for researchers in social neuroscience to focus on basic emotions has left many other (arguably more social) aspects of emotion under-researched, such as pride, guilt, and jealousy.

MOTIVATION: REWARDS AND PUNISHMENT, PLEASURE AND PAIN

Motivation makes one work to obtain a reward, or work to avoid a punishment (Rolls, 2005). A motivational state is one in which a goal is *desired*, whereas an emotional state is elicited when a goal is *obtained* (or not). For example, hunger is a motivational state (related to the goal of eating) and happiness and disgust are emotional states that may be an outcome of eating. Unlike eating, many of our goals have a social dimension to them, such as a need for love and group affiliation. One influential attempt to categorize different aspects of motivation is Maslow's (1943) **hierarchy of needs**. At the lowest level are physiological needs such as food and sex. At the top of the hierarchy is our need to realize the full potential of our abilities (self-actualization). In between, Maslow lists safety needs (e.g. financial security, good health), social needs (e.g. for friendship, family), and self-esteem (our need

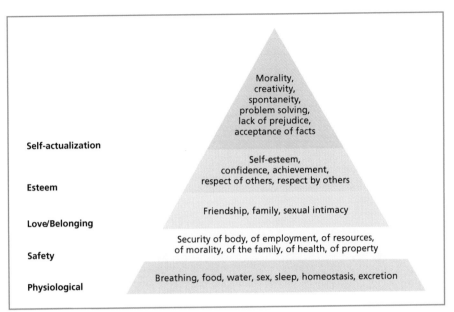

Self-actualization

Morality,
creativity,
spontaneity,
problem solving,
lack of prejudice,
acceptance of facts

Esteem

Self-esteem,
confidence, achievement,
respect of others, respect by others

Love/Belonging

Friendship, family, sexual intimacy

Safety

Security of body, of employment, of resources,
of morality, of the family, of health, of property

Physiological

Breathing, food, water, sex, sleep, homeostasis, excretion

Maslow argued that there were different kinds of motivation and he also suggested that they could be arranged hierarchically. Although it is reasonable to suggest different forms of motivation, there is little evidence for this organization.

to be liked and valued by others). This is reminiscent of Freud's hierarchy of id, ego, and super-ego. There is little evidence that such needs are organized hierarchically (Wahba & Bridgewell, 1976) but it provides a useful description of the various motivators of behavior.

Innate versus conditioned likes and dislikes

As noted earlier, the emotional characteristics of stimuli, in terms of whether they are rewarding or punishing, can either be innate or learned. A reinforcer is a stimulus that increases or decreases a particular pattern of behavior. **Primary reinforcers** act as rewards or punishers without any learning. **Secondary reinforcers** act as rewards or punishers as a result of learning. For example, in learning that a particular tone predicts a shock (as in fear conditioning) the secondary reinforcer is the tone, which was previously paired with a primary reinforcer (a painful shock). As already discussed, the amygdala is considered important for learning the secondary reinforcement properties of aversive stimuli (e.g. fear conditioning; Le Doux, 2000) and in some aspects of learning about rewarding stimuli (e.g. the amygdala is involved in learning that a pleasant secondary reinforcer has been devalued but not in the initial Pavlovian conditioning of neutral stimulus with reward; Baxter & Murray, 2002).

Primary reinforcers consist of certain tastes (e.g. sweet is rewarding, bitter is punishing), smells (e.g. putrefying is a punisher; pheromones are rewarding), touch (e.g. pain is a punisher, stroking is rewarding), and so on. Within the social domain, Rolls (2005) provides a list of possible reward-related primary reinforcers in humans, including attachment (e.g. to one's parent, partner, and child), cooperation (or reciprocal altruism), group acceptance, and 'mind-reading'. To this list, one could add social exclusion as a possible punishment-related primary reinforcer (see Chapter 8) and imitation as a possible reward-related primary reinforcer (see Chapter 3).

Facial expressions also act as primary reinforcers. If human infants are given a novel object, their behavior will be influenced by the response of their primary caregiver – a phenomenon termed **social referencing** (e.g. Klinnert, Campos, & Source,

1983). If the caregiver displays disgust, then the object will be avoided, but if the caregiver smiles, then the child will interact with the object. The object itself has acquired an emotional value and is now classed as a secondary reinforcer. In adults, facial expressions of fear (Mineka & Cook, 1993), sadness (Blair, 1995), and disgust (Rozin, Haidt, & McCauley, 1993) can all be used as negative reinforcers. Happy

Facial expressions can act as primary reinforcers in that they modulate behavior towards an (initially) affectively neutral stimulus (unconditioned stimulus), which may then act as a secondary reinforcer (conditioned stimulus).

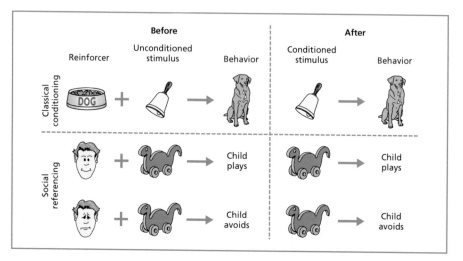

THE FRONTAL LOBES AND COGNITIVE CONTROL OVER EMOTIONS

Although emotional processing is often considered to be automatic, it can be controlled given an appropriate task or context. This involves an interplay between lateral prefrontal cortex (PFC) involved in cognitive control and 'executive functions', the ventromedial/orbital parts of the PFC (involved in emotional experience and contextualizing emotions), and regions such as the amygdala.

Ochsner, Bunge, Gross, and Gabrieli (2002) presented negative images (e.g. of someone in traction in a hospital) to participants in one of two conditions: either passively viewing them or a 'cognitive' condition in which they were instructed to reappraise each image 'so that it no longer elicited a negative response'. Their analysis revealed a trade-off between activity in the lateral PFC (high when reappraising) and the ventromedial PFC and amygdala (high during passive looking). When participants are asked to reappraise the stimulus negatively (i.e. making it worse than it looks) then this also engenders a similar network in the lateral PFC but tends not to dampen activity in the ventromedial PFC and amygdala (Ochsner et al., 2004). Similar results are found when comparing passive viewing of negative images and explicitly describing/labeling the images (Hariri, Bookheimer, & Mazziotta, 2000; Lieberman et al., 2007). These studies highlight the different role of the major distinctions within the PFC but, of equal importance, they indicate how they might be coordinated to regulate (and contextualize) emotions.

expressions can be used as positive reinforcers (Matthews & Wells, 1999). Disgust expressions are often employed in the context of food. Sickness itself can also be a very powerful reinforcer for food. Novel food eaten during a period of illness, for example during chemotherapy (Fredrikson et al., 1993), may elicit a highly durable subsequent avoidance of that food – called conditioned taste aversion. In this instance, the sickness is the primary reinforcer and the food paired with the sickness becomes a learned secondary reinforcer. Anger, by contrast, tends not to act as an unconditioned stimulus but, rather, is used to curtail ongoing behavior by implying violation of social norms (Blair & Cipolotti, 2000).

The orbitofrontal cortex computes the motivational value of rewards

The most basic anatomical division within the prefrontal cortex is that between the three different cortical surfaces: lateral, medial, and orbital. The lateral prefrontal cortex is more closely associated with sensory inputs than the orbitofrontal cortex. It receives visual, somatosensory, and auditory information, as well as receiving inputs from multi-modal regions that integrate across senses. In contrast, the medial and orbital prefrontal cortex is more closely connected with medial temporal lobe structures critical for long-term memory and processing of emotion.

The orbitofrontal cortex (OFC) consists of the ventral surface of the frontal lobes above the eye sockets (the orbits). It consists of Brodmann's areas 10, 11, 12, 13, and 14 (Walker, 1940), with area 12 being co-extensive with area 47 and areas 11 and 14 extending into the medial surface. The ventral part of the medial prefrontal cortex (VMPFC), including areas 25 and 32, has similar connections to the orbital surface and is often similarly affected by strokes. These two regions (orbital and ventromedial) tend to act as a functional network (Öngür & Price, 2000).

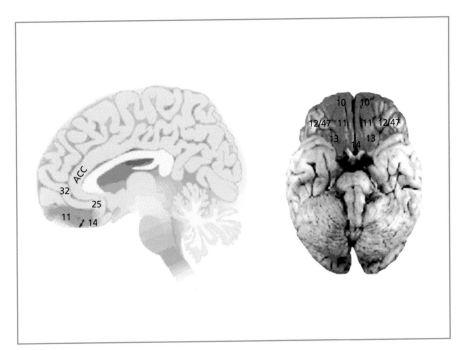

Different regions of the orbitofrontal cortex (and neighboring ventromedial frontal cortex), displayed in terms of Brodmann's areas. The left image is a medial view, and the right image is viewed from the underside of the brain. ACC = anterior cingulate cortex.

The same stimulus can elicit pleasure or aversion depending on context (e.g. the person's motivational state). Chocolate is normally pleasant, but if you have just eaten two bars of it you probably do not want any more. The orbitofrontal cortex computes the current emotional status of a stimulus (i.e. whether it is *currently* desired or not), thus enabling flexible behavior. Other regions in the brain may code the long-term value of a stimulus (i.e. whether it is *normally* desired or not).

One general function of the OFC is in computing the *current* value of a stimulus (i.e. how rewarding the stimulus is within the current context). For example, chocolate may be a rewarding stimulus but it may not be *currently rewarding* if one is full-up or if eating it may incur the anger of someone else. Small, Zatorre, Dagher, Evans, and Jones-Gotman (2001) asked participants to eat chocolate during several blocks of PET scanning. Initially, the chocolate was rated as pleasant and participants were motivated to eat it, but the more they ate the less pleasant it became and they were less motivated to eat it. This change in behavior was linked to changes in activity in orbitofrontal regions. Specifically, there was a shift in activity from medial regions (pleasant/wanting) to lateral regions (unpleasant/not-wanting). Other studies are consistent with different regions of OFC coding rewards and punishments (e.g. for a review see Kringelbach, 2005). For instance, activation of lateral OFC is found when a rewarding smile is expected but an angry face is presented instead (Kringelbach & Rolls, 2003) and is correlated with amount of monetary loss on a trial (O'Doherty, Kringelbach, Rolls, Hornak, & Andrews, 2001).

The OFC may enable flexible changes in behavior to stimuli that are normally rewarding (or recently rewarding) but suddenly cease to be. This can account for its role in **reversal learning** (in which rewarded and non-rewarded stimuli are reversed) and **extinction** (in which a rewarded stimulus is no longer rewarded). Eating chocolate until it is no longer pleasant can be regarded as a form of extinction. Lesions in these regions in humans lead to difficulties on these tasks, and the amount of difficulty in reversal learning correlates with the level of socially inappropriate behavior of the patients (Rolls, Hornak, Wade, & McGrath, 1994). Many patients with damage to these regions are inappropriately joking or flirtatious rather than aggressive (Damasio et al., 1990; Grafman et al., 1996). This is consistent with the notion that they struggle to 'devalue' these positive behaviors according to social context or feedback (e.g. that other people are uncomfortable with their behavior). The **Iowa Gambling Task** requires participants to

KEY TERMS

Reversal learning
Learning that the reward values of two stimuli have been swapped.

Extinction
Learning that a previously rewarded stimulus is no longer rewarded.

Iowa Gambling Task
A task involving selection between four options, each associated with different levels of risk; most participants learn to avoid choosing high-risk options (with some high rewards but resulting in a net loss).

avoid choosing high-risk options (which have some high rewards but result in a net loss) and patients with lesions in the OFC/VMPFC have difficulties on this task (Bechara, Damasio, Damasio, & Anderson, 1994).

Blair et al. (2006) show that activity in the OFC/VMPFC is related to the magnitude of rewards, but interestingly not the magnitude of punishments (which was more closely related to anterior cingulate activity). It was also related to the magnitude of foregone as well as chosen rewards. Thus, a choice between $500 and $400 will elicit greater activity in this region relative to $500 versus $100, even though the actual reward (i.e. $500) is the same in both. In a similar vein, Coricelli et al. (2005) argue that the OFC/VMPFC is important for the emotion of **regret** – the emotion that occurs when an outcome is worse than one would have experienced if one had made a different choice. Participants were asked to make a monetary gamble by choosing one of two options. In one condition, they were given feedback about what would have happened if they had chosen the other option (resulting in feelings of regret on some trials) and in another condition they were not given feedback (with no opportunity for regret). The ventral striatum was active when participants won money, but the medial OFC was only active when they were given feedback and this was correlated with the amount of regret (modeled here as the difference between chosen and not-chosen gains).

Activity in the OFC has been linked to participants' subjective reports of pleasantness to stimuli such as taste (McClure et al., 2004) and music (Blood & Zatorre, 2001). Importantly, these ratings of pleasantness are not just affected by the stimulus itself but also by the participants' beliefs about the product (which can be construed as a motivational bias). Being told the price of a wine affects the ratings of pleasantness upon tasting it – more expensive wines taste nicer – and perceived pleasantness was again related to activity in the medial part of the OFC (Plassmann, O'Doherty, Shiv, & Rangel, 2008). Of course, the experimenters administered some of the same wines twice and gave the participants different prices, so the stimuli were physically identical but their beliefs about the quality of the wine were not identical. Presumably our beliefs about the valence of people and social situations may operate along similar principles to that documented for taste.

The anterior cingulate cortex: Cognitive and affective evaluation of responses

The anterior cingulate is typically divided into two sections – a dorsal and ventral region – serving different functions (Bush, Luu, & Posner, 2000). The ventral region is considered an 'affective division'. Lesions in the *ventral* anterior cingulate in humans can produce symptoms comparable to that found after orbitofrontal lesions, including reduced subjective frequency of emotional experiences and changes in social behavior (Hornak et al., 2003). Individual differences in gray matter density in this more ventral region (from within the normal population) correlate with subjective social status on a 'social ladder' even when potential confounds such as age, sex, income,

The anterior cingulate cortex lies above the corpus callosum on the medial surface of each hemisphere. It has been suggested that there are two broad divisions: a dorsal region (blue) and a ventral region (green). The ventral region is generally agreed to be involved in processing of affective/social stimuli. But is the function of the dorsal region purely 'cognitive' or is it involved in evaluating rewards and punishments too?

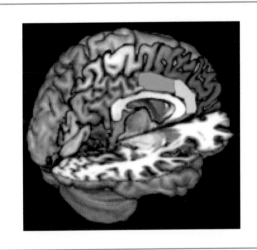

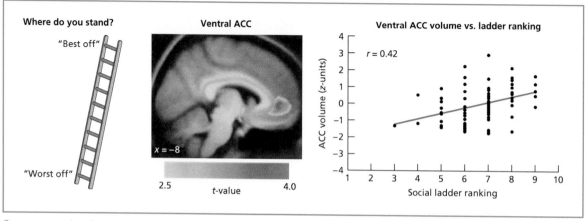

Gray matter density in the ventral portion of the anterior cingulate correlates with subjective social standing on a 'social ladder', even after more objective measures of social standing (income, education level) are controlled for. Decreased density may reflect a reduced control mechanism for social aspects of stress. From Gianaros et al. (2007). Reproduced with permission from the authors.

and education level are taken into consideration (Gianaros et al., 2007). The question of whether the *dorsal* anterior cingulate is purely cognitive (e.g. Bush et al., 2000) or is involved in reward/punishment (Bush et al., 2002; Rushworth, Behrens, Rudebeck, & Walton, 2007) is discussed below.

The dorsal anterior cingulate is regarded as being particularly important in the detection of errors and in the monitoring of responses in which errors are likely to occur, for instance, when a stimulus is compatible with several responses (e.g. Botvinick, Braver, Barch, Carter, & Cohen, 2001). In support of this, anterior cingulate activity, measured with fMRI, is greater on errorful trials than non-errorful trials (Kerns et al., 2004) and there is an ERP deflection, called error-related negativity, occurring within 100 ms of an error (Gehring, Goss, Coles, Meyer, & Donchin, 1993) that has its origins in the anterior cingulate (Dehaene, Posner, & Tucker, 1994). The classic example of **response conflict** is provided by the **Stroop test** (Stroop, 1935). In this task, participants must name the color of the ink and ignore reading the word (which also happens to be a color name). The standard explanation for the response conflict generated by this task is that reading of the word occurs automatically and can generate a response that is incompatible with that required (e.g. MacLoed & MacDonald, 2000). Both functional imaging studies (e.g. Carter et al., 2000) and lesion studies (Stuss, Floden, Alexander, Levine, & Katz, 2001) highlight the role of the anterior cingulate. One widely accepted model of the anterior cingulate, based on these and similar findings, is that it generates a 'warning signal' when responses are likely to err, and that other regions of the brain (e.g. in the lateral prefrontal cortex) act on this signal by, for example, being slower and more cautious (e.g. Botvinick et al., 2001).

There is, however, evidence that the anterior cingulate is involved in evaluating social and emotional stimuli. Rushworth et al. (2007) argue that the function of the anterior cingulate is to assess the value of responses (i.e. whether an *action* is likely to elicit a reward or punishment). This may differ from the function of the OFC, which computes whether a given *stimulus* is currently rewarded or punished. Rats with anterior cingulate lesions are more likely to choose the 'laziest' of two options when given a small reward in a nearby chamber or a larger reward in a

KEY TERMS

Response conflict
Situations in which the desired response is not the easiest response.

Stroop test
A task in which participants must name the color of the ink and ignore reading the word (which also happens to be a color name).

chamber that requires jumping over a wall – they choose the nearer response with least effort (Walton, Bannerman, Alterescu, & Rushworth, 2003). If two responses require the same effort but one requires waiting longer for a larger reward, then the rats with anterior cingulate lesions will wait for the larger reward whereas rats with orbitofrontal lesions behave impulsively by favoring the quick reward over the delayed larger reward (Rudebeck, Walton, Smyth, Bannerman, & Rushworth, 2006). Male monkeys with anterior cingulate lesions fail to adjust their responses, when reaching for food, when simultaneously shown a dominant male or a female in oestrus whereas most control monkeys will pay close attention to these social stimuli, and hence take longer to respond to the food (Rudebeck, Buckley, Walton, & Rushworth, 2006).

Are the cognitive-based accounts (based on errors and response conflict) and the action-value account (based on evaluating the reward/punishment value of a response) compatible with each other? Potentially yes, given that both accounts emphasize a role in *response evaluation* but differ in the extent to which they emphasize social versus purely cognitive situations. For example, in a human fMRI study, activity in the dorsal anterior cingulate increases when the difference between rewards or punishments decreases (Blair et al., 2006). Thus it is more active when deciding between a loss of 50 or a loss of 100 (or gain of 50 and gain of 100) relative to a loss of 50 or a loss of 300 (or gain of 50 and gain of 300). This is consistent with greater response conflict when the difference between rewards/punishments is small.

The ventral striatum and reward

One important method for discovering the 'reward centers' of the brain has been to use electrical **self-stimulation** in animals (Olds & Milner, 1954). Electrodes are implanted at a specific location in

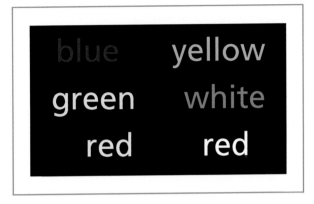

The Stroop test involves naming the color of the ink and ignoring the written color name (i.e. the correct response is 'red, green, yellow, blue, yellow, white').

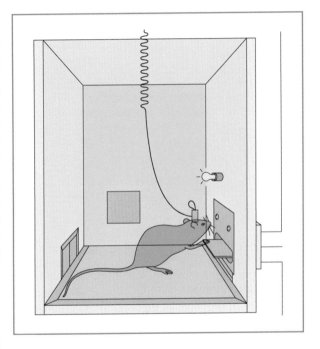

The rat is pressing a lever in order to obtain brain stimulation. This method has been used to identify regions of the brain that are associated with rewards (i.e. stimuli that the animal is motivated to work to obtain). These regions can also be considered the 'pleasure centers of the brain'. The nature of the reward differs according to the region stimulated and the current motivational state of the animal (e.g. whether it is hungry, lonely, etc.). From Olds, J. (1956). Pleasure centers of the brain. *Scientific American, 195*, 105–116. Copyright © 1956 Scientific American, Inc. All rights reserved. Reproduced with permission.

KEY TERM

Self-stimulation
In animals, the tendency to work (e.g. lever press) in order to receive direct stimulation of certain brain regions.

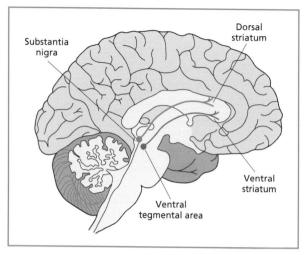

The striatum is located bilaterally in the basal ganglia, towards the front of the brain. The dorsal striatum and ventral striatum receive different dopaminergic inputs from mid-brain regions, namely the substantia nigra (implicated in Parkinson's disease) and the **ventral tegmental area**, respectively.

the brain and a tiny current occurs when the animal produces a response, such as a lever press. If the site of stimulation is rewarding then the animal will carry on repeating this action. (Recall that the behavioral definition of a reward is a stimulus that an animal is willing to work to obtain; a non-behavioral definition could be stimuli that an animal considers pleasant.) If the stimulation site is not rewarding (i.e. neutral or punishing) then the animal does not self-stimulate. Self-stimulation does not occur for electrodes implanted in most parts of the brain, including sensory and motor cortices, but it is found in areas such as the OFC, the lateral hypothalamus, the amygdala, and – the area considered in this section – the nucleus accumbens located in the **ventral striatum** (for an overview see Rolls, 2005). Whereas the OFC and amygdala are important for learning about the emotional value of stimuli, and changes to the value of stimuli, the ventral striatum is concerned with learning the emotional value of an action, such as a lever press that delivers food or some other reward (Cardinal, Parkinson, Hall, & Everitt, 2002), and also for learning the reward value of a decision (e.g. Hare, O'Doherty, Camerer, Schultz, & Rangel, 2008).

The ventral striatum is part of the basal ganglia. These consist of subcortical gray matter structures, including the caudate nucleus, the putamen, and the globus pallidus. The caudate nucleus and putamen are collectively known as the **striatum**. The dorsal region of the striatum has more sensorimotor properties (e.g. involved in **habit** formation) whereas the ventral region may be more specialized for emotions, although the distinction is relative and not absolute (Voorn, Vanderschuren, Groenewegen, Robbins, & Pennartz, 2004). There are several loops that connect regions within the frontal cortex to the basal ganglia and on to the thalamus before returning to the frontal cortex. Each loop targets different regions of the frontal

---

KEY TERMS

Ventral striatum
Part of the basal ganglia that includes the nucleus accumbens; Involved in a 'limbic circuit' connecting the orbitofrontal cortex, basal ganglia and thalamus.

Ventral tegmental area
A mid-brain structure in which the reward-related dopaminergic system originates.

Striatum
Collectively, the caudate nucleus and the putamen (parts of the basal ganglia).

Habit
An action that tends to be repeated, perhaps because it has previously been rewarded.

cortex and passes through different structures within the basal ganglia and thalamus (e.g. Alexander & Crutcher, 1990). The loops modulate brain activity within these frontal structures and, hence, increase or decrease the probability of a particular behavior. For example, there is a loop concerned with voluntary movement ('the motor circuit'), starting and ending in the supplementary motor area, and damage to various portions of this loop passing through the basal ganglia can lead to motor deficits such as Parkinson's disease (associated with rigidity and lack of voluntary movement) or Huntington's disease (associated with uncontrolled movements). However, the loop that is of particular relevance to reward-based learning (the 'limbic circuit') starts and ends in the OFC and limbic regions (amygdala, hippocampus, anterior cingulate) and connects to the ventral striatum (including the nucleus accumbens and head of caudate nucleus), the ventral globus pallidus (also part of the basal ganglia) and the mediodorsal nucleus of the thalamus. As such, the ventral striatum has been described as a 'limbic–motor interface' (Mogenson, Jones, & Yim, 1980).

Neurons containing the neurotransmitter dopamine project from the mid-brain to the nucleus accumbens, and psychomotor stimulants such as amphetamine and cocaine may exert their effects on action (i.e. by increased drug taking) via this system (Koob, 1992). Initial drug taking, driven by their pleasant/rewarding nature, may transform itself into compulsive, habit-based drug taking (in which the rewards are less immediately apparent) and this may be associated with shift from the ventral striatum (emotion based) to dorsal striatum (sensorimotor based) (Everitt & Robbins, 2005). Other rewarding stimuli activate this region. Dopamine release in the nucleus accumbens of male rats increases when a female is introduced to the cage, and increases further if they have sex (Pfaus et al., 1990). Secondary reinforcers, previously associated with food, increase the release of dopamine in the nucleus accumbens of rats (e.g. Robbins, Cador, Taylor, & Everitt, 1989). In humans, an fMRI study shows that the greater the monetary reward that could be obtained in a task, the larger the activity in the ventral striatum (Knutson, Adams, Fong, & Hommer, 2001) and, to give a more social example, activity in the ventral striatum correlates with male participants' desire to give a punishment to someone who has cheated (Singer et al., 2006).

One contemporary idea is that these dopaminergic neurons are not encoding reward per se but the difference between the *predicted* reward and actual reward (e.g. Schultz, Dayan, & Montague, 1997). After training to perform an action when presented with a light or tone cue, dopaminergic neurons in monkeys eventually respond to the conditioned cue itself rather than the subsequent reward (Ljungberg, Apicella, & Schultz, 1992; Schultz, Apicella, Scarnati, & Ljungberg, 1992). If no subsequent reward appears then their activity drops below baseline, indicating that a reward was expected. Some fMRI studies of decision making in humans also suggest that activity in the ventral striatum is greater when a reward is better than expected, rather than when a reward is high per se (Hare et al., 2008). Self-reported

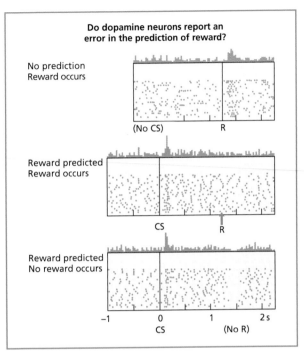

Single-cell recordings of dopamine neurons in the ventral striatum of monkeys show that the neuron responds when an unexpected reward of fruit juice is given (top), but if the reward is predicted by a cue (the conditioned stimulus) then the neuron responds to the cue and not the reward (middle). If an expected reward is omitted (bottom) the firing of the neuron falls below baseline. The results suggest that these neurons code the difference between the predicted reward and actual reward, rather than reward itself. From Schultz et al. (1997). Copyright © 1997 American Association for the Advancement of Science. Reproduced with permission.

lonely people show less activity in the ventral striatum when shown photos of social scenes (relative to non-lonely people), arguably because they predict them to be less rewarding (Cacioppo, Norris, Decety, Monteleone, & Nusbaum, 2009).

Evaluation

This section on motivation has highlighted the importance of three brain regions that are inter-connected: the ventral striatum, the OFC (and ventromedial frontal cortex), and the anterior cingulate cortex. The dorsal section of the anterior cingulate is important for assessing the rewards and risks of an action/response. The ventral striatum responds to rewards and the anticipation of rewards (e.g. responding to cues that may predict a subsequent reward). The OFC integrates reward-related information (e.g. from the ventral striatum) and emotional associations (e.g. from the amygdala) with other contextual information to ascertain the current motivational status of a stimulus (i.e. how much it is wanted). It facilitates behavioral flexibility enabling people to, for instance, stop responding when a previously rewarded stimulus becomes aversive.

SUMMARY AND KEY POINTS OF THE CHAPTER

- Emotions are states associated with stimuli that are rewarding (i.e. that we seek) or punishing (i.e. that we avoid) and they are multi-faceted in nature, consisting of emotional expressions, internal bodily responses, and subjective liking/disliking. They enable certain stimuli in the environment to be prioritized.
- Ekman has argued that there are six basic emotions that are defined on the basis of their having innate facial expressions, dedicated neural substrates, and that have evolved to deal with specific situations (e.g. disgust related to contamination). One problem with this approach is that many emotions appear to be 'basic' in one sense but not another.
- The amygdala has been specifically linked to the processing of fear (e.g. fear conditioning, recognition of fear on faces). The amygdala also has a role to play for processing other emotions (e.g. appetitive conditioning) but its role in other emotions is not necessarily the same as that for fear.
- A motivational state is one in which a goal is *desired*, whereas an emotional state is elicited when a goal is *obtained* (or not).
- The orbitofrontal cortex is involved in computing the motivational value of a stimulus (i.e. how much it is currently desired). It links emotion with current context in order to guide behavior.
- The dorsal anterior cingulate cortex is involved in computing whether an action is rewarded or punished (including detecting errors).
- The ventral striatum connects to the orbitofrontal cortex and is involved in reward-based learning. It is involved in the prediction and anticipation of reward rather than reward per se.

EXAMPLE ESSAY QUESTIONS

- What are 'basic emotions' and does current evidence support their existence?
- Is the amygdala the fear center of the brain?
- Contrast the roles of the orbitofrontal cortex and anterior cingulate cortex in emotions and decision making?
- Does damage to the emotional circuitry of the brain lead to impaired social functioning?

RECOMMENDED FURTHER READING

- Fox, E. (2008). *Emotion science*. New York: Palgrave Macmillan. Clear, comprehensive, and up-to-date.
- Le Doux, J. (1998). *The emotional brain*. New York: Simon & Schuster. Although this book is now showing its age, it is still a very clear and accessible overview and a good place to start.

CHAPTER 5

CONTENTS

Perceiving faces   101

Perceiving bodies   111

Joint attention: From perception to intention   115

Trait inferences from faces and bodies   120

Summary and key points of the chapter   126

Example essay questions   127

Recommended further reading   127

Reading faces and bodies

5

Our social interactions exist between other members of our species, so-called **conspecifics**. As such we need an effective system of keeping track of who is who. We need to remember what people look like and what their typical behaviors are. Facial and bodily appearances provide only superficial clues as to a person's inner state. But given that we cannot directly observe inner states but we can observe faces and bodies, there is a strong incentive to extract whatever information we can from a face or body. We need to know whether someone is likely to cooperate or cheat. Skilled basketball players, for instance, learn to detect fake passes from body language alone (Sebanz & Shiffrar, 2009). We need to know whether someone is happy or sad, or angry and likely to use force. Faces and bodies (together with voice cues) provide an important source of such information. Recognizing someone's expression involves making inferences about someone's *current* state; they are smiling therefore they are happy. However, there is a natural tendency to go beyond this. Many people believe that we can read character traits, such as trustworthiness and aggression, from faces even when they have neutral facial expressions (Hassin & Trope, 2000). Indeed, people tend to vote for political candidates whose faces are judged to be associated with greater competency. Todorov, Mandisodza, Goren, and Hall (2005) presented participants with pairs of photographs of faces of the winner and runner-up of seats in the US congressional election. Participants were not informed about how the stimuli were constructed and were asked to note if they recognized any of the faces in the pair. (These pairs were then discarded and only those judged unfamiliar were analyzed.) Judging which of the faces was more competent predicted the overall result better than chance (68.8% relative to 50% chance), and was correlated with the margin of victory.

This chapter starts by considering the basic mechanisms of recognizing a face from both a cognitive and neural perspective. Particular consideration is given to the issue of whether or not recognizing an emotional expression involves different mechanisms from recognizing familiar faces, or reading other dynamic cues in a face (such as gaze direction). Recent research in the less-studied area of body perception is then evaluated. The second half of the chapter considers how perceivers go beyond the raw information provided in order to infer other peoples' intentions from faces and bodies, and to infer their stable personality traits.

PERCEIVING FACES

Face perception has several goals. In some instances, the goal may be to recognize a particular individual (e.g. 'that is my wife'). In other instances, the goal may be to extract other types of socially relevant information such as whether the person is happy, attractive, old, where they are looking, etc. To some degree, these different aspects of face perception reflect different cognitive and neural mechanisms.

Todorov et al. (2005) asked participants to judge which of two faces looked more competent. The faces were unknown to the participants making the rating, but both had in fact been contenders in an election. Faces rated as more competent were likely to win with an increased majority of votes. From Todorov et al. (2005). Copyright © 2005 American Association for the Advancement of Science. Reproduced with permission.

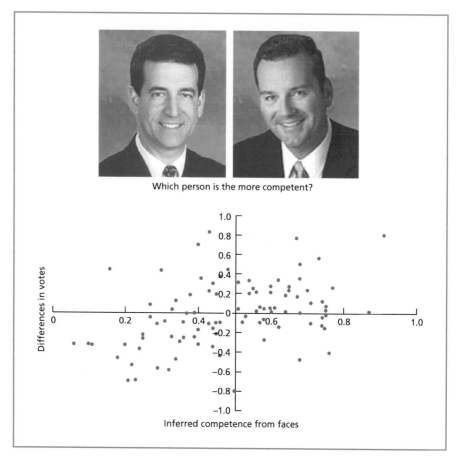

Which person is the more competent?

A cognitive model

Bruce and Young (1986) proposed a cognitive model of face recognition that has largely stood the test of time and is illustrated below. They assume that the earliest level of processing involves structural encoding of the face by detecting shading and curvature of surfaces and detection of edges. Following this, a distinction is made between the processing of familiar and unfamiliar faces. Recognizing familiar faces is assumed to involve matching the visual description of the face with a stored memory representation of a face. There is assumed to be a store of all known faces, and each face is said to have its own '**face recognition unit**'. In neural terms, single-cell recordings of primates suggest that there is a class of neurons that respond to faces but not objects and, moreover, that these neurons respond to certain faces more than other faces (e.g. Rolls & Tovee, 1995). However, neurons that respond to one (and only one) face tend not to be found (Quiroga et al., 2005). As such, face recognition units are likely to be coded in terms of the activity of a set of neurons rather than having a single neuron responding to each known face. For familiar faces, once the face has been recognized then other information may become available, such as their occupation or name. The semantic level of description (which relates to conceptual knowledge of people, rather than the face per se) is termed the '**person identity node**'. The process of familiar face recognition involves matching a face seen from one particular viewpoint and one particular lighting condition to a memory representation that stores the three-dimensional structure of the face (enabling it to be recognized from any view). A separate route (termed directed visual processing) was postulated to deal with unfamiliar faces, for example in order to match them across different views or lighting conditions. Aside from the mechanism for recognizing familiar faces, there are a number of other pathways on the Bruce and Young (1986) model that apply equally to familiar and unfamiliar faces. These include detecting the emotional expression of faces, and also using lip-reading cues from faces.

Evidence from patients with acquired neurological impairments lends support to this basic model. Impairments of face processing that do not reflect difficulties in early visual analysis are termed **prosopagnosia** (Bodamer, 1947). The term prosopagnosia is also sometimes used specifically to refer to an inability to recognize previously familiar faces. As such, care must be taken to describe the putative cognitive mechanism that is impaired rather than relying on simple labeling. The case study reported by De Renzi (1986) had profound difficulties in recognizing the faces of people close to him, including his family, but could recognize them by their voices or other non-facial information. He once remarked to his wife: 'Are you [wife's name]? I guess you are my wife because there are no other women at home, but I want to be reassured.' The patient's ability to recognize and name other objects was spared. Within the Bruce and Young (1986) model his deficit would be located at the face recognition unit stage. Patients with acquired prosopagnosia still retain the ability to recognize other socially salient information from (familiar and unfamiliar) faces, including sex, age, and emotional expressions (Tranel, Damasio, & Damasio, 1988). They can also use lip-reading cues to aid in speech perception (Campbell, Landis, & Regard, 1986). More recently, congenital prosopagnosia has been documented, which has similar characteristics to acquired prosopagnosia but is present throughout the lifespan with no known external cause (Duchaine & Nakayama, 2006).

The Bruce and Young (1986) model of face processing postulates a number of separate mechanisms that reflect the different goals of the perceiver, including: recognizing a familiar face; recognizing a facial expression; and matching two images of an unfamiliar face.

Neural basis of face perception

Beyond the earliest stages of cortical visual processing (in the primary visual cortex), neurons become increasingly specialized in their response properties. In primates, cells in the inferotemporal cortex respond to specific shapes but not where the shape is presented in space (Gross, Rocha-Miranda, & Bender, 1972). This suggests that these neurons are concerned with the appearance of objects rather than their location. A distinction is drawn between two visual streams – a **ventral visual stream** that

KEY TERMS

Dorsal visual stream
Runs from the occipital to the parietal lobes and is concerned with locating and acting on objects.

Occipital face area (OFA)
A region in the occipital cortex that responds to faces more than objects but does not process facial identity.

Fusiform face area (FFA)
A region in the fusiform cortex that responds to faces more than objects and is responsive to facial identity.

is concerned with identifying objects, largely irrespective of where they are, and a **dorsal visual stream** that is concerned with locating objects, largely irrespective of what they are (Ungerleider & Mishkin, 1982). Face perception depends primarily on the ventral visual stream.

In this section, three regions of the human brain are considered in detail: the so-called **occipital face area (OFA), fusiform face area (FFA)**, and the superior temporal sulcus (STS). The label 'face area' has attracted controversy due to the claim that it may be a domain-specific region that processes only one kind of information, namely faces (e.g. Gauthier, Tarr, Anderson, Skudlarski, & Gore, 1999). The model of Haxby, Hoffman, and Gobbini (2000) presents a neuro-anatomically inspired model of face perception that contrasts with the purely cognitive account offered by Bruce and Young (1986). In their model, Haxby et al. (2000) consider the core regions involved in face perception to lie in the fusiform gyrus in humans (corresponding to the inferotemporal cortex identified in primates) and the STS. They also identify an 'extended system' to denote other areas of the brain that receive inputs from the core face perception system but are not essential for face perception (e.g. regions supporting semantic knowledge of people).

Occipital face area

The OFA is located in the inferior occipital gyrus. It is considered to be an early stage in perceptual analysis of faces that sends inputs to fusiform and superior temporal regions (e.g. Haxby et al., 2000). Like the 'fusiform face area' it is defined on the basis of showing a greater fMRI BOLD response to faces relative to other categories. However, it differs from the FFA in a number of key respects. For example, the OFA responds to both upright and inverted faces whereas the FFA responds more to upright faces (Yovel & Kanwisher, 2005). Fox, Moon, Iaria, and Barton (2009) used an fMRI adaptation paradigm to compare activity in the OFA with other face-sensitive regions. This method relies on the finding that the BOLD

The model of Haxby et al. (2000) divides the neural substrates of face processing into a number of core mechanisms (relatively specialized for faces) and an extended system in which face processing makes contact with more general cognitive mechanisms (e.g. concerning emotion, language, action).

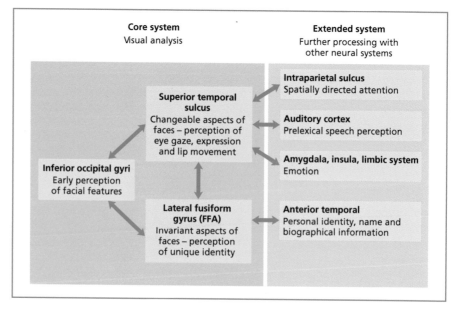

Occipital face area (OFA)	Fusiform face area (FFA)	Superior temporal sulcus (STS)
• Relatively specialized for faces (not bodies or objects) • Codes the physical aspects of facial stimuli	• Relatively specialized for faces (not bodies or objects) • Important for computing an invariant facial identity	• Responds to faces and bodies • Important for action perception and dynamic stimuli (e.g. lip movements) • Integrates visual and auditory information

signal is reduced when the same stimulus is presented twice. However, one can vary how 'same' is defined to reveal different properties of different regions (e.g. physically same, same person, same expression). The OFA activity is sensitive to any physical change in the face stimulus consistent with a role in the early perception of facial structure. Other regions (including FFA and STS) show a more complex pattern that is not related to physical changes in the stimulus, but to whether or not the participant actually perceives a change in identity or expression. Rotshtein, Henson, Treves, Driver, and Dolan (2005) presented participants with morphed images of two famous people (e.g. Marilyn Monroe, Margaret Thatcher) during fMRI. Morphs of 0% Maggie/100% Marilyn and 30% Maggie/70% Marilyn tend to be perceived as 'Marilyn'. In contrasts, morphs of 40% Maggie/60% Marilyn and 70% Maggie/30% Marilyn tend to be perceived as different to each other (as 'Marilyn' and 'Maggie', respectively). Crucially, both pairs of images are equally different physically. Whereas the FFA responded only when the identity was perceived to change, the OFA responded to physical changes in general.

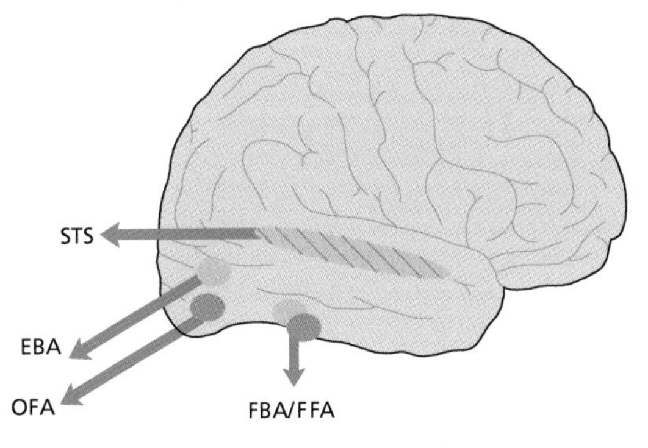

Approximate locations of face- and body-sensitive visual regions, shown here in the right hemisphere. The fusiform body and face areas are located on the under side of the brain, but are shown here projected onto a side view. Other regions are located laterally. EBA = extrastriate body area; FBA = fusiform body area.

Fusiform face area

The FFA responds to faces more than other stimuli, including bodies, and may be particularly important for recognizing known faces (Kanwisher, McDermott, & Chun, 1997; Kanwisher & Yovel, 2006). The FFA is found bilaterally, with a generally more robust BOLD response on the right. The degree of left/right asymmetry differs between individuals and is related to individual differences in visual field asymmetry for identifying faces (Yovel, Tambini, & Brandman, 2008). Thus, a greater ability at identifying a face in the left visual field is associated with greater activity in the right FFA relative to left FFA. Functional imaging shows that the FFA shows fMRI adaptation when the same face is repeated even if physical aspects of the images change (see Kanwisher & Yovel, 2006).

The main rival claim to the suggestion that the FFA is specialized for faces is that this region is sensitive to expert within-category visual discriminations but not to faces per se (e.g. Gauthier et al., 1999). These accounts have two key elements: that faces require discrimination within a category (between one face and another), whereas most other object recognition requires a superordinate level of discrimination (e.g. between a cup and comb); and consequently that we become 'visual experts' at making these fine within-category distinctions through

FAMILY 1   FAMILY 2   FAMILY 3   FAMILY 4   FAMILY 5

Greebles can be grouped into two genders and come from various families. To what extent does discriminating amongst Greebles resemble discriminating across faces? Images provided courtesy of Michael J. Tarr (Carnegie Mellon University, Pittsburgh, PA; see www.tarrlab.org).

prolonged experience with thousands of exemplars. The evidence for this theory comes from training participants to become visual experts at making within-category discriminations of non-face objects, called 'Greebles'. As participants become experts they move from part-based to holistic processing, as has often been proposed for faces (Gauthier & Tarr, 1997). In addition, they have shown that Greeble experts activate the FFA (Gauthier et al., 1999) and similar findings have been reported for experts on natural categories such as birds and cars (Gauthier, Skudlarski, Gore, & Anderson, 2000).

There is some evidence from prosopagnosia that supports the face specificity account against the visual expertise alternative. Sergent and Signoret (1992) reported a prosopagnosic patient, RM, who had a collection of over 5000 miniature cars. He was unable to identify any of 300 famous faces, or the face of himself or his wife, or match unfamiliar faces across viewpoints. Nevertheless, when shown 210 pictures of miniature cars he was able to give the company name, and for 172 he could give the model and approximate year of manufacture. This suggests that face perception and visual within-category expertise are not necessarily the same thing (see also McNeil & Warrington, 1993).

Superior temporal sulcus

According to the model of Haxby et al. (2000), the STS responds to the changeable aspects of a face (e.g. particular poses, gaze directions) whereas the FFA responds to the stable aspects of a face (i.e. the person's identity). The changeable aspects of a face are particularly important for extracting social cues that are likely to be fleeting (Allison, Puce, & McCarthy, 2000). In support of the distinction between STS/FFA made by Haxby et al. (2000), functional imaging studies show that when participants are asked to make judgments about eye gaze (deciding whether the face is looking in the same direction as the last face) then activity is increased in the STS but not in the FFA (Hoffman & Haxby, 2000). In contrast, when participants are asked to make judgments about face identity (deciding whether the face is the same as the last one presented) then activity is increased in the FFA but not in the STS.

The STS region responds to bodies as well as faces, for example when observing people walk (e.g. Grossman et al., 2000). The STS may also be important for linking the ventral visual stream with the dorsal stream. This may enable dynamic visual percepts of faces and bodies (in the STS) to be processed motorically (by parietal and frontal lobe systems) – a social mirroring of perception and action. The STS receives greater multi-sensory inputs than the FFA. Functional imaging studies show that it responds to both seen speech (i.e. facial lip-reading) and heard speech, and that the response is greater when the two correspond in terms of content and timing (Calvert, Hansen, Iversen, & Brammer, 2001). Single-cell recordings from monkeys show neurons that respond to both the sight and sound of certain actions such as lip-smacking or threat noises (Barraclough, Xiao, Baker, Oram, & Perrett, 2005).

The role of the STS in detecting biological motion is discussed in the section on body perception, and the role of STS in initiating joint attention is considered later.

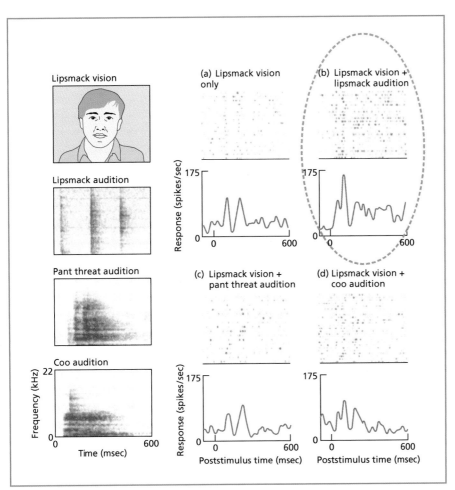

Lipsmack vision

(a) Lipsmack vision only

(b) Lipsmack vision + lipsmack audition

Lipsmack audition

Response (spikes/sec)

175

0

0   600

175

0

0   600

Pant threat audition

(c) Lipsmack vision + pant threat audition

(d) Lipsmack vision + coo audition

Coo audition

Frequency (kHz)

22

0

0   600

Time (msec)

Response (spikes/sec)

175

0

0   600

Poststimulus time (msec)

175

0

0   600

Poststimulus time (msec)

Single-cell recordings in the monkey STS show increased firing when the same vocalization is both seen and heard. This suggests that the region is not a purely visual one, but integrates across hearing and vision. From Barraclough et al. (2005). Copyright © 2005 by the Massachusetts Institute of Technology. Reproduced with permission.

SOCIAL DISPLAY RULES

Although basic emotions are considered to be universal across cultures, culture can exert an effect in terms of social **display rules** – the extent to which one regulates emotional expressions in the presence of others. Ekman (1972) found that when both Japanese and American participants watched a stressful film in isolation there were no cultural differences in facial expression. However, in the presence of a high status experimenter the Japanese participants were more inclined to hide a display of negative emotion with a smile. A recent study carried out in 32 different cultures asked participants questions about their attitudes towards displays of emotion (Matsumoto et al., 2008). The results revealed that individualistic cultures (e.g. USA, Australia) were more likely to endorse the display of happiness than collectivist cultures (e.g. Indonesia and Hong Kong). Moreover, participants from all cultures reported a greater emotional display towards their ingroup than an outgroup (considered here in terms of their nationality).

KEY TERM

Display rules
The extent to which one regulates emotional expressions in the presence of others.

Perceiving emotion from faces

This section will consider two different issues: the question of whether there is a specialized route for recognizing facial emotions relative to recognizing familiar faces (as in Bruce & Young, 1986) or whether recognizing facial expressions is part of a more general system for detecting dynamic changes in faces (as in Haxby et al., 2000); and the question of whether familiar faces (even with a neutral facial expression) have an emotional signature and how this might go awry in a fascinating symptom called Capgras delusion.

How does facial expression recognition relate to other aspects of face perception?

Both the cognitive model of Bruce and Young (1986) and the neural model by Haxby et al. (2000) make a distinction between recognizing familiar faces (facial identity) and recognizing emotional expressions in faces. Whilst there is good evidence for this basic idea, Calder and Young (2005) argue that neither model offers a satisfactory account of the evidence. Although prosopagnosic patients have been reported who are better at recognizing facial expressions than facial identity (e.g. Tranel et al., 1988), Calder and Young (2005) argue that the dissociation is not absolute. Both facial expression and facial identity tend to be impaired in prosopagnosia relative to controls, even though recognition of facial expressions is less impaired. They suggest that the structural encoding stage is important for both expressions and identity (including both the OFA and FFA) but the latter fares worse after brain damage because facial recognition is more demanding. There is evidence from fMRI that the FFA responds to changes in expression as well as identity (Ganel, Valyear, Goshen-Gottstein, & Goodale, 2005).

The Bruce and Young (1986) model assumes that there is a separate route for analyzing facial expressions. If this is so, then one might be able to find brain-damaged patients who are unable to recognize facial expressions but can recognize facial identity. Insofar as such patients exist, they seem to be associated with orbital and ventromedial frontal lesions (Heberlein, Padon, Gillihan, Farah, & Fellows, 2008; Hornak, Rolls, & Wade, 1996) or somatosensory regions (Adolphs, Damasio, Tranel, Cooper, & Damasio, 2000) but *not* the STS. However, rather than postulating a single route for dealing with all emotions, Calder and Young (2005) favor the idea that each emotion is dealt with separately (e.g. the amygdala for fear, the insula for disgust) and is part of the extended system, to borrow the terminology of Haxby et al. (2000), rather than the core system of face processing.

Although not specifically discussed by Calder and Young (2005) or Haxby et al. (2000), there is one candidate mechanism that could serve as a general system for recognizing expressions but not identity – namely in terms of sensorimotor simulation (e.g. Heberlein & Adolphs, 2007). **Simulation theory** will be encountered many times during this book; it actually consists of a collection of somewhat different theories based around a unifying idea – namely that we come to understand others (their emotions, actions, mental states) by vicariously producing their current state on ourselves – and is considered at length in Chapter 6. With regard to emotions, the claim is that when we see someone smiling we also activate our own affective pathways for happiness. Moreover, we may activate the motor program needed to make us smile (this may make us smile back, or it may prepare a smile response) and we may simulate what this might feel like in terms of its sensory consequences (e.g. muscle stretch and tactile sensations on the face). As such, one could possibly recognize emotions such as happiness, fear, and disgust not just in terms of their visual appearance but in terms of the way that they activate the sensorimotor program of the perceiver.

There is evidence from electromyographic studies that viewing a facial expression produces corresponding tiny changes in our own facial musculature, even if the face is viewed briefly so as to be unconsciously perceived (Dimberg et al., 2000). However, this does not necessarily imply that this is used to recognize expressions. To address this, Oberman, Winkielman, and Ramachandran (2007) report that biting a pen lengthways uses many of the same muscles involved in smiling. They subsequently showed that the bite task selectively disrupts the recognition of happiness. Lesion studies also suggest a direct contribution of simulation mechanisms to recognizing emotional expressions. Adolphs et al. (2000) tested the critical lesion sites in 108 patients asked to identify facial expressions from the Ekman categories. Damage to sensorimotor areas, including the right somatosensory cortex and Broca's area, were found to predict poor performance. Pitcher, Garrido, Walsh, and Duchaine (2008) applied TMS to the right somatosensory cortex of healthy participants and noted that recognition of facial expressions, but not facial identity, was disrupted by TMS in this region. The effects occurred relatively early (when TMS was applied within 170 ms of stimulus onset), suggesting that simulation pathways may be activated in parallel with visual mechanisms of expression recognition.

Do familiar faces have an emotional signature?

A rather different issue concerns the extent to which a familiar face (with a neutral expression) generates an emotional response. In several studies, the emotional response has been measured in terms of the skin conductance response (SCR). This response occurs within seconds of an appropriate stimulus and is considered to be automatic (insofar as participants cannot control its absence or presence) and unconscious (insofar as participants are generally unaware of it).

In the neurologically typical population, familiar faces generate a greater SCR than unfamiliar faces (Tranel, Fowles, & Damasio, 1985). Intriguingly, patients with acquired prosopagnosia show evidence of the same trend (e.g. Tranel, Damasio, & Damasio, 1995) – that is, they generate a greater SCR to familiar faces than unfamiliar ones despite not knowing who the familiar people are or, indeed, being unable

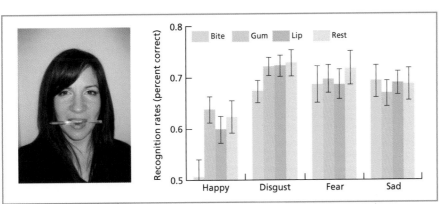

Placing a pen in the mouth horizontally and holding it with the teeth uses many of the same muscles as smiling. Performing this task can also disrupt recognition of facial expressions of happiness. Data from Oberman et al. (2007).

to classify them as familiar. There are several possible explanations for this. One is that both the conscious recognition of familiar faces and the unconscious (emotional) recognition of familiar faces use the same damaged mechanism (e.g. visual processes in the fusiform). These mechanisms are presumably degraded rather than absent, but obtaining an emotional signature is less demanding and can still occur with partial damage. Another explanation is that there are separate processes at early visual stages that are linked to these different outcomes: generating an emotional response versus consciously identifying the face. This would be an interesting question to explore with functional imaging.

A dissociation between the emotional content of faces and the recognition of facial identity has been put forward to account for a delusional symptom called **Capgras syndrome**. In the Capgras syndrome, people report that their acquaintances (spouse, family, friends, and so on) have been replaced by 'body doubles' (Capgras & Reboul-Lachaux, 1923; Ellis & Lewis, 2001). They will acknowledge that their husband/wife looks like their husband/wife. Indeed, they are able to pick out their husband/wife from a line-up while maintaining that he/she is an imposter. To account for this, Ellis and Young (1990) suggest that these patients are the opposite of prosopagnosia. Thus, they can consciously recognize the person/face but they lack an emotional response to them. As such, the person/face is interpreted as an imposter. This explains why the people who are doubled are those closest to the patient, as these would be expected to produce the largest emotional reaction. This theory 'makes the clear prediction that Capgras patients will not show the normally appropriate skin

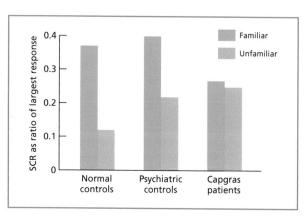

Most people produce a greater skin conductance response (SCR) to personally familiar people relative to unfamiliar ones, but patients with Capgras delusion do not. From Ellis and Lewis (2001). Copyright © 2001 Elsevier. Reproduced with permission.

conductance responses to familiar faces' (Ellis & Young, 1990, p. 244). Subsequent research has confirmed this prediction (Ellis, Young, Quayle, & DePauw, 1997). However, the findings of Tranel et al. (1995) are problematic. Their patients with damage to the ventromedial frontal lobes had abolished skin conductance to familiar faces but reported no signs of Capgras delusion. Thus, a lack of emotional response may well be necessary but it is unlikely to be sufficient. Perhaps a mechanism involved in decision making is compromised in addition to facial emotional processing.

Evaluation

There is good evidence that face processing can be divided into different operations serving different functions and with different neural substrates. The models of Bruce and Young (1986) and Haxby et al. (2000) offer useful, and complementary, accounts of these different mechanisms. However, neither account offers a full explanation of the current evidence. For example, there appear to be multiple routes for recognizing facial expressions: one based on sensorimotor simulation that is not specialized for particular emotions, and others that make contact with core regions of emotion processing (e.g. amygdala, insula). In addition, there are conscious and unconscious correlates of emotion processing of faces – the latter may be conveniently measured by SCR.

RECOGNIZING ONE'S OWN FACE USES A SPECIAL MECHANISM

Seeing our own face activates a circuit of regions in the frontal and parietal lobes that are not found for other personally familiar faces (Uddin, Iacoboni, Lange, & Keenan, 2007). Of course, we never directly see our own faces except via mirrors or external images. Recognizing one's own reflection in a mirror requires more than just face recognition (a basic-level knowledge), it requires appreciating that the face out there corresponds to the person 'in here'. **Mirror self-recognition** (MSR) has been considered by some to be an important test of self-awareness in other species (for reviews see de Veer & Van den Bos, 1999; Schilhab, 2004). Gallup (1970) noted that, during development, both humans and chimpanzees alter their behavior towards mirrors. At early stages of development they behave as if the mirror-image is another animal and show social behavior towards it. When older (18–24 months in humans) they show self-directed behaviors, such as using the mirror to explore unseen body parts. Of particular interest is their behavior on the 'mark test' (Gallup, 1970). If a red mark is placed on the forehead during anesthesia, then the animal will attempt to groom it on encountering its reflection in a mirror. This behavior is found in humans, common and pygmy chimpanzees, and orang-utans, but not in gorillas or monkeys (Schilhab, 2004). Does this behavior unequivocally demonstrate self-awareness? At the very least it is consistent with *bodily* self-awareness or an ability to form a link between an external image and one's own body.

PERCEIVING BODIES

There is far less research on the visual perception of bodies compared to faces. There are no long-standing cognitive models of body perception that are equivalent to Bruce and Young's (1986) model. Whilst brain regions have been identified that are important for body perception, their precise functions remain relatively under-specified relative to comparable regions in the face domain.

Visual perception of bodies

Downing, Jiang, Shuman, and Kanwisher (2001) reported an area in visual cortex that responded more to whole bodies and body parts than faces and objects. This was termed the **extrastriate body area (EBA)**. It appears to code a rather abstract description of a body plan insofar as it responds strongly not only to real photographs but also to line drawings, stick figures, and body parts (Downing et al., 2001). It shows a graded response to other types of bodies such that it has the greatest response to humans, then mammals, then fish and birds, and lastly objects (Downing, Chan, Peelen, Dodds, & Kanwisher, 2006). TMS over this region affects the speed of deciding whether two body parts are the same, but not an equivalent task for faces or motorcycle parts (Urgesi, Berlucchi, & Aglioti, 2004), and brain lesions in the EBA region prevent successful performance on this task (Moro et al., 2008).

More recently, a second body-sensitive region was discovered that spatially overlaps with the fusiform face area, and has been termed the **fusiform body area (FBA)** (Peelen & Downing, 2005). The different contributions of the EBA and FBA are not well understood, but the FBA is noted to respond relatively more to whole bodies than body parts versus the EBA (Taylor, Wiggett, & Downing, 2007), which is consistent with its location further along the ventral stream.

The processing of body movement, so-called **biological motion**, may be more a function of the STS, whereas the processing of body configuration is a function of the EBA and FBA. The perception of biological motion is assessed by attaching light-emitting diodes (LEDs) to the joints and then recording someone walking/running in the dark. When only the LEDs are viewed, most people are still able to detect bodily movement relative to a control condition in which these moving lights are presented jumbled up. A comparison of BOLD response to biological motion relative to jumbled motion reveals activity in a posterior region of the STS (Grossman et al., 2000). This region is different from the main region responsible for visual movement perception, known as **V5 (or MT)**, and there may be separate pathways into STS (for biological motion) versus V5/MT (for visual motion in general). For example, a patient with a bilateral lesion to V5/MT was unable to detect visual motion and perceived the world in terms of jerky snapshots (Zihl, Von Cramon, & Mai, 1983). Nonetheless, she was able to discriminate biological from non-biological motion

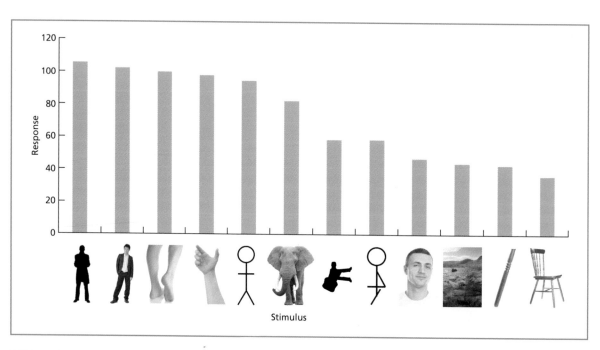

BOLD activity in the extrastriate body area (EBA) in response to different kinds of stimuli (scaled according to the response to body parts). From left to right: body silhouettes; whole bodies; various body parts; hands; stick figures; mammals; scrambled silhouettes; scrambled stick figures; faces; scenes; object parts; and whole objects. Adapted from Peelen and Downing (2007).

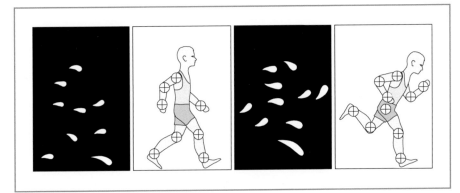

When this array of dots is set in motion, most people can distinguish between motion from biological forms (e.g. the human body) and non-biological motion (e.g. the same dots scrambled).

(McLeod, Dittrich, Driver, Perrett, & Zihl, 1996). Single-cell recordings in monkey STS show cells that respond not only to a figure walking but also the direction of motion (towards or away from the monkey), whether the figure has his back or front to the monkey, and also whether the figure was near or far from the monkey in the testing room (Jellema, Maassen, & Perrett, 2004).

Perceiving emotion from bodies

Bodies can convey somewhat different emotional information from a face. For example, whilst a facial expression might signal fear, the body might additionally convey whether the person is likely to stay or run.

Fearful body language elicits activation in the amygdala and fusiform gyrus relative to happy and neutral body language (Hadjikhani & de Gelder, 2003). (In these stimuli the face is blurred out so that one can be sure that it is the body itself that conveys the expression.) In order to study how emotional information from faces and bodies is combined, Meeren, van Heijnsbergen, and de Gelder (2005) created composite images in which angry/fearful faces were superimposed on angry/fearful bodies, thus generating both congruent (e.g. fear face + fear body) and incongruent (e.g. fear face + angry body) combinations. Participants asked to categorize the faces as angry or fearful were affected by the emotional status of the body, suggesting that this information is hard to ignore. An ERP study using the same stimuli suggests that the brain detects an emotional mismatch between face and body quickly, at around 100 ms (Meeren et al., 2005).

De Gelder (2006) put forward a model of emotional body language perception. The model contains two circuits and it is noteworthy that both routes have a visual perception component and a motor-based component. This raises the possibility that emotional body language, like facial expressions, could be recognized on the basis of visual appearance, or on the basis of motoric simulation of the body language, or both. Observers do not necessarily have to reproduce an action themselves, but the suggestion is that watching emotional body language prepares one for an action and that this preparatory signal may itself convey information about the other person.

1. *Reflex-like perception of emotional body language.* This involves a rapid and unconscious evaluation of the visual stimulus using mainly subcortical visual pathways (e.g. from the pulvinar and superior colliculus to the amygdala). This may initiate over-learned motor responses through a subcortical basal-ganglia

When asked to judge facial expressions of anger or fear, the emotional body expression interferes with this decision. This implies that emotional body language is recognized automatically. From de Gelder (2006). Copyright © 2006 Nature Publishing Group. Reproduced with permission.

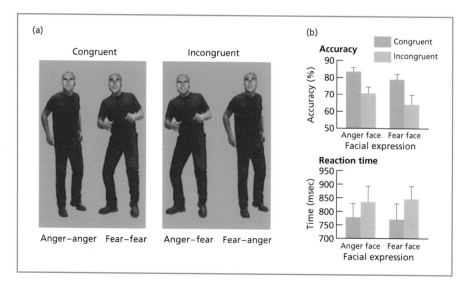

motor circuit. Lesions to the basal ganglia, such as those found in Parkinson's disease, lead to better performance at producing posed emotional expressions than spontaneous ones (Smith, Smith, & Ellgring, 1996). Patients with Parkinson's disease are often noted to have a 'mask-like' blank facial expression.

2. *Visuo-motor perception of emotional body language.* This involves a slower and more deliberative assessment that uses predominantly cortical structures. Visual analysis is assumed to occur in the fusiform and STS regions, which connect to emotional centers (e.g. the amygdala), and also regions containing mirror neurons in the intraparietal sulcus and premotor cortex.

This model is yet to be tested in detail. However, there is good evidence of social mirroring of body language, which is a key idea in the model. Chartrand and Bargh (1999) asked two participants to cooperate in a task. However, one participant was actually an experimenter who was instructed to surreptitiously perform one of two body movements – nose rubbing or foot shaking. Different participants saw different actions so that the unperformed action in the pair served as a control. Participants who saw foot shaking were more likely to produce foot shaking than nose rubbing and the reverse was true for those who saw nose rubbing. In a second study, Chartrand and Bargh (1999) asked the confederate to surreptitiously copy the body language of the participant (or not, as a control condition). Participants whose body language was imitated reported liking the confederate more and judged that the task was completed more smoothly. This important result suggests that mirroring of body language serves an important social function that goes beyond simulation for the sake of identifying a person's emotional state, as implied by de Gelder's model.

Evaluation

In terms of early visual processing, body perception relies on different neural substrates to face perception (the EBA is particularly important). However, beyond these earlier stages there is evidence of convergence. Bodies, like faces, activate

VOICES CONTAIN IMPORTANT SOCIAL CUES TOO

Voices, like faces and bodies, convey a large amount of socially relevant information about the people around us. It is possible to infer someone's sex, size, age, and mood from their voice. Physical changes related to sex, size, and age affect the vocal apparatus in systematic ways. Larger bodies have longer vocal tracts and this leads to greater dispersion of certain frequencies (the formants found, for example, in human vowels and dog growls are more dispersed in larger animals). Adult men have larger vocal folds (17–25 mm) than adult women (12.5–17.5 mm), resulting in a lower pitched male voice. One can also even infer the current emotional state (angry, sad, etc.) from a voice even in an unfamiliar language (Scherer, Banse, & Wallbott, 2001). Individual differences in the shape and size of the vocal apparatus (teeth, lips, etc.) and resonators (e.g. nasal cavity), together with learned speaking style (e.g. accent), create a unique voice signature. As with faces, there is evidence that identifying a speaker uses different cognitive and neural resources from identifying emotional states from a voice (Belin, Zatorre, Lafaille, Ahad, & Pike, 2000).

the STS when they are dynamic (biological motion) and emotional body language activates regions of the brain specialized for emotion. Body perception is important for imitation, which promotes both social cohesion and social learning.

JOINT ATTENTION: FROM PERCEPTION TO INTENTION

Joint attention refers to the process by which attention is oriented to a particular object/location in response to another person's attention. Direction of eye gaze is important, but head and body orientation, or pointing, can elicit joint attention too. This may provide the foundations for making inferences about other people's intentions and actions.

Eye gaze detection

Making eye contact can be important for establishing one-to-one communication (dyadic communication), and the direction of gaze can be important for orienting attention to critical objects in the environment. Direct eye contact, in many primates, can be sufficient to initiate emotional behaviors. Macaques are more likely to show appeasement behaviors when shown a direct gaze relative to indirect or averted gazes (Perrett & Mistlin, 1990). Moreover, dominance struggles are often initiated with a mutual gaze and terminated when one animal averts its gaze (Chance, 1967). The discrimination of gaze direction in humans may be easier than in other animals due to the smaller dark region (pupil and iris) surrounded by the

KEY TERM

Joint attention
The process by which attention is oriented to a particular object/ location in response to another person's attention.

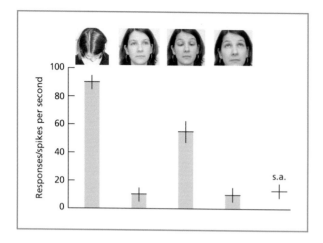

This neuron responds when gaze is oriented downwards. The activity of the neuron (spikes per second) is shown when presented with four faces and during spontaneous activity (s.a.). Adapted from Perrett et al. (1992).

Gaze cueing is modulated by head orientation. In this picture, attention is cued to the right because the person is looking rightwards (from his perspective), even though he looks straight ahead from our perspective.

white sclera. One suggestion is that the white sclera evolved specifically to facilitate joint attention and enable cooperation (e.g. Tomasello, Hare, Lehmann, & Call, 2007).

Baron-Cohen argues that an eye direction detector is an innate and distinct component of human cognition (Baron-Cohen, 1995a; Baron-Cohen & Cross, 1992). Babies are able to detect eye contact from birth, suggesting that it is not a learned response (Farroni, Csibra, Simion, & Johnson, 2002). This ability is likely to be important for the development of social competence because the eyes code relational properties between objects and agents (e.g. 'mummy sees daddy', 'mummy sees the box').

The region that appears to be particularly important for eliciting joint attention from faces and bodies is the STS. Functional imaging shows that other regions such as the intraparietal sulcus are important for directing attention more generally from both social (e.g. eyes) and non-social (e.g. arrows) cues, but the STS responds particularly to social attention cues (Hooker et al., 2003). The effect is greater when the eyes look at an object rather than empty space (Pelphrey, Singerman, Allison, & McCarthy, 2003). The STS of monkeys contains many cells that respond to eye direction (Perrett et al., 1985) and lesions in this area can impair the ability to detect gaze direction (Campbell, Heywood, Cowey, Regard, & Landis, 1990). Cells in the STS are not only sensitive to the direction of the eyes but also to the orientation of the head and the body. Perrett, Hietanen, Oram, and Benson (1992) report that the response rate of single cells depends mostly on eye gaze, then head orientation, followed by body posture. Some neurons may respond, for example, to a downward oriented gaze irrespective of whether the head is pointing down or whether the eyes are pointing down with the head full frontal. This suggests that these particular neurons are indeed coding the locus of *attention* rather than the orientation of particular body parts.

In reaction time experiments, the direction of eye gaze appears to automatically orient participants to a particular location – termed **gaze cueing** (e.g. Frischen, Bayliss, & Tipper, 2007). Thus, eyes seen looking to the left will facilitate detection of a visual target presented on the left, relative to the right. Similar effects are found when participants view head stimuli that are oriented to the left or right (Langton & Bruce, 1999). However, when head and eye cues are independently manipulated the results are more complex (Hietanen, 1999). Thus a head oriented towards the left but looking towards the right can be an effective cue to a target on the right. In this example, the person is looking towards *his* right side even though the eyes are looking directly at the observer.

Finally, the direction of gaze and the processing of facial expression can interact with each other. A facial expression of anger is recognized faster if the gaze is direct, and a facial expression of fear is recognized faster if gaze is averted (Adams & Kleck, 2003). This is unlikely to be a perceptual level of interference between gaze and expression, but rather a more conceptual level of interference concerned with inferring behavioral intentions of approach or avoidance. Anger and direct gaze

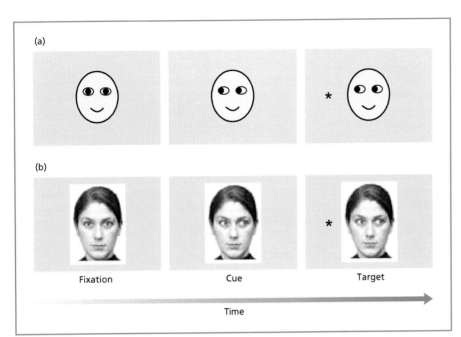

(a)

(b)

Fixation   Cue   Target

Time

In gaze cueing experiments, participants must press a button when they see a particular target (e.g. an asterisk). Performance is speeded up if gaze is directed towards the target (top) and slowed when gaze is directed away (bottom). From Frischen et al. (2007). Copyright © 2007 American Psychological Association. Reproduced with permission.

both signal approach whereas fear and averted gaze both signal avoidance, and when these two cues are mismatched the behavioral intention is more ambiguous. Mismatched gaze and expression signals are reflected in greater amygdala activity than matched signals (Adams, Gordon, Baird, Ambady, & Kleck, 2003).

WHY DO AUTISTIC PEOPLE OFTEN FAIL TO MAKE EYE CONTACT?

The social interactions of autistic people are characterized by an absence of joint attention and a failure to make direct eye contact (Sigman, Mundy, Ungerer, & Sherman, 1986). This could be due to either a difficulty in detecting where people are looking or a failure to understand the social significance of this behavior. In order to distinguish between these two possibilities, Baron-Cohen, Campbell, Karmiloff-Smith, Grant, and Walker (1995) devised a number of tests. In one test, children with autism were asked whether the eyes of another person are directed at them. They were unimpaired at this. They do, however, have difficulties in using gaze information to predict behavior or infer desire. In the four-sweets task, a cartoon face of Charlie directs his gaze to one of the sweets (Baron-Cohen et al., 1995). Children with autism are unable to decide: 'which chocolate will Charlie take?' or 'which one does Charlie want?'.

Children with autism are able to detect which person is looking at them (top), but are unable to infer behavior or desires from eye direction (bottom). For example, they are impaired when asked 'which chocolate will Charlie take?' or 'which one does Charlie want?'. Top photo from Baron-Cohen and Cross (1992). Copyright © 1992 John Wiley and Sons. Reprinted with permission. Bottom panel from Baron-Cohen et al. (1995). Reproduced with permission from *British Journal of Developmental Psychology* © British Psychological Society.

Pointing and reaching

Pointing is a seemingly simple behavior, but very few animals do it. Most non-human animals do not understand it – they look at the finger tip, not at where it is aimed. Wild chimpanzees do not point, but captive ones raised with human contact do to a limited extent (Leavens, Hopkins, & Bard, 2005). It has even been suggested that dogs, domesticated through human contact, can outperform chimpanzees on understanding pointing (Hare, Brown, Williamson, & Tomasello, 2002).

Researchers distinguish between at least two types of pointing behavior: **proto-imperative pointing** is related to wanting (e.g. meaning 'give me that!') whereas **proto-declarative pointing** elicits joint attention for its own sake (e.g. meaning 'look at that!'). The latter, in particular, is considered to contain an element of 'mind reading' because it requires computing what the other person can and cannot see, whereas the former could be acquired from reward-based learning in the same way as a child learns to put his or her arms in the air to be picked up. A failure to engage in proto-declarative pointing at 18 months is an early behavioral marker of autism, several years before theory-of-mind tests are administered (Baird et al., 2000).

Materna, Dicke, and Thier (2008) asked participants to move their eyes to where the person is looking and contrasted this with the same task elicited by watching a person pointing. Functional imaging reveals that both tasks engage the same region in the posterior STS, and suggests that the region has a more general role in social orienting of attention.

Although it is questionable whether monkeys point, or understand human pointing, their brains contain cells in STS that support the visual decoding of such behavior. Other mechanisms, absent or less developed in monkeys, presumably support the decoding of behavior in terms of mental states. Jellema, Baker, Wicker, and Perrett (2000) reported cells in the STS that respond to the sight of the arm reaching when the person directs attention to the same location. When the same reaching movement is performed but the person looks away then the response is significantly lower – the

KEY TERMS

Proto-imperative pointing
Pointing that implies wanting (e.g. meaning 'give me that!').

Proto-declarative pointing
Pointing that elicits joint attention for its own sake (e.g. meaning 'look at that!').

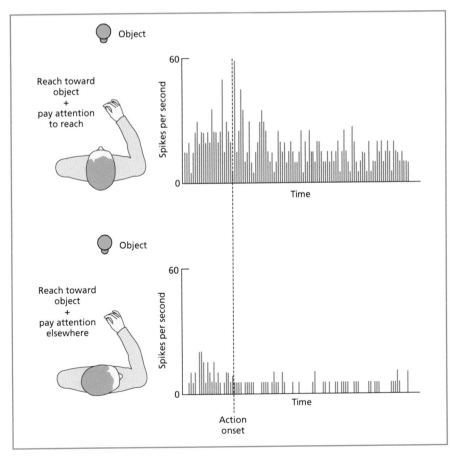

Some neurons respond only when the monkey watches an action performed when the actor also attends to the action. This kind of mechanism may distinguish intentional from accidental actions. From Jellema and Perrett (2005).

neurons did not respond strongly to head orientation per se, but only to the conjunction of head orientation and reaching. They argue that cells such as these provide the building blocks for understanding actions that are intended and goal directed, as opposed to accidental. Arm movements that are accompanied by attention are more likely to be intentional. Pointing would be one specific example of such an action.

In summary, gaze detection and pointing or acting (particularly when accompanied by gaze) provide clues about the intentions of others – such as what they want, whether an action was deliberate, and whether they want to engage you in an activity. The STS is important for the visual decoding of this information but other regions of the brain are implicated in making mentalizing inferences based on this information (see Chapter 6).

TRAIT INFERENCES FROM FACES AND BODIES

Traits, in psychology, refer to long-term dispositions to behave or think in a particular way – for example, being extrovert, caring, a worrier, a risk-taker, and so on. Our **personality** can be viewed as a collection of such traits.

The suggestion that our face may be a window into our traits has a long history. The 'science' of physiognomy attempted to match facial characteristics onto traits such as criminal behavior but with no real success (Alley, 1988). However, contemporary researchers have a better handle on what the core set of traits might consist of (criminal behavior is not one of them) and rely on image-based configural techniques (e.g. that compute the relative separation of features) rather than measures of face parts such as nose length.

The sections below will consider evidence that people make trait inferences on the basis of superficial information, and that people tend to agree (more than chance) on which traits belong with which people. Of course, proving an above-chance agreement between people does not make the association objectively true. Even if it were objectively true at the group level (i.e. in terms of statistical reliability), it does not mean that it holds true for each and every individual. In the discussion below, several possibilities should be borne in mind:

- *'A grain of truth'*. It is possible that there are real associations between certain facial characteristics and certain personality traits. For example, testosterone affects masculine facial development and is also known to influence certain behaviors (i.e. correlated development of traits and physical characteristics).
- *Self-fulfilling prophecies*. People with certain facial characteristics may tend to be treated in particular ways that reinforce a particular behavioral outcome. For example, attractive people may receive more positive social interactions whereas people with facial abnormalities may tend to be shy or withdrawn.
- *Using expression cues to make trait inferences*. A smile may be used to infer friendliness, but even in neutral expressions people may infer traits from structural features that resemble expressions, such as upturning lips (Said, Sebe, & Todorov, 2009).
- *Culturally generated stereotypes with little or no objective basis*.

In all these cases, there is clearly an over-generalization. However, the accounts differ in terms of the source of the signal used for this over-generalization and the objective reliability of that signal.

Beautiful = good

The factors that make a face attractive are now fairly well understood. In a recent review, Rhodes (2006) lists three factors:

1. *Averageness.* Langlois and Roggman (1990) studied composite images of faces and found that as more faces were averaged together then attractiveness increased (up to a composite of about 16 faces). This factor is related to symmetry (averaging faces makes them more symmetrical) but is not identical to it. For example, averageness remains an important predictor of attractiveness when profiled faces are used (Valentine, Darling, & Donnelly, 2004).
2. *Symmetry.* In non-human animals, increased asymmetry has been related to difficulties in withstanding physical stress during development, poor nutrition, and inbreeding. Perfectly symmetrical faces created by morphing an image with its mirror image are judged to be more attractive, even when averageness is excluded (Rhodes, Sumich, & Byatt, 1999).
3. *Sexual dimorphism.* At puberty, increased testosterone in males stimulates growth of the jaw, cheekbones, brow ridges, and nose. In females, growth of these traits is inhibited by estrogen, which may also lead to the growth of fuller lips. There is evidence that females with enhanced feminine features are rated as more attractive than averaged female faces (Perrett, May, & Yoshikawa, 1994). The effects for enhanced masculine features in males tends to be weaker (see Rhodes, 2006).

KEY TERM

Halo effect
A person who is rated positively in one dimension tends to be rated positively in other dimensions.

In a classic study called 'What is beautiful is good', Dion, Berscheid, and Walster (1972) found that photographs of attractive faces were judged to be more likely to have socially desirable personality traits. Each person was given three photographs to rate on 27 personality traits (e.g. exciting vs. dull). The photos were of attractive, neutral, and unattractive faces (either three male or three female photos). The tendency to associate positive traits with attractive people was found irrespective of the sex of the rater or the sex of the face. This effect is related to the **halo effect** in which a person who is rated positively in one dimension tends to be rated positively in other dimensions. For example, in a mock videotaped interview featuring a potential teacher to the students, the teacher was either warm and friendly or cold and distant (Nisbett & Wilson, 1977). Participants rated the interviewer's accent and physical appearance more negatively in the cold condition even though these aspects were the same in both interviews.

The hypothesis that 'what is beautiful is good' implies that we assign positive personality traits to attractive faces. However, some theories in evolutionary psychology would turn this on its head by arguing that traits that we consider desirable (e.g. physically strong males, youthful females) have determined what we consider as beautiful (e.g. Buss, 1989). Indeed, cultures tend to agree not only on face

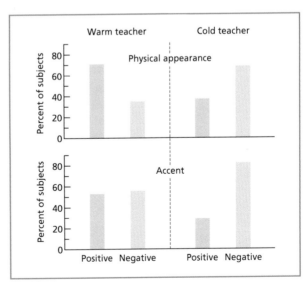

If a person behaves in a warm or cold manner then irrelevant attributes (such as appearance, accent) become judged in a positive or negative light. From Nisbett and Wilson (1977). Copyright © 1977 American Psychological Association. Reproduced with permission.

Which of these four faces do you find most attractive? Your answer to the question depends on the traits you desire in a partner. People who desire partners who are not easy going tend to choose (a) and those desiring easy-going partners choose (b). Those desiring non-assertive and assertive partners choose (c) and (d), respectively. From Little et al. (2006). Copyright © 2006 Elsevier. Reproduced with permission.

attractiveness but also on what long-term traits are desired in male and female partners (e.g. Buss, 1989). Nevertheless, within each culture there are individual differences in terms of what traits are desired – for example, women differ in the extent to which they desire a dominant or cooperative partner. As such, one might predict that facial attractiveness is related to the personality traits that people desire in a partner rather than (or in addition to) general factors such as symmetry and averageness. Little, Burt, and Perrett (2006) provided evidence for this prediction. For instance, women who find masculine personality traits attractive would tend to find men with masculine facial characteristics attractive, whereas those favoring more easy-going and less-assertive traits in a man would find baby-faced men attractive. The results of this study suggest that attractiveness is not just a structural property of a face but is also a projection of the traits that we value in a partner.

The 'Big Five' personality traits

Contemporary research into personality has attempted to address the question of how many traits are needed to describe the range of personalities in a population. According to Costa and McCrae (1985), a five-factor model offers the most satisfactory solution and the corresponding traits are known as the **Big Five**: openness to experience; conscientiousness; extraversion; agreeableness; and neuroticism (or emotional stability). Collectively, they spell the acronym OCEAN.

Penton-Voak, Pound, Little, and Perrett (2006) obtained black-and-white face photos of 294 people and each of these people rated their own personality by questionnaire based on the Big Five. These photos were then shown to an independent group of raters who were asked to rate the photographs using the same questionnaire. Male and female faces were analyzed separately because personality is known to differ across sexes: for instance, women rate themselves as more neurotic and more agreeable (Costa, Terracciano, & McCrae, 2001). As such, raters could be expected

THE BIG FIVE PERSONALITY TRAITS

- *Openness to experience*: appreciation for art, emotion, curiosity, and unusual ideas and activities.
- *Conscientiousness*: a tendency to show self-discipline, be meticulous, and aim for achievement.
- *Extraversion*: a tendency to seek stimulation and the company of others.
- *Agreeableness*: a tendency to be compassionate and cooperative towards others.
- *Neuroticism*: emotionally reactive and vulnerable to stress.

to produce above-chance results based on gender stereotypes rather than facial characteristics. For male faces, there was a significant correlation between self-reported personality and independent ratings based on faces alone for three traits: extraversion, neuroticism, and openness to experience. For female faces, only extraversion came out as significant. In a second study, Penton-Voak et al. (2006) produced composite faces based upon the top and bottom 10% rating of self-reported traits: blending the faces of the top 10% most extraverted people to create an 'extrovert face'; blending the faces of the 10% least extraverted faces to create an 'introvert face'; and so on. Independent raters tended to rate the composite images as more attractive when they had been derived from socially desirable personality traits (e.g. high agreeableness, low neuroticism, etc.).

Aggressive face versus baby face

Aggressive behavior is defined as any act that is intended to harm another individual who is motivated to avoid the behavior (Baron & Richardson, 1994).

Carre and McCormick (2008) investigated the facial width-to-height ratio (measured between lip and brow) of professional and amateur ice hockey players. This ratio is believed to be independent of body size and may be related to facial growth at puberty stimulated by testosterone. A lower width-to-height ratio is associated with a rounder, more baby-like face in males. In order to assess actual aggressive behavior they counted up the penalty minutes that each ice hockey player had accrued for behavior such as slashing, elbowing, and fighting. This objective measure of aggression correlates with facial width-to-height ratio.

Oosterhof and Todorov (2008) adopted a somewhat different procedure by obtaining fourteen different trait ratings for a large number of faces and then using computerized statistical techniques to determine what the underlying structure is. They

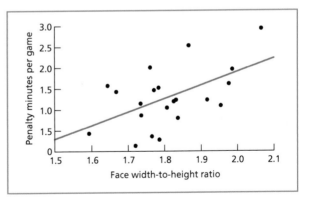

The facial width-to-height ratio predicts aggressive behavior in ice hockey players. Why might testosterone be a mediating factor? From Carre, J. M., & McCormick, C. M. (2008). In your face: Facial metrics predict aggressive behaviour in the laboratory and in varsity and professional hockey players. *Proceedings of the Royal Society of London, Series B, 275*(1651), 2651–2656. Copyright © 2008 The Royal Society. Reproduced with permission.

Oosterhof and Todorov (2008) found that trait ratings from faces can be described in terms of two independent dimensions: how dominant the face appears (aggressive looking vs. baby-faced) and how positive or negative the face is judged (which relates to perceived trustworthiness). From Todorov, Said, Engell, and Oosterhof (2008). Copyright © 2008 Elsevier. Reproduced with permission.

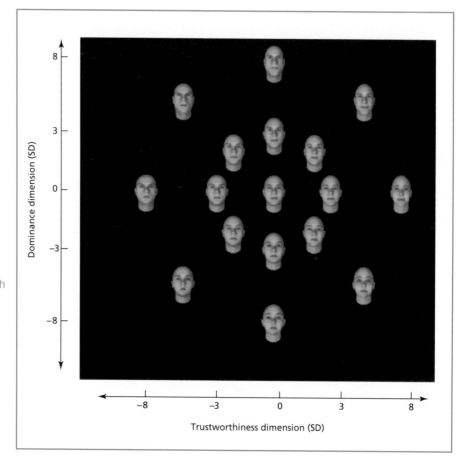

found that faces could be described in terms of two separate traits: how dominant the face appears (this is linked to how aggressive or baby-faced it appears) and how positive/negative the face is judged. Trustworthiness was the trait that was most strongly linked with this positive/negative component. The authors noted a tendency for the trustworthy faces to have mouths resembling a smile (up-turned lips) and untrustworthy faces to have mouths resembling anger (down-turned lips). Thus, even though neutral facial expressions were used, participants might still use subtle differences relating to expression detection to infer traits.

Berry and Brownlow (1989) found that men who were independently rated to have 'baby faces' rate themselves as being less aggressive, more approachable, and warmer than non-baby-faced men, whereas women who were independently rated to have 'baby faces' rate themselves as having low physical power and low assertiveness. Interestingly, baby-faced adolescents from low socio-economic backgrounds are *more* likely to enter into delinquency, which suggests that these individuals may actively react against the social stereotype of low dominance that is applied to them (Zebrowitz, Andreoletti, Collins, Lee, & Blumenthal, 1998). Baby-faced men are more likely to receive lighter sentences for crimes because they are probably judged (correctly or incorrectly) to have lower trait aggression (Zebrowitz & McDonald, 1991).

Trustworthiness

One possible metric that people may use to evaluate trustworthiness from a face is to determine whether the face looks similar to oneself. One may tend to trust people who look like oneself because such people may include family members. DeBruine (2005) tested this theory by morphing faces of strangers to be closer to the participant's own face and showed that such faces are judged as more trustworthy. Interestingly, they are judged as *less* attractive, arguably as an evolutionary mechanism against incest (i.e. to prevent sexual attraction to possible family members). Whatever the explanation, it suggests that trustworthiness and attractiveness are not the same thing.

Although self-similarity might be one mechanism for assessing trust, it is unlikely to be the only mechanism given that different individuals tend to agree on which faces are trustworthy (e.g. Oosterhof & Todorov, 2008). In one of the earliest functional imaging studies of trustworthiness, Winston, Strange, O'Doherty, and Dolan (2002) found that activation in the amygdala increased with increasing levels of untrustworthiness. Lesions to the amygdala impair evaluation of trustworthiness, such that untrustworthy faces are deemed trustworthy (Adolphs, Tranel, & Damasio, 1998). But congenital prosopagnosics who are unable to recognize familiar faces are still able to attribute trustworthiness in similar ways to those without face recognition difficulties (Todorov & Duchaine, 2008). Similarly, people with Asperger's syndrome are able to evaluate trustworthiness in faces (White, Hill, Winston, & Frith, 2006). According to Todorov and Engell (2008) the role of the amygdala in trustworthiness judgments may be related to categorizing stimuli according to positive and negative valence, but is unlikely to be involved in social trait attribution per se. For instance, they found that amygdala activity in fMRI was related to trustworthiness ratings but not to dominance ratings (using the aggressive vs. baby-faced stimuli discussed above). Dominance is less easy to categorize in terms of being 'good' or 'bad' than levels of trust.

Trait inferences from bodies

There is very little evidence concerning trait inferences from bodies, especially from a neuroscience angle. Swami et al. (2008) asked participants to make trait inferences to different images of male and female bodies that vary in body size (the body mass index, BMI). Increasing BMI was associated with the perceived trait of laziness for both male and female bodies. There was also evidence that people attribute loneliness to certain body sizes, specifically for overweight people of either sex and for underweight women. Kramer, Arend, and Ward (2010) extracted biological motion cues (displayed to participants using point light displays) of Barack Obama and John McCain during presidential debates. (A second experiment used UK political leaders.) Participants were told only that the movies were from people giving a public speech and they were asked to rate them on attractiveness, trustworthiness, caring, dominance, leadership, anxiety, depression, and physical health. There was some agreement across participants on trait inferences. For instance, Obama was rated as more trustworthy and dominant, whereas McCain was rated as more anxious. After the ratings were collected participants were asked who they would vote for (again they did not know who the people were). In terms of voting behavior, perceived physical health was the strongest predictor.

Evaluation

The start of this section considered various possible explanations for a link between faces and traits, and it would be interesting to return to this in light of the evidence. The strongest evidence for the 'grain of truth' explanation comes from the link between aggression (measured through both self-report and objectively) and independently rated 'aggressive looking' faces. With regard to trustworthiness, the judgment appears to be an over-generalization of expression-relevant mechanisms to trait judgements – neutral expressions with a smile-like facial configuration are rated as trustworthy. It is unclear whether there is a grain of truth in this. Attractive faces are rated as having attractive traits, and it is argued that this reflects our desire to seek out certain traits in potential mates.

SUMMARY AND KEY POINTS OF THE CHAPTER

- Face perception involves a number of different mechanisms with somewhat different neural substrates. The occipital face area (OFA) may compute structural properties of a face whereas the fusiform face area (FFA) computes facial identity.
- To some extent, recognizing facial expressions is separable from recognizing facial identity, but there are several candidate mechanisms for recognizing expressions: using dynamic information (e.g. STS); mapping faces onto regions specialized for emotional stimuli (e.g. amygdala, insula); or by simulating the expression motorically (i.e. the mirror system).
- The superior temporal sulcus (STS) is important for action perception for both faces and bodies, including lip reading, biological motion, eye gaze, and pointing. Eye gaze and pointing are important for establishing joint attention and inferring intentions.
- Bodies and voices provide other cues to socially relevant information. As with faces, there is evidence for some degree of separation between identifying people and recognizing their emotional state.
- There is a tendency for people to infer stable characteristics (i.e. traits) from the faces of others. Attractive faces tend to be linked to attractive traits. Similarly, faces judged to be aggressive or baby-faced tend to be linked with matching behavior (i.e. dominant vs. non-dominant behavior). These reflect a tendency to over-generalize and reflect the desire to infer the inner world of people around us based on sparse information.

EXAMPLE ESSAY QUESTIONS

- Is recognizing a facial expression different from recognizing other properties of a face?
- What kinds of social information might be conveyed in body perception and how is this related to face perception?
- What is the role of the superior temporal sulcus (STS) in face and body perception?
- Do we use facial information to infer the character of people around us? Is this information accurate?

RECOMMENDED FURTHER READING

There are no known books that cover faces and bodies predominantly from a social neuroscience perspective. As such, the following review papers are recommended:

- Calder, A. J., & Young, A. W. (2005). Understanding the recognition of facial identity and facial expression. *Nature Reviews Neuroscience*, 6(8), 641–651.

- Haxby, J. V., Hoffman, E. A., & Gobbini, M. I. (2000). The distributed human neural system for face perception. *Trends in Cognitive Sciences*, 4(6), 223–233.

- Peelen, M. V., & Downing, P. E. (2007). The neural basis of visual body perception. *Nature Reviews Neuroscience*, 8(8), 636–648.

- Todorov, A., Said, C. P., Engell, A. D., & Oosterhof, N. N. (2008). Understanding evaluation of faces on social dimensions. *Trends in Cognitive Sciences*, 12(12), 455–460.

CHAPTER 6

CONTENTS

Empathy and simulation theory   130

Theory of mind and reasoning about mental states   139

Explaining autism   145

Summary and key points of the chapter   154

Example essay questions   154

Recommended further reading   155

Understanding others

If you see someone yawning do you yawn too? Most people probably do to some extent. Some behavior, such as laughing and yawning, is socially contagious. But can any wider significance be attached to such findings? One study of contagious yawning in chimpanzees speculates that 'contagious yawning in chimpanzees provides further evidence that these apes possess advanced self-awareness and empathic abilities' (Anderson, Myowa-Yamakoshi, & Matsuzawa, 2004). Another study, this time on humans, administered tests requiring reasoning about the mental states of other people (e.g. beliefs, knowledge) as well as measuring yawn contagion, and concluded that 'contagious yawning may be associated with empathic aspects of mental state attribution' (Platek, Critton, Myers, & Gallup, 2003). Of course, there is unlikely to be anything special about yawning itself. There might be a general tendency to *simulate* the behavior of others on ourselves (internally in our minds and brains) even if we do not overtly *reproduce* it (as observable behavior on our bodies). Thus, we may understand others by creating a similar response in our brain to that found in the other person's brain. Contagious yawning, under this account, is one extreme example of this more general and, normally, more subtle tendency. This chapter will attempt to unpick these claims and place them alongside traditional concepts in social and cognitive psychology, such as empathy and theory of mind. The chapter will also consider how these processes may be disrupted after brain injury and in people with autism.

The overarching question of the chapter is how do we understand the mental states of others? **Mental states** consist of knowledge, beliefs, feelings, intentions, and desires. The process of making this inference has more generally been referred to as **mentalizing**. The term is generally used in a theory-neutral way, insofar as it is used by researchers from a wide spectrum of views. It could be contrasted with the

KEY TERMS

Mental states
Knowledge, beliefs, feelings, intentions and desires.

Mentalizing
The process of inferring or attributing mental states to others.

It just takes one yawn to start other yawns off. How does this kind of simple contagion mechanism relate to empathy and theory of mind?

term 'theory of mind', which has essentially the same meaning but has tended to be adopted by those advocating a particular position, namely the notion that there is a special mechanism for inferring mental states. According to some researchers, this theory-of-mind mechanism cannot be reduced to general cognitive functions such as language and reasoning, or those involved in imitating. These arguments lie at the heart of the social neuroscience enterprise in that they raise important and divisive issues about the nature of the mental and neural processes that support social behavior and the extent to which they are related to other aspects of cognition.

WHAT IS SIMULATION THEORY?

Simulation theory is not strictly a single theory but a collection of theories proposed by various individuals (e.g. Gallese, 2001; Goldman, 2006; Hurley, Clark, & Kiverstein, 2008; Preston & de Waal, 2002). However, common to them all is the basic assumption that we understand other people's behavior by recreating the mental processes on ourselves that, if carried out, would reproduce their behavior – that is, we use our own recreated (or simulated) mental states to understand, and empathically share, the mental state of others. Within this framework there are various ways in which this could occur. Gallagher (2007) broadly distinguishes between two: one could create an explicit, narrative-like simulation of another person's situation and behavior in order to understand it; or when we see someone else's behavior (e.g. their action, emotional expression) we may automatically, and perhaps unconsciously, activate the corresponding circuits for producing this behavior in our own brain. These latter versions of simulation theory tend to be intimately linked to the idea of mirror systems in which perception is tightly coupled with action.

EMPATHY AND SIMULATION THEORY

The word **empathy** is relatively modern, being little more than 100 years old. It was coined by Titchener (1909) from the German word *einfühlung* (Lipps, 1903) and originally referred to putting oneself in someone else's situation. This would also go under the contemporary name of **perspective taking**. This section will first consider the various different ways in which the term empathy is used today, which reveals potentially important differences in the way that it may be accounted for.

Empathy as a multi-faceted concept

If one starts with the working definition of empathy introduced above ('putting oneself in someone else's situation') it is clear that there are subtle, but potentially crucial, different ways in which this could be understood. Some of these are listed below and are an abridged version from Batson (2009):

1. Knowing another person's internal state, including his or her thoughts and feelings.
2. Adopting the posture or matching the neural response of an observed other.

3. Having an emotional reaction to someone else's situation, although it need not be the same reaction.

4. Imagining how I would feel/react in that situation (i.e. given *my* personal history, traits, knowledge, beliefs).

5. Imagining how the other person would feel/react in that situation (i.e. given *their* personal history, traits, knowledge, beliefs).

The first three scenarios differ with respect to whether the knowledge/feeling is the same in self and other. Knowing about another person's internal state need not necessarily imply that the observer shares that state. This important consideration lies at the heart of some tests of theory of mind, specifically **false belief** tasks, but they are relevant to some conceptions of empathy too. The second sense in which empathy is used ('adopting the posture or matching the neural response of an observed other') is the one most closely linked with mirror systems, imitation, and contagion (emotional contagion, yawning contagion, etc.). For example, one might feel **personal distress** in response to someone else's suffering. The third sense in which the term empathy may be used differs from the second in that the person's response is not matched. For instance, one might feel a sense of **pity** to another's situation or **sympathy** towards someone who is suffering. These reactions are directed outwards (other-oriented) rather than being self-oriented (as in personal distress), and the response of the perceiver does not match that of the other person. The fourth and fifth notions of empathy relate more directly to the idea of perspective taking, but they differ in the degree to which they are self-oriented versus other-oriented. The fourth scenario ('imagining how I would feel/react in that situation') could be construed as a shallow attempt to empathize, in which the level of success is dependent on self–other similarity rather than a true understanding of the other.

Given these somewhat different conceptions of empathy, it is not surprising that there is no single agreed-upon measure of empathy. Theory-of-mind tests, discussed in detail below, normally involve assessments based on linguistic reasoning of the sort 'If X believes Y then how will he/she behave in situation Z?' Others use neural or bodily responses to seeing others in pain, for example, as a measure of empathy (e.g. Bufalari, Aprile, Avenanti, Di Russo, & Aglioti, 2007; Jackson, Meltzoff, & Decety, 2005). Of

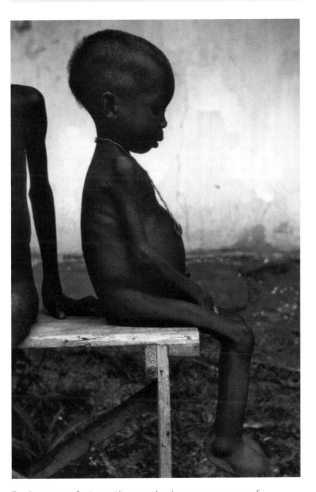

Do images of starvation evoke in you a sense of personal distress (self-focused) or a sense of pity or sympathy (other-focused)? Different individuals may have different reactions, although both can be broadly construed as empathic.

course, this presupposes a certain idea of what empathy is (i.e. that it can be measured solely in physiological ways). There are various questionnaire measures of empathy, such as the Interpersonal Reactivity Index (IRI; Davis, 1980) and the Empathy Quotient (EQ; Baron-Cohen & Wheelwright, 2004), which touch upon some of the distinctions discussed above. For example, the IRI contains separate subscales such as personal distress (items such as 'I tend to lose control during emergencies'), perspective taking (items such as 'Before criticizing somebody, I try to imagine how I would feel if I were in their place'), and empathic concern (items such as 'I often have tender, concerned feelings for people less fortunate than me'). One current trend is to incorporate questionnaire measures in functional imaging experiments. For example, watching someone drinking a pleasant or disgusting drink may activate the gustatory (taste) regions of the perceiver (Jabbi, Swart, & Keysers, 2007). Moreover, the extent to which this occurs may be greater in those people who report higher empathy on questionnaire measures (Jabbi et al., 2007). Findings such as these are often used to argue that the different concepts of empathy are related or, at least, share a common core (perhaps based upon simulation). Finally, one could potentially measure the ability to *accurately* empathize (i.e. to accurately state what another person is thinking or feeling) rather than the extent to which the person may report the motivation to empathize (i.e. most questionnaire measures) or to simulate that state themselves (which need not be linked to the ability to consciously report that state). As an example of such a test, the 'reading the mind in the eyes' test requires participants to match expressions in the eye region of faces to labels denoting mental states such as bored, sorry, or interested (Baron-Cohen, Wheelwright, Hill, Raste, & Plumb, 2001). Another test requires two participants to work together in a scenario that is video recorded. Each participant can then watch it back and report their own internal states as well as attempting to infer that of the other participant, thus enabling the experimenter to cross-reference the responses together in order to infer empathic accuracy (e.g. Ickes, 1993; Ickes, Gesn, & Graham, 2000). A recent functional imaging study based on this method found that empathic accuracy was related to a network of regions including the medial prefrontal cortex, implicated in mentalizing/theory of mind (although not the temporo-parietal junction), and the premotor cortex, which has been associated with mirror systems (Zaki, Weber, Bolger, & Ochsner, 2009).

Playful — Comforting — Irritated — Bored

The extent to which people can accurately detect the mental states of others (also called empathic accuracy) may differ from the extent to which they try to empathize or take perspectives. One test along these lines is the 'reading the mind in the eyes' test. From Baron-Cohen et al. (2001). Copyright © 2001 Association for Child Psychology and Psychiatry. Reproduced with permission from Wiley-Blackwell.

From imitation to empathy?

A link between imitation and empathy receives some support from social psychology. These studies generally use unconscious imitation in which the participant engages in a task with another person (a confederate) and the extent to which the participant imitates the confederate is measured. The participant is unaware of the true nature of the study (i.e. that his/her imitative behavior is being assessed). Participants who imitate more (based on blind scoring of their actions) whilst performing a cooperative task with a confederate tend to rate themselves as higher in trait empathy (Chartrand & Bargh, 1999). When the confederate deliberately imitates the participant in a cooperative task, then he/she is liked more by the participant than in a control condition in which imitation is avoided (Chartrand & Bargh, 1999). Van Baaren, Holland, Kawakami, and van Knippenberg (2004) showed that being imitated increases the chances of helping behavior when a confederate drops something. However, the effects are quite general. The person who has been imitated is not just more likely to help the imitator but they are more likely to help others too. It also increases the amount of money that the participant opts to donate to charity at the end of the experiment.

Iacoboni (2009) has argued that the mirror system for action may be co-opted by other regions of the brain to support empathy. Mirror neurons respond both when an animal performs an action and when it observes another performing the same (or similar) action – they act as a neural 'bridge' between self and other. They respond not just to the motor properties of an action but to the goal of the action. For example, it has been shown that neurons that code grasping actions respond in different ways to the sight of the grasp according to whether a container is present or absent (Fogassi et al., 2005). In this study the presence of the container was reliably associated with one particular goal, placing a piece of food inside it, whereas the absence was associated with another goal, eating it. In this example, the action is the same (grasp) but the subsequent goal is not and the mirror neurons (in the parietal lobe) respond according to the implied goal. Umilta et al. (2008) have shown that neurons that respond to grasping will also respond when pliers are used to grasp, even when a different action is required. In this example, the action is different but the goal is the same and the neural response is determined by the goal. Studies such as these have been used to argue that mirror neurons enable understanding of at least one mental state: intentions.

Carr, Iacoboni, Dubeau, Mazziotta, and Lenzi (2003) examined more directly a possible link between empathy and imitation using fMRI in humans. They showed

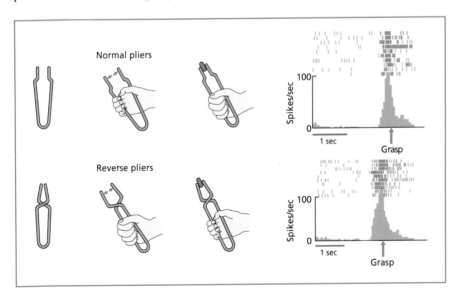

Mirror neurons respond to the same goal rather than the same action. Mirror neurons in monkeys responded similarly after training with both normal pliers and reverse pliers (maximum responding at point of grasping the food), even though both required different actions. From Umilta et al. (2008). Copyright © 2008 National Academy of Sciences, USA. Reproduced with permission.

participants emotional facial expressions under two conditions: observation versus deliberate imitation. (Note that this is different from the social psychology studies above, in which imitation was spontaneous rather than instructed.) They found increased activation for the imitation condition relative to observation in classical mirror system areas such as the premotor cortex. In addition, they found increased activation in areas involved in emotion, such as the amygdala and insula. Their claim was that imitation activates shared motor representations between self and other but, crucially, there is a second step in which this information is relayed to limbic areas via the insula. This action-to-emotion route was hypothesized to underpin empathy. Other studies have reported a positive correlation between questionnaire-based empathy scores and activation in the premotor region when observing actions (Kaplan & Iacoboni, 2006) or listening to actions (Gazzola, Aziz-Zadeh, & Keysers, 2006).

The model proposed by Carr et al. (2003) and Iacoboni (2009) is simple, but it is also perhaps simplistic. The assumption that limbic = emotion is an over-simplification (Le Doux, 1996), as is the claim that emotion imitation = empathy. As argued above, empathy is a broader concept than this. Recall also from Chapter 3 that the link between mirror neurons themselves and imitation is by no means uncontroversial. For example, monkeys (who possess mirror neurons) do not imitate tool use without extensive training.

It is possible to imagine alternative scenarios to the imitation-to-empathy model within a general simulation theory framework. For example, de Vignemont and Singer (2006) suggest that it may be possible to have simulation of emotions (and empathy for emotions) without having action/motor representations as a linking step. Singer et al. (2004b) investigated empathy for pain in humans using fMRI. The brain was scanned when anticipating and watching a loved-one suffer a mild electric shock. There was an overlap between regions activated by expectancy of another person's pain and experiencing pain oneself, including the anterior cingulate cortex and the insula. This provides evidence for a mirror system for pain – a system that responds to pain in self and other. However, there was little evidence that this system depends on the 'classic' mirror system for actions/goals that may support imitation.

Empathy beyond simulation

Some theories of empathy propose a variety of different mechanisms of which simulation is only one. In such models, simulation may either be a junior or senior partner.

As noted above, watching someone in pain activates certain parts of our own pain circuitry. This offers clear support for simulation theories. However, our beliefs about the person in pain can modulate or over-ride this mechanism. Singer et al. (2006) had participants in an fMRI scanner play a game with someone who plays fairly (a 'Goodie') and someone else who plays unfairly (a 'Baddie'). Mild electric shocks were then delivered to the Goodie and Baddie (who, of course, were only virtual characters but the participant did not know this). Participants empathically activated their own pain regions when watching the Goodie receive the electric shock. However, this response was attenuated when they saw the Baddie receiving the shock. In fact, male participants often activated their pleasure and reward circuits (such as the nucleus accumbens) when watching the Baddie receive the shock, which is the exact opposite of simulation theory. This brain activity correlated with their reported desire for revenge, which suggests that although simulation may tend to operate automatically it is not protected from our higher order beliefs.

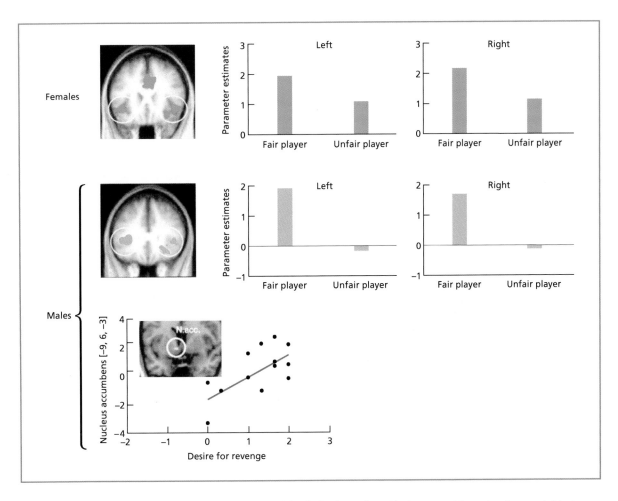

Females (pink) and males (blue) show reduced activity in brain regions that respond to pain when watching an unfair player receive a shock (shown here for the insula). In males, activity in the nucleus accumbens, measured whilst the unfair player received a shock, correlates with their self-reported desire for revenge. From Singer et al. (2006). Copyright © 2006 Nature Publishing Group. Reproduced with permission.

The findings of this study have implications for conditions associated with a lack of empathy, such as autism and psychopathy. It suggests that there are multiple reasons why empathy might fail – because of a failure to simulate the emotions of others or because of personally or socially constructed beliefs about who is 'good' and who is 'bad'. The eminent social psychologist Bandura (2002) argues that simulation has a relatively minor role to play in empathy, arguing that if it did it would lead to emotional exhaustion, which would debilitate everyday functioning. Moreover, Bandura (2002) argues that acts of inhumanity, such as genocide, depend on our ability to self-regulate and dissociate self from other. Although genocide is an extreme example, displaying lack of empathy towards socially marginalized groups (e.g. illegal immigrants, welfare cheats) could be regarded as a typical facet of human behavior.

Other studies support this view. Although doctors may be expected to show empathy for their patients, it would be unhelpful for them to experience personal distress when performing painful procedures. Indeed acupuncturists show less

It may be important for doctors performing painful procedures to switch off their empathic tendencies. What kind of mechanisms in the brain might support this?

activity, measured by fMRI, in the pain network (including the anterior insula and anterior cingulate) when watching needles inserted into someone, relative to controls (Cheng et al., 2007). Lamm, Batson, and Decety (2007) found that activity in these pain-related regions, induced by watching painful facial expressions induced by medical treatment, was modulated by the observer's beliefs about whether the treatment was successful or not (more activity in pain-processing regions when less successful). It was also related to whether the participants were instructed to imagine the feelings of the patient or to imagine themselves to be in that situation (more activity in pain-processing regions when imagining self). This suggests that the tendency to simulate is moderated by cognitive control (e.g. based on our beliefs) and also our efforts to take different perspectives.

Studies of imitation also show that the extent to which two people imitate each other depends on the characteristics of the imitator and the person being imitated, as well as characteristics of the social situation (van Baaren, Janssen, Chartrand, & Dijksterhuis, 2009). This suggests that imitation-based simulation is flexible and context sensitive, taking into account information beyond perception–action links. For example, imitation is less likely when the confederate has a social stigma such as a facial scar or is heavily obese (Johnston, 2002). Similarly, non-deliberate imitation of facial expressions is greater for one's ethnic ingroup relative to an outgroup (Bourgeois & Hess, 2008).

Some models of empathy propose a divide between so-called cognitive empathy and affective empathy (e.g. Baron-Cohen & Wheelwright, 2004; Shamay-Tsoory, Aharon-Peretz, & Perry, 2009). For example, in the experiment of Singer et al. (2006) the tendency to simulate another's pain would be part of the affective empathy system, and the representation of the other's intentions (to deceive or cooperate) would be part of the cognitive empathy system (which is often linked to a theory of mind in general). The ability to regulate (e.g. inhibit) the affective responses evoked by seeing another in pain would also be linked to this system. The terms 'cognitive' and 'affective' require some clarification. Many researchers would not regard emotions as existing outside of cognition (Lazarus, 1984; Phelps, 2006). A better terminology might be affective and non-affective empathy, as this stresses the different informational content. Patients with acquired brain damage to the orbital and ventromedial prefrontal cortex have difficulties in recognizing emotions in others (Hornak et al., 1996) as well as reporting feeling less emotions in themselves (Hornak et al., 2003). These patients may fail tests of theory of mind based on affective information but not on non-affective information (Shamay-Tsoory, Tibi-Elhanany, & Aharon-Peretz, 2006). This provides some support for the affective/non-affective ('cognitive') distinction. However, strictly speaking it does not prove that there is a separate affective theory-of-mind 'module', only that this kind of affective reasoning task depends on the integrity of regions that give rise to our own emotional feelings.

Most simulation theories do not fit squarely in either of the putative 'cognitive' or affective divisions. For example, emotion contagion would be an example of simulation based on affective information, whereas studies on action and mirror neurons suggest that it is possible to simulate goals and intentions, which are 'cognitive' (i.e. non-affective) mental states. Mirror neurons themselves are non-affective insofar as their response does not differ between actions that result in a reward (e.g. grasping food) and those that do not (e.g. grasping an object) (Rizzolatti & Craighero, 2004).

The model of empathy proposed by Decety and Jackson (2004, 2006) argues for a distinction between mechanisms based on simulation and other types of mechanism, but does not draw a sharp line between affective and non-affective processes. It brings together many of the strands discussed already. Decety and Jackson (2004) argue that there are three components of empathy:

1. *Shared representations between self and other, based on perception–action coupling.* This would include mechanisms for action understanding and imitation, emotional contagion, and pain processing. However, Decety and Jackson (2004) suggest that these are widely distributed throughout the brain rather than all loading on some core regions (such as premotor cortex).
2. *An awareness of self–other as similar but separate.* This is related to mechanisms of self-awareness (see Chapter 9) that enable us to attribute our own thoughts and actions as self-generated. Decety and Jackson (2004) suggest that one important brain region for this process is the right temporo-parietal junction (rTPJ). For instance, this region responds more when watching a moving dot controlled by someone else's action relative to self-generated action (Farrer & Frith, 2002) and responds more when participants are asked to imagine someone else's feelings and beliefs compared to their own (Ruby & Decety, 2004).
3. *A capacity for mental flexibility to enable shifts in perspective and self-regulation.* Decety and Jackson (2004) suggest that this is a candidate for a uniquely human component of empathy. It involves deliberate perspective taking of another's situation, which may also involve inhibiting one's own beliefs and self-referential knowledge. People with high self-reported personal distress may tend to over-rely on emotional contagion rather than cognitive control. Eisenberg et al. (1994) have shown that individual differences in personal distress are related to ability to control and shift attention, and Spinella (2005) reports negative correlation between behavioral measures of executive function and personal distress. Decety and Jackson (2004) suggest that regions in the prefrontal cortex responsible for the control of emotions (ventromedial and orbital regions) and the control of thought and action (lateral regions) are important. A region in the medial prefrontal cortex (considered below and in Chapter 9) responds to self-referential perspective relative to other perspective.

As such, this model offers a good account of the multi-faceted nature of empathy both in terms of cognitive mechanisms, social influences, and neural substrates. It also offers one way of connecting the literature on empathy with the other main topic of this chapter: theory of mind.

Evaluation

Empathy should perhaps best be regarded as a multi-faceted concept, and is likely to be explained via several interacting mechanisms rather than a single one. One possible division is between affective and cognitive (or non-affective) empathy, in which the former is based on emotion simulation and the latter on mental state reasoning.

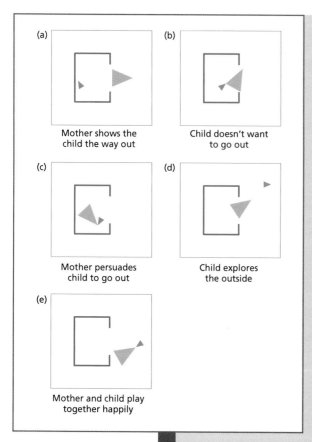

(a) Mother shows the child the way out

(b) Child doesn't want to go out

(c) Mother persuades child to go out

(d) Child explores the outside

(e) Mother and child play together happily

Mental states (e.g. want), behaviors (e.g. play), and other human characteristics (e.g. mother, child) are readily attributed to animated geometric shapes. Watching these animations, during functional imaging, activates a network of regions implicated in theory of mind. The captions were not presented in the studies, but are shown here for clarification. From Castelli et al. (2000). Copyright © 2000 Elsevier. Reproduced with permission.

PROJECTING MENTAL STATES EVERYWHERE – THE ORIGINS OF ANTHROPOMORPHISM?

Anthropomorphism refers to the attribution of human characteristics to non-human animals, objects, or other concepts. This could reflect a natural tendency to attribute mental states externally, and not just to other humans who are 'like me'. Living objects are commonplace in our popular culture – think of Pixar's bouncing lamp. It has also been suggested that a belief in God is a result of the tendency to attribute mental states externally (Guthrie, 1993).

To some extent, the tendency to anthropomorphize may depend on whether something looks like us – an angry dog shows its teeth like an angry human. Movement as well as appearance is important. Heider and Simmel (1944) found that people readily ascribe mental states to animations of two interacting geometric objects, such as 'the blue triangle wanted to surprise the red one'. In a functional imaging study that compared these kinds of animations with aimless movements, it was found that these moving shapes activated a network of regions that are typically activated in theory-of-mind tasks (Castelli, Happe, Frith, & Frith, 2000). They argued that this supports the idea that intentions tend to be inferred from actions, even in situations in which participants know that the objects are not capable of having mental states.

Although anthropomorphism may be a universal tendency, some people may do it more and others may do it less. One study found that this tendency, measured in terms of mental state ratings for gadgets or terms used to describe pets, is greater in lonely people (Epley, Akalis, Waytz, & Cacioppo, 2008). This suggests that it may be a compensatory mechanism for social isolation. In contrast, people with autism use less mental state terms to describe the moving geometric shape stimuli and show less activity in regions linked to theory of mind when watching these animations (Castelli, Frith, Happe, & Frith, 2002).

The model of Decety and Jackson proposes a set of different mechanisms that underpin empathy, but without evoking a dichotomy between cognitive/affective empathy. The idea of simulation is likely to remain an important component of models of empathy for the foreseeable future, but whether or not it is the main or core component of empathy in real-life social situations remains to be determined. Certainly, there is evidence that behaviors related to simulation, such as emotion contagion, are modulated by social biases, beliefs, and deliberate attempts at cognitive control (e.g. when deliberately adopting the other perspective).

THEORY OF MIND AND REASONING ABOUT MENTAL STATES

This section distinguishes itself from the previous one by considering in detail a certain kind of task: namely deliberate attempts to reason about mental states, and deliberate attempts to attribute mental states to others. To some extent these sorts of mechanisms are linked to those involving empathy, as discussed previously. However, the tasks used in the theory-of-mind literature are typically quite different from those considered previously in the section on empathy. The stimuli themselves are typically narratives or sequences of events, rather than observation of a particular state (e.g. pain). The tasks also typically require an overt response (e.g. what does Sally think or do?) whereas studies on empathy often do not (e.g. a typical measure could be degree of imitative behavior or subtle contraction of facial muscles). We may be able to tell from someone's face or voice that they are being thoughtful, but knowing what they are thinking may involve a different computation.

The term 'theory of mind' derived originally from research on primate cognition. Premack and Woodruff (1978) conducted a number of studies on a chimpanzee to see if it understood an experimenter's intentions. For example, the chimp might point to a picture of a key when an experimenter was locked in a cage, the inference being 'he wants to get out'. A number of criticisms were leveled at the study. For instance, it may reflect knowledge of object associations (e.g. between key and lock) rather than mental states. In a reply to the article, Dennett (1978) suggested that one way of testing for theory of mind would be to consider false beliefs, in which someone else may hold a mental state (e.g. a belief) that differs from one's own belief and from the current state of reality. In developmental psychology, the paradigmatic false belief test is the object transfer task, such as the Sally–Anne task (Baron-Cohen, Leslie, & Frith, 1985; Wimmer & Perner, 1983). Sally puts a marble in a basket so that Anne can see. Sally then leaves the room, and Anne moves the marble to a box. When Sally enters the room, the participant is asked 'where will Sally look for the marble?' or 'where does Sally think the marble is?' A correct answer ('in the basket') is typically taken to indicate the presence of a theory of mind. An incorrect answer is potentially more problematic to interpret. It could imply a lack of theory of mind. However, one also has to rule out other factors such as language comprehension difficulties or a failure to inhibit a more dominant response (one's own belief). False beliefs are harder to accommodate within simulation theories because one's own belief is at odds with that attributed to the other person. This cannot be done by straightforward simulation involving shared self–other representations. It requires taking one's own mental states 'offline' and creating a hypothetical scenario different to current reality. So-called meta-representation and pretense is often regarded as a hallmark of theory-of-mind ability (Leslie, 1987).

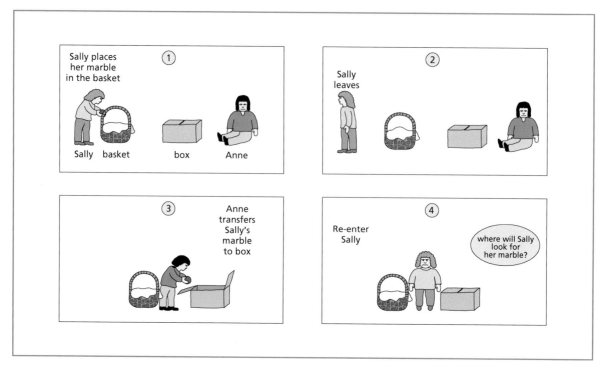

The Sally–Anne task requires an understanding of false belief and an attribution of second-order intentionality. Adapted from Wimmer and Perner (1983).

Social psychologists use the term **attribution** to refer to the process of inferring the causes of people's behavior. The philosopher Dennett (1983) uses his own term of **intentional stance** to refer to our tendency to explain behavior in terms of mental states, which could otherwise be considered synonymous with mentalizing or theory of mind. However, Dennett (1983) has a particularly useful way of describing different levels of intentionality that might be used to account for behavior. For example, an observer might have to evoke zero-order intentionality to explain the behavior of an object, first-order intentionality to explain the behavior of some animals, and second-order intentionality to explain some human behavior.

KEY TERMS

Attribution
In social psychology, the process of inferring the causes of people's behavior.

Intentional stance
The tendency to explain or predict the behavior of others using intentional states (e.g. wanting, liking).

First-order intentionality
An agent possesses beliefs and desires, but not beliefs about beliefs.

Second-order intentionality
An agent possesses beliefs about other people's beliefs.

- *Zero-order intentionality*. The assumption that an agent possesses no beliefs and desires. It responds to stimuli reflexively, such as producing a scream when frightened or running to evade a predator.
- *First-order intentionality*. The inference that an agent possesses beliefs and desires, but not beliefs about beliefs. It may produce a scream because it *believes* a predator is present or *wants* others to run away.
- *Second-order intentionality*. The inference that an agent possesses beliefs about other people's beliefs. It may produce a scream because it wants

others to believe that a predator is nearby. False belief tests operate at this level
(e.g. 'I think that Sally thinks that the marble is in the box'). .

- *Third-order intentionality.* An agent possesses beliefs about other people's
beliefs concerning beliefs about other people, such as 'I think that John thinks
that Sally doesn't know where the marble is'.

In this taxonomy, first-order intentionality and above would constitute 'mentalizing',
taking an 'intentional stance' or theory of mind (depending on one's preferred term).
Second-order intentionality does not have a special status (from a theoretical point
of view), but it has acquired a special status by virtue of the fact that most tests of
theory of mind operate at this level because they are more stringent and cannot be
solved by stating one's own beliefs.

Domain-general versus domain-specific accounts of theory of mind

Domain specificity is linked to the notion of modularity (Fodor, 1983). A cognitive
mechanism, or brain region, can be said to be domain specific if it is specialized to
process only one kind of information. Thus, a domain-specific theory-of-mind mechanism would be a process that is specialized for attributing mental states (Leslie,
1987). There are two dominant lines of evidence that have been brought to bear on
this. Firstly, there is the question of whether there is a specific region of the brain that
responds to reasoning about mental states but not other kinds of things. It is possible
that such a mechanism could be distributed in several locations, or that only one of
the regions in that network is truly domain specific. Secondly, one can look to see if
there are specific impairments in mental state attribution but not in other domains.
Most evidence related to this question has come from the developmental condition
of autism (e.g. Baron-Cohen, 1995b) but other lines of research have addressed this
question from the perspective of acquired brain damage (e.g. Samson, 2009).

Historically, explanations of theory of mind have fallen into two camps that are
termed **theory-theory** and simulation theory. Theory-theory argues that we store, as
explicit knowledge, a set of principles relating to mental states and how these states
govern behavior (e.g. Gopnik & Wellman, 1992). In this sense, the 'theory' in theory
of mind is like a mental rule-book for understanding others. This can be contrasted
with simulation theory, which in one form would argue that perceptual-motor systems
(rather than thinking and theorizing) are all that is needed for understanding others
(e.g. Gallese & Goldman, 1998). When phrased in this way, it is reasonable to say that
theory-theory makes more domain-specific assumptions whereas simulation theory
can be considered a domain-general account. However, one needs to be cautious in
dividing explanations into black and white dichotomies. For example, some versions
of simulation theory argue that we do reason about mental states (rather than it being
solely an outcome of perceptual-motor processes) but these versions are distinguished
from theory-theory by making the claim that our own mental states form the foundation for understanding others (e.g. Mitchell, Banaji, & Macrae, 2005a). In a review
of the neuroimaging literature on theory of mind, Apperly (2008) concludes that
the strong division between simulation theory and theory-theory is no longer useful.
Apperly (2008) argues instead that many of the concepts from social neuroscience
research are likely to be more fruitful for understanding theory of mind, including: an
understanding of how processing of self-related and other-related information is carried

out; how both conscious beliefs and unconscious intuitions drive behavior; and so on. However, what such a 'third way' explanation will look like remains to be seen.

Stone and Gerrans (2006) argue against the notion of a domain-specific theory-of-mind mechanism and propose instead that the available data are more consistent with the notion of theory of mind arising out of the interaction of several different mechanisms (and not theory-theory either). This kind of explanation is in the spirit of the models of empathy discussed previously (Decety & Jackson, 2004). It is to be noted that Stone and Gerrans (2006) do not reject the idea of domain specificity per se. They claim that there are domain-specific mechanisms for detecting eye gaze, for example, and claim that deficits here could contribute to problems in theory of mind. Whilst the idea of a domain-specific mechanism for theory of mind is controversial, the idea that theory of mind requires basic competency in a number of domain-general mechanisms such as executive functions is not controversial, and a basic competency in language may be required for many tasks.

Language ability in typically developing children predicts success on a false belief task independently of age (Dunn & Brophy, 2005), and deaf children whose parents are non-native signers are delayed in passing such a task (Peterson & Siegal, 1995). This suggests that language is important for the development of theory of mind. Language may serve several functions: both a social, communicative role and also the acquisition of semantic knowledge of mental state words such as 'want' and 'think'. For example, children have to learn that these words denote concepts that are privately held (Wellman & Lagattuta, 2000). However, once a normal theory of mind is established it may not be dependent solely on language. Evidence for this assertion comes from brain-damaged patients with acquired **aphasia**. Apperly, Samson, Carroll, Hussain, and Humphreys (2006) report a single case study of a man with left hemisphere stroke who was impaired in many aspects of language, including syntax comprehension, but showed no impairments on non-verbal tests of theory of mind, including second-order inferences (X thinks that Y thinks).

Having sketched out the battle lines, the next section will go on to consider the neural substrates for theory of mind as evidenced from functional imaging (of neurologically normal adults) and neuropsychology (of brain-damaged adults). The following section will then consider autism in detail. Developmental issues will be covered specifically in Chapter 11.

KEY TERMS

Aphasia
Deficits in spoken language comprehension or production, typically acquired as a result of brain damage.

Schema
An organized cluster of different information (e.g. describing the subroutines of a complex action).

Neural substrates of theory of mind

Evidence for the neural basis of theory of mind has come from two main sources: functional imaging studies of normal participants and behavioral studies of patients with brain lesions. Numerous tasks have been used, including directly inferring mental states from stories (e.g. Fletcher et al., 1995), from cartoons (e.g. Gallagher et al., 2000), or when interacting with another person (e.g. McCabe, Houser, Ryan, Smith, & Trouard, 2001a). A review and meta-analysis of the functional imaging literature was provided by Frith and Frith (2003), who identified three key regions involved in mentalizing.

Temporal poles

This region is normally activated in tasks of language and semantic memory. Frith and Frith (2003) suggest that this region is involved with generating **schemas** that specify the current social or emotional context, as well as in semantics more generally. Zahn et al. (2007) report an fMRI study suggesting that this region responds to comparisons between social concepts (e.g. brave–honorable) more than matched non-social concepts (e.g. nutritious–useful). Also, not all the tests of mentalizing that activated this region involved linguistic stimuli. For example, one study used triangles that appeared to interact by, say, chasing or encouraging each other (Castelli et al., 2000).

Brain damage to the temporal poles is a feature of the degenerative disorder known as **semantic dementia** (Mummery et al., 2000). Patients with semantic dementia lose their conceptual knowledge of words and objects and show difficulties in language comprehension and production. However, there is little evidence from these patients that social concepts are selectively impaired. In general, although the temporal poles are important for theory of mind, there is no convincing support that it is domain specific for this kind of information.

KEY TERM

Semantic dementia
A neurological condition associated with progressive deterioration in the meaning of objects and words.

Medial prefrontal cortex (mPFC)

Frith and Frith (2003) reported that this region is activated in all functional imaging tasks of mentalizing to that date. Saxe (2006) argues that a sub-region of this area is involved in 'uniquely human' aspects of social cognition. This region lies in front of, but extends into, the ventral region of the anterior cingulate, labeled by Bush et al. (2000) as the affective division. Functional imaging studies reliably show that this region responds more to: thinking about people than thinking about other entities such as computers or dogs (e.g. Mitchell, Banaji, & Macrae, 2005a; Mitchell, Heatherton, & Macrae, 2002); thinking about the *minds* of people than thinking about their other attributes, such as their physical characteristics (Mitchell et al., 2005b); and thinking about the minds of certain people compared to others, such as similar people to ourselves (Mitchell et al., 2005b) and those who are most humanized relative to dehumanized (Harris & Fiske, 2006).

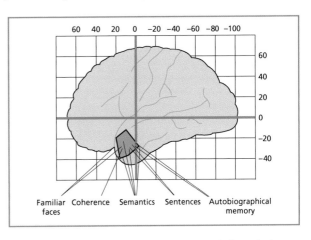

The temporal poles may support semantic knowledge, including of social concepts. Numbers indicate Talairach coordinates. Adapted from Frith and Frith (2003).

Some studies of patients with frontal lobe damage have suggested that the medial regions are necessary for theory of mind (e.g. Stuss, Gallup, & Alexander, 2001), but by no means all (e.g. Bird, Casteli, Malik, Frith, & Husain, 2004). This region also seems to be implicated in the pragmatics of language, such as irony ('Peter is well read. He has even heard of Shakespeare') and metaphor ('your room is a pigsty') (Bottini et al., 1994). Interestingly, people with autism have difficulties with this aspect of language (Happe, 1995). In such instances, the speaker's *intention* must be derived from the ambiguous surface properties of the words (e.g. the room is not literally a pigsty). Functional imaging suggests that this region is involved both in theory of mind and in establishing the pragmatic coherence between ideas/sentences, including those that do not involve mentalizing (Ferstl & von Cramon, 2002).

Can a generic function be ascribed to this region? If so, how does it relate to theory of mind? Amodio and Frith (2006) argue that the function of this region is in reflecting on feelings and intentions, which they label a 'meeting of minds'. One

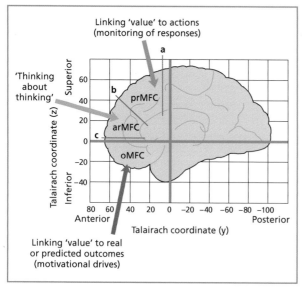

The medial frontal cortex (and adjacent regions of cingulate cortex) may contain three sub-regions with different functional specializations. Amodio and Frith (2006) regard the anterior rostral region (arMFC) as involved in 'thinking about thinking', or meta-cognition. This region is typically activated in tests of theory of mind. The orbital region (oMFC) is involved in linking value (positive or negative reinforcement) to outcomes, whereas the posterior rostral (or dorsal) region (prMFC) is involved in linking value to actions. These latter two regions are considered in Chapter 3. Adapted from Amodio and Frith (2006).

intriguing finding concerning this region is that it can be activated when a person believes they are playing a computer game against another person relative to when they think they are playing against a computer (Rilling, Sanfey, Aronson, Nystrom, & Cohen, 2004). Even though the situation is physically identical (the participant always played the computer), the act of cooperating with another person/mind engenders activity in this region. A more recent explanation of the function of this region is similar, but different, to that of Amodio and Frith (2006). Krueger, Barbey, and Grafman (2009) argue that the function of this region is to bind together different kinds of information (actions, agents, goals, objects, beliefs) to create what they term a 'social event'. They note that within this region some sub-regions respond more when participants make judgments about themselves and also about others who are considered to be similar to themselves (this is discussed in detail in Chapter 9). This suggests that this region is not attributing mental states per se, but is considering the self in relation to others (e.g. when playing a game against a human rather than a computer). It is also consistent with some versions of simulation theory in which participants understand others via deliberate perspective taking (e.g. Mitchell et al., 2005b). The notion of creating internal social events could also explain some of the findings of the role of this region in linking ideas in story comprehension (Ferstl & von Cramon, 2002).

Temporo-parietal junction (TPJ)

This region tends to be activated not only in tests of mentalizing but also in studies of the perception of biological motion, eye gaze, moving mouths, and living things in general. These skills are clearly important for detecting other 'agents' and processing their observable actions. Some simulation theories argue that mentalizing need not involve anything over and above action perception. It is also conceivable that this region goes beyond the processing of observable actions, and is also concerned with representing mental states and perhaps even the mental states of others over and above one's own mental states. Congenitally blind people activate essentially the same network of regions identified by Frith and Frith (2003) when they perform theory-of-mind tasks (Bedny, Pascual-Leone, & Saxe, 2009). This suggests that the computations of these regions are, at least partially, independent from visual perception of agents.

The TPJ region was previously highlighted in the discussion on empathy because it responds more when participants are asked to imagine how someone else would feel relative to how they would feel (e.g. Ruby & Decety, 2004). Patients with brain lesions in this region fail theory-of-mind tasks that cannot be accounted for by difficulties in body perception (Samson, Apperly, Chiavarino, & Humphreys, 2004). Saxe and Kanwisher (2003) found activity in this region, on the right, when comparing false belief tasks (requiring mentalizing) with false photograph tasks (not requiring mentalizing but entailing a conflict with reality). A false photograph may involve taking a picture of an apple on the tree, and then the apple falling down. In this scenario, there is a conflict between reality and a representation of reality. The result was also found when the false photograph involved people and actions, consistent with a role in mentalizing beyond any role in action/person perception. The region responds to false beliefs more than false maps or signs, which differ in an important way from a false photograph in that they are designed to represent *current* reality (Perner, Aichhorn, Kronbichler, Staffen, & Ladurner, 2006). Saxe and colleagues do not dismiss the fact that this region has a role to play in recognizing people and actions, but they claim that there may be different sub-regions within it, with one sub-region specialized for the attribution of mental states (Scholz, Triantafyllou, Whitfield-Gabrieli, Brown, & Saxe, 2009). Moreover, Saxe (2006) argues that it is

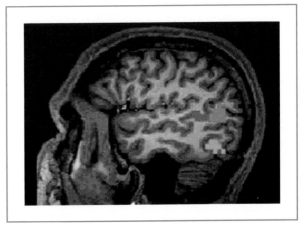

The TPJ region may contain separate sub-regions for dealing with theory of mind (shown here in blue) and recognizing actions and expressions (shown here in purple). For comparison, the position of the extrastriate body area (in green) is shown, which is involved in body perception. From Saxe (2006). Copyright © 2006 Elsevier. Reproduced with permission.

uniquely human in doing so. It is important to note that this region is not specialized for false belief per se. It responds to true beliefs and other types of mental state (Saxe & Wexler, 2005). In other words, it responds to attributions of first-order intentionality as well as higher order intentionality (in Dennett's terms). Saxe and Powell (2006) have shown that this region responds to attribution of contentful mental states (such as thoughts and beliefs) rather than subjective states (such as hunger or tiredness). This suggests that it may have a role over and above 'thinking about others'. However, it is important to mention that one should be cautious in making strong claims about relative differences in BOLD signal. The differences can reflect different functional specialization (Saxe's claim) but they can also reflect the different difficulty of tasks, and the attention or strategy deployed to solve them. One could defend the claim of functional specialization by noting that other regions that respond to theory of mind do not show the same selective responses as the TPJ (Saxe & Wexler, 2005).

Evaluation

Functional imaging studies of the general population and, to a lesser extent, studies of people with acquired brain damage have helped to reveal the key regions involved in theory of mind and their somewhat different functions. There remains no consensus as to whether there is a domain-specific mechanism for theory of mind (i.e. a particular neural region that is dedicated to attributing mental states), but the strongest candidate region for domain specificity has shifted away from the medial prefrontal area to the TPJ region.

EXPLAINING AUTISM

He wandered about smiling, making stereotyped movements with his fingers, crossing them about in the air. He shook his head from side to side, whispering or humming the same three-note tune. He spun with great pleasure anything he could seize upon to spin ... When taken into a room, he completely disregarded the people and instantly went for objects, preferably those that could be spun ... He angrily shoved away the hand that was in his way or the foot that stepped on one of his blocks.

(This description of Donald, aged 5, was given by Leo Kanner (1943), who also coined the term autism. The disorder was independently noted by Hans Asperger (1944), whose name now denotes a variant of autism.)

Autism has been formally defined as 'the presence of markedly abnormal or impaired development in social interaction and communication and a markedly restricted repertoire of activities and interests' (American Psychiatric Association, 1994). It is a severe developmental condition that is evident before 3 years of age and lasts throughout life. There are a number of difficulties in diagnosing autism. First, it is defined according to behavior because no specific biological markers are known (for a review, see Hill & Frith, 2003). Second, the profile and severity may be modified during the course of development. It can be influenced by external factors (e.g. education, temperament) and may be accompanied by other disorders (e.g. attention deficit and hyperactivity disorder, psychiatric disorders). As such, autism is now viewed as a spectrum of conditions spanning all degrees of severity. It is currently believed to affect 1.2% of the childhood population, and is three times as common in males (Baird et al., 2006). **Asperger's syndrome** falls within this spectrum, and is often considered a special sub-group. The diagnosis of Asperger's syndrome requires that there is no significant delay in early language and cognitive development, although the term is also used to denote people with autism who fall within the normal range of intelligence. Learning disability, defined as an IQ lower than 70, is present in around half of all cases of autism (Baird et al., 2006).

Much of the behavioral data has been obtained from high-functioning individuals in an attempt to isolate a specific core of deficits. On a purely theoretical level, one reason why researchers have been interested in the study of autism is the belief that it might reveal something fundamental about social interactions more generally.

Autism as mind blindness

One candidate deficit is the ability to represent mental states, or theory of mind (e.g. Baron-Cohen, 1995b; Fodor, 1992). The first empirical evidence in favor of this hypothesis came with the development of a test of false belief devised by Wimmer and Perner (1983) and tested on autistic children by Baron-Cohen et al. (1985) as the Sally–Anne task (described above). Autistic children tend to fail the task whereas normally developing children (from 4 years on) pass the test, as do control participants with learning disability matched in IQ to the autistic children. The erroneous reply is not due to a failure of memory, because the children can remember the initial location. It is as if they fail to understand that Sally has a belief that differs from physical reality – that is, a failure to represent mental states. This has also been called 'mind-blindness' (Baron-Cohen, 1995b). Autistic children are still impaired when the false belief was initially their own. For example, in one task, the child initially expects to find candy in a candy packet and is surprised to find a pencil, but

The child initially expects to find candy in a tube of Smarties and is surprised to find a pencil. When asked what other people will think is in the packet, autistic children reply 'pencil' whereas typically developing children reply 'candy'.

when asked what other people will think is in the packet the child replies 'pencil' (Perner, Frith, Leslie, & Leekam, 1989).

Passing false belief tasks requires the ability to form meta-representations (i.e. representations of representations: in this instance, beliefs about beliefs). It was originally suggested that a failure of meta-representation may account for impaired theory of mind in autism (Baron-Cohen et al., 1985). However, other studies suggest that autistic people can form meta-representations in order to reason about false photographs in which the information depicted on the photograph differs from current reality (Leekam & Perner, 1991). If their deficit really is related to mental state representations rather than physical representations, then this offers support for the domain-specific account. A number of other studies have pointed to selective difficulties in mentalizing compared to carefully controlled conditions. For example, people with autism can sequence behavioral pictures but not mentalistic pictures (Baron-Cohen, Leslie, & Frith, 1986); they are good at sabotage but not deception – they tend to think that everyone tells the truth (Sodian & Frith, 1992); and they tend to use desire and emotion words but not belief and idea words (Tager-Flusberg, 1992). In all instances, the performance of people with autism is compared to mental-age controls to establish that the effects are related to autism and not to general level of functioning.

Functional imaging studies of autistic people carrying out theory of mind (Happe et al., 1996) or related tasks (Castelli et al., 2002) have shown reduced activity in the network of regions commonly activated by controls.

Finally, it may be necessary to make a distinction between implicit mentalizing (intuitive, reflexive) and more explicit forms of mentalizing (based on reasoning). Whilst the latter tend to be measured by overt predictions of behavior, the former may be measured by non-declarative means (e.g. monitoring of eye movements). For example, some high-functioning people with autism pass standard theory-of-mind measures but may still lack an intuitive understanding of others and may still show abnormal performance on other measures (e.g. eye movements to a location consistent with a false belief; Senju, Southgate, White, & Frith, 2009). By contrast, children under the age of 4 years show some implicit understanding of false beliefs (based on the same measure) despite failing on explicit measures (Onishi & Baillargeon, 2005). This is considered in detail in Chapter 11.

Autism as executive dysfunction

The mentalizing or theory-of-mind account of autism has not been without its critics. These criticisms generally take two forms: that other explanations can account for the data without postulating a difficulty in mentalizing (e.g. Russell, 1997); or that a difficulty with mentalizing is necessary but insufficient to explain all of the available evidence (e.g. Frith, 1989). A number of studies have argued that the primary deficit in autism is one of executive functioning (Hughes, Russell, & Robbins, 1994; Ozonoff, Pennington, & Rogers, 1991; Russell, 1997). **Executive functions** refer to control processes that are needed to coordinate the operation of more specialized components of the brain, thus enabling us to switch attention from one task to another, to give priority to certain kinds of information, or to develop novel solutions, which would include inhibiting familiar solutions (e.g. Goldberg, 2001). For example, the incorrect answer might be chosen on false belief tasks because of a failure to suppress the strongly activated 'physical reality' alternative. Some patients with brain damage in prefrontal regions do this when given false belief tasks (Samson, 2009). However, it is not clear that this explanation can account for all the studies relating to mentalizing (e.g. picture sequencing). Moreover, high-functioning autistic people often have normal executive functions (e.g. Baron-Cohen, Wheelwright, Stone, & Rutherford, 1999) and early brain lesions can selectively disrupt theory-of-mind abilities without impairing executive functions (e.g. Fine, Lumsden, & Blair, 2001).

Rather than difficulties in executive function explaining impairment on theory-of-mind tasks, Baron-Cohen (2009) speculates that the opposite could be true – namely, autistic people may develop, and stick to, their own rule system rather than the 'correct' one as determined by another person, the experimenter. An experiment is, in effect, a social contract. One study found that autistic people show the greatest impairment on open-ended tasks of executive function (in which participants may induce their own rules), rather than those that require the following of simple, stated rules (White, Burgess, & Hill, 2009). On some tests of executive function autistic people show differences in the medial prefrontal region, which is implicated in mentalizing (Gilbert, Bird, Brindley, Frith, & Burgess, 2008). This again suggests that difficulties on some aspects of executive functions could be related to their social difficulties.

Autism as weak central coherence

One difficulty with the theory-of-mind explanation is that it fails to account for cognitive strengths as well as weaknesses. One popular notion of autistic people is that they have unusual gifts or 'savant' skills, as in the film *Rain Man*. In reality, these skills are found only in around 10% of the autistic population (Hill & Frith, 2003). Nevertheless, some account of them is needed for a full explanation of autism. The unusual skills of some autistic people may be partly an outcome of their limited range of interests. Perhaps one reason why some individuals are good at memorizing dates is that they practice it almost all the time. However, there is also evidence for more basic differences in processing style. For example, people on the autistic spectrum are superior at detecting embedded figures (Shah & Frith, 1983) and searching for a target in an array of objects (for a review see Mitchell & Ropar, 2004). One explanation for this is in terms of 'weak central coherence' (Frith, 1989; Happe, 1999). This is a cognitive style, assumed to be present in autism, in which processing of parts (or local features) takes precedence over processing of wholes (or global features).

What would cause such a pattern? One study has suggested that weak central coherence is linked to differences in brain size and connectivity (White, O'Reilly, & Frith, 2009). However, it is also possible that differences in social cognition in autism cause differences in the style of perceptual processing, rather than vice versa. For example, cultures that regard themselves as socially inter-dependent (i.e. strongly connected with the people around them in terms of shared goals and identity) show more global processing than those who construe themselves more socially independent (Davidoff, Fonteneau, & Fagot, 2008; Lin & Han, 2009; Nisbett, Peng, Choi, & Norenzayan, 2001). People with autism could be regarded as lying at one extreme end of this normal scale.

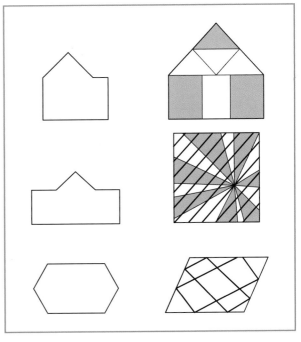

People with autism may be faster at spotting embedded figures such as the ones shown here (the figures on the left are embedded within those on the right).

Autism as an extreme form of the male brain

Baron-Cohen (2002, 2009) argues that the characteristics of all individuals can be classified according to two dimensions: 'empathizing' and 'systemizing'. Empathizing allows one to predict a person's behavior and to care about how others feel. Systemizing requires an understanding of lawful, rule-based systems and requires an attention to detail. Males tend to have a brain type that is biased towards systemizing (S > E) and females tend to have a brain type that is biased towards empathizing (E > S). However, not all men and women have the 'male type' and 'female type', respectively. Autistic people appear to have an extreme male type (S >> E), characterized by a lack of empathizing (which would account for the mentalizing difficulties) and a high degree of systemizing (which would account for their preserved abilities and unusual interests). Questionnaire studies suggest that these distinctions hold true (Baron-Cohen, Richler, Bisarya, Gurunathan, & Wheelwright, 2003; Baron-Cohen & Wheelwright, 2004). However, it remains to be shown whether these distinctions are merely descriptive or indeed do reflect two real underlying mechanisms at the cognitive or neural level.

How does the extreme male brain hypothesis relate to other theories of autism? Baron-Cohen (2002, 2009) regards this explanation as an extension of the earlier mind-blindness theory, which has the advantage of being able to incorporate additional data. Specifically, it accounts for some of the non-social differences found in autism and it offers an explanation for why autism is more common in men (i.e. because men are more likely to have S>E type brains). However, there are at least

SYSTEMIZING IN CLASSIC AUTISM AND/OR ASPERGER'S SYNDROME

Type of systemizing	Classic autism	Asperger's syndrome
sensory systemizing	tapping surfaces or letting sand run through one's fingers	insisting on the same foods each day
motoric systemizing	spinning round and round, or rocking back and forth	learning knitting patterns or a tennis technique
collectible systemizing	collecting leaves or football stickers	making lists and catalogues
numerical systemizing	obsessions with calendars or train timetables	solving maths problems
motion systemizing	watching washing machines spin round and round	analysing exactly when a specific event occurs in a repeating cycle
spatial systemizing	obsessions with routes	developing drawing techniques
environmental systemizing	insisting on toy bricks being lined up in an invariant order	insisting that nothing is moved from its usual position in the room
social systemizing	saying the first half of a phrase or sentence and waiting for the other person to complete it	insisting on playing the same game whenever a child comes to play
natural systemizing	asking over and over again what the weather will be today	learning the Latin names of every plant and their optimal growing conditions
mechanical systemizing	learning to operate the VCR	fixing bicycles or taking apart gadgets and reassembling them
vocal/auditory/verbal systemizing	echoing sounds	collecting words and word meanings
systemizing action sequences	watching the same video over and over again	analysing dance techniques

Source: Baron-Cohen, S., Ashwin, E., Ashwin, C., Tavassoli, T., & Chakrabarti, B. (2009). Talent in autism: Hyper-systemizing, hyper-attention to detail and sensory hypersensitivity. *Philosophical Transactions of the Royal Society of London, Series B,* 364(1522), 1377–1383. Copyright © 2009 The Royal Society. Reproduced with permission.

two ways in which these different ideas (mind blindness vs extreme male brain) could be related: that an inability to engage with others (due to a theory-of-mind deficit) leads to systemizing as a kind of compensatory strategy; or that an unusual interest or ability in systemizing leads to a lack of interest and understanding of social behavior. A third possibility is that both are true – that whatever it is that causes high systemizing also causes low empathizing. Possible mechanisms include fetal testosterone levels (e.g. Auyeung et al., 2009) or sex-related genetic differences (e.g. Creswell & Skuse, 1999). Although the extreme male brain theory predicts an autistic advantage for understanding systems, it differs from the weak central coherence theory by not making predictions about a difference between local versus global information. Finally, some research has tried to suggest a link between the extreme male brain theory and the broken mirror theory (discussed below), noting that there are sex differences (within the non-autistic population) in white/gray matter density in regions associated with the mirror system, with females showing greater density (Cheng et al., 2009). An EEG signature linked to functioning of the mirror system, termed **mu suppression**, also shows a sex difference, with females showing greater suppression (Cheng et al., 2008).

The broken mirror theory of autism

The **broken mirror theory** of autism argues that the social difficulties linked to autism are a consequence of mirror system dysfunction (Iacoboni & Dapretto, 2006; Oberman & Ramachandran, 2007; Ramachandran & Oberman, 2006; Rizzolatti & Fabbri-Destro, 2010). Hadjikhani, Joseph, Snyder, and Tager-Flusberg (2006) examined, using structural MRI, the anatomical differences between the brains of autistic individuals and matched controls. The autistic individuals had reduced gray matter in several regions linked to the mirror system, including the inferior frontal gyrus (Broca's region), the inferior parietal lobule, and the superior temporal sulcus. Although these were not the only regions where differences were found, the degree of thinning in these regions correlated with autistic symptom severity.

EEG, fMRI, and TMS data also suggest differences in mirror system functioning during certain tasks. Oberman et al. (2005) used EEG to record mu waves over the motor cortex of high-functioning autistic children and controls. **Mu waves** occur at a

KEY TERMS

Broken mirror theory
An account of autism in which the social difficulties are considered as a consequence of mirror system dysfunction.

Mu waves
EEG oscillations at a particular frequency (8-13 Hz) that are greatest when participants are at rest.

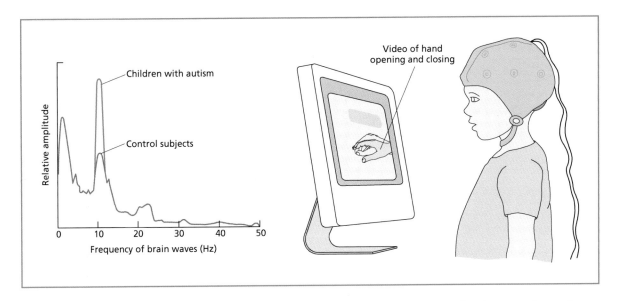

Mu waves are EEG oscillations in the 8–13 Hz range that are reduced both when performing an action and when watching someone else perform an action (relative to rest). As such, they may provide a neural signature for human mirror neurons. Autistic children show less mu suppression when watching others perform a hand action, which provides evidence in support of broken mirror theory. From Ramachandran and Oberman (2006). Reproduced with permission from Lucy Reading-Ikkanda for *Scientific American* Magazine.

particular frequency (8–13 Hz) and are greatest when participants are doing nothing. However, when they perform an action there is a decrease in the number of mu waves, a phenomenon termed mu suppression. Importantly, in typical controls mu suppression also occurs when people *observe* actions and, as such, it has been regarded by some as a measure of mirror system activity (e.g. Pineda, 2005). Oberman et al. (2005) found that the autistic children failed to show as much mu suppression as controls during action observation (watching someone else make a pincer movement) but did so in the control condition of action execution (they themselves make a pincer movement).

Similar findings have been obtained with fMRI. Dapretto et al. (2006) conducted a study in which autistic children and matched controls either observed or imitated emotional expressions. The imitation condition produced less activity in the inferior frontal gyrus of the autistic children relative to controls, and this was correlated with symptom severity. Differences in regions linked to face recognition (fusiform gyrus) and emotion recognition (amygdala) did not differ between groups.

Finally, watching someone perform an action increases one's own motor excitability, measured as a motor-evoked potential (MEP) on the body when TMS is applied to the motor cortex. However, this effect is reduced in autistic people even though their motor cortex behaves normally in other contexts (Theoret et al., 2005).

The broken mirror theory makes some novel predictions about what people with autism might be impaired at, such as imitation and understanding the goals of others based on action observation. Boria et al. (2009) compared children with autism against typically developing controls in which actions were either

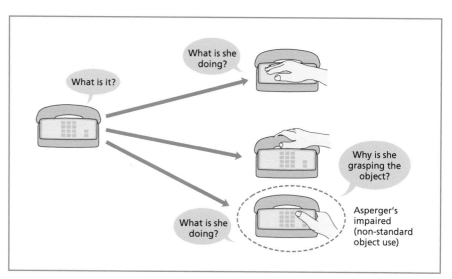

People with autism perform worse than controls at inferring intentions for non-standard actions. In this example, they are more likely than controls to say that the person intends to make a call than to say that they are moving the phone. Figure based on Boria et al. (2009).

consistent with typical use of a phone (e.g. making a call) or not (e.g. picking it up to move it). They found that the autistic children were more likely to base their understanding of actions based on the object rather than action. In this example, they are more likely to answer 'making a call' when the object is being moved.

Although deficits in imitation are found in autism (Williams, Whiten, & Singh, 2004), these may be more apparent in spontaneous imitation than instructed imitation (e.g. Hamilton, Brindley, & Frith, 2007). This suggests that autistic people have a poor intuitive understanding of when and what to imitate (i.e. social rules) rather than in perceptual-motor interactions (at the heart of the broken mirror theory). The broken mirror theory has its critics (Dinstein, Thomas, Behrmann, & Heeger, 2008; Southgate & Hamilton, 2008). In general, the criticism takes two forms. Firstly, it does not account for all the unusual behavior found in autism (e.g. embedded figures; interest in systems). Defendants of the theory argue that it is not trying to explain all the features of autism (i.e. it is not a theory of autism but a theory of certain characteristics of autism). The second general criticism surrounds the extent to which empathy and imitation are linked to mirror systems. Earlier in the chapter, many examples were given of how both imitation and empathy are modulated by social rules, deliberate attempts at perspective taking, and so on. A core deficit elsewhere (e.g. in representing mental states) could nevertheless affect the functioning of the mirror system, and perhaps even lead to structural changes within that system. Heyes (2010) argues that the properties of mirror neurons may be learned as a result of social interactions. Thus, impoverished social interactions may cause mirror system dysfunction, as well as vice versa.

Evaluation

For many years the dominant explanation of autism has been a failure to represent the mental states of others. This has been termed mind blindness and has tended to have been regarded as a failure to develop a theory of mind (although not necessarily with commitment to the idea that this exists as a domain-specific module). Other theories, such as weak central coherence theory and extreme male brain theory, maintain this basic idea but adopt a wider perspective in order to explain other features of autism. The most significant challenge to this idea previously came from the notion of executive dysfunction in autism, but now comes in the form of broken mirror theory. There is good evidence of mirror neuron dysfunction in autism, but it is less clear whether this dysfunction is a core feature of autism or a by-product of other deficits – given that mirror systems in general are modulated by beliefs, social knowledge, and cognitive control.

SUMMARY AND KEY POINTS OF THE CHAPTER

- Simulation theory argues that we understand the mental states (thoughts, feelings, beliefs, etc.) of others by activating our own mechanisms for producing that behavior. To some extent, we literally share the experiences of the people around us. As such, simulation theory is an appealing way of explaining empathy.
- Empathy is a broad concept that may include simulation, but it is unlikely to be limited to it. It also involves perspective taking (either automatically or deliberately) and cognitive control, which may inhibit the tendency to simulate.
- Both empathy and theory of mind (or mentalizing) involve understanding the mental states of other, but the latter is typically assessed via conscious attempts to reason about mental states, such as in false belief tasks.
- Functional imaging of the normal population reveals a network of regions that are consistently activated by tests of theory of mind, and the two regions that have provoked the most research interest are the temporo-parietal junction region and a medial prefrontal cortex region. However, it remains controversial whether either region can be classed as domain specific for attributing mental states.
- People with autism often fail theory-of-mind tasks, leading to the theory that they have a specific impairment in representing mental states. Their difficulty is not well explained by difficulties in executive function alone or difficulties in meta-representation per se.
- There is good evidence of mirror neuron dysfunction in autism, but it is less clear whether this dysfunction is a core feature of autism or a by-product of other deficits (given that mirror systems in general are modulated by beliefs, social knowledge, and cognitive control).

EXAMPLE ESSAY QUESTIONS

- What is the evidence for and against simulation theories of empathy?
- How is empathy related to theory of mind, and in what ways are they different?
- Is there a theory-of-mind module in the human brain?
- How can the social behavior of people with autism be explained?

RECOMMENDED FURTHER READING

- Decety, J., & Ickes, W. (2009). *The social neuroscience of empathy*. Cambridge, MA: MIT Press. An excellent collection of papers on empathy.

- Hill, E. L., & Frith, U. (2004). *Autism: Mind and brain*. Oxford: Oxford University Press.

- Saxe, R., & Baron-Cohen, S. (2006). *Theory of mind*. New York: Psychology Press. A very good collection of papers on all aspects of theory of mind.

CHAPTER 7

CONTENTS

Altruism and helping behavior   158

Game theory and social decision making   166

Summary and key points of the chapter   177

Example essay questions   177

Recommended further reading   177

Interacting with others

7

Interactions have been defined as 'dyadic behavior in which the participants' actions are interdependent such that each actor's behavior is both a response to and a stimulus for the other participant's behavior' (Rubin, Bukowski, & Parker, 2006). This chapter is about two kinds of interaction: cooperation and competition. Cooperation entails sharing of commodities (e.g. food) and knowledge, and providing helping behavior (e.g. if someone is injured). This type of behavior is also termed **altruism**, but with the added twist that altruism is often described as 'selfless' in that no personal gain is obtained. Non-cooperation entails keeping commodities and knowledge for oneself and not providing help to others. For most people, Darwin's theory of natural selection is synonymous with competition, as exemplified in the phrase (not actually used by Darwin) 'survival of the fittest'. When put this way, cooperation seems like a puzzle. Being social and cooperative compromises one's own time and resources. If my genes (and my traits) are to survive then they have to be of benefit to me, not to you. However, whether a system of interactions is competitive ('selfish') or cooperative does not depend entirely on the cost of being cooperative. It depends on whether the benefits of cooperation exceed the costs. Short-term interests have to be balanced against the longer term gains to be had through group living. Individuals working together in groups may increase chances of survival by, for instance, hunting as a group and sharing knowledge and skills.

In humans, at least, cooperative interactions between individuals are predicated upon **trust** (i.e. the belief that others will treat you fairly). Knowing who to trust and when to trust involves a complex decision-making process that will be discussed throughout the chapter. It will be dependent on the particular situation and prior knowledge of how others have behaved in the past. However, the basic fact that we are capable of trust at all may depend on our ability to understand that others have similar mental states to our own and on our ability to form shared goals between individuals. People who receive the benefits of cooperation but do not contribute to the group themselves are termed **free loaders (or free riders)**, and groups typically impose sanctions on those who free load, such as social exclusion, physical punishment, or fines. Such sanctions require norms to regulate or enforce cooperation, and these norms require consensual agreement as to what is 'fair' or 'right'.

In his book, 'Why we Cooperate', Tomasello (2009) outlines a number of different answers to that question:

1. *There is an intrinsic desire to help.* In humans at least, helping others is personally rewarding. This may relate to the capacity for empathy, which was discussed in detail in the previous chapter and will be returned to here.
2. *The benefits of reciprocity* – 'you scratch my back and I'll scratch yours'. There may be certain things that I cannot do for myself now but that you can do for me; in the future the situation may be reversed.
3. *Punishment for non-cooperation.* This has also been termed **altruistic punishment** (Fehr & Gachter, 2002). It is altruistic in the sense of being 'selfless': it has no direct benefit to the punisher, but comes at a cost to the punisher (in terms of time

Why do people engage in altruistic acts such as helping the elderly, giving to charity, or performing favors? Given that altruism entails a cost, what kind of benefits accrue from altruism to ensure that it survives within a group?

and effort, and in terms of risk of reprisal). This factor is more likely to explain why cooperation is maintained rather than why it exists at all.

4. *A desire to conform*, by sharing in a group-level identity. Groups and identity are primarily considered in a subsequent chapter.

The evidence for these different motivations for cooperation will be outlined below. This chapter will not only consider *why* we cooperate, but also *how* we cooperate in terms of the cognitive and neural mechanisms that support this kind of behavior.

ALTRUISM AND HELPING BEHAVIOR

Evolutionary biological approaches

From an evolutionary perspective, the problem of altruism lies in explaining how it is possible to improve the survival chances (or 'fitness') of altruists, given that helping others necessarily entails a personal cost to those who help. Evolutionary biology has come up with two main mechanisms to explain the evolution of altruism: **kin selection** and **reciprocal altruism (or reciprocity)**. Kin selection assumes that we help others who are related to us. Reciprocal altruism assumes that we provide help to others in order to obtain help from others in the future.

In proposing kin selection theory, Hamilton's (1964) significant insight was to realize that it is the survival chances of a trait (in this case, helping behavior) rather than an individual that matters – it is survival of the fittest *trait* not the fittest *person*. (This is essentially the same point made by Dawkins (1976) in his famous 'selfish gene' theory.) If an individual helps their kin then there is a greater chance that the helping trait will survive, because there is a greater chance that their kin also carry this same trait. However, this still only applies if the benefits of helping are greater than the costs. Specifically, helping behavior can spread through a population if the cost to the organism's own reproduction (C) is offset by the benefit to the reproduction of its kin member (B) multiplied by the probability that the kin member inherits the same helping behavior (r). Thus, helping traits will tend to spread when $C < r \times B$.

There are several examples from the natural world that are consistent with kin selection. Ground squirrels are able to discriminate full siblings from half-siblings and unrelated individuals (Holmes & Sherman, 1982), and are more likely to give an alarm call if close relatives are nearby (Sherman, 1977). Kinship predicts whether vampire bats will share food with another individual (Wilkinson, 1988). It also explains why sterile worker insects forego their own fertility for their kin (Trivers & Hare, 1975).

Kin selection does not offer an explanation of altruism towards unrelated individuals, except as an error in detecting kin from non-kin. However, such altruism does exist. For instance, unrelated chimpanzees will come to each other's aid when threatened (De Waal & Luttrell, 1988). To give another example, a major predictor of whether a vampire bat will share food with a non-relative is whether the non-relative shared with that individual in the past (Wilkinson, 1988). Williams (1966) and Trivers (1971) proposed the alternative mechanism of reciprocal altruism (or 'reciprocity' for short) to account for this. Reciprocal altruism is based on the economic concept of trade (e.g. exchanging food for protection). The exchange is most often delayed, so a cost to an individual initially will be a benefit to the individual at some later point. What kinds of cognitive processes are needed for reciprocal altruism? Minimally, it requires an ability to distinguish between conspecifics and to remember their previous behavior. However, the decision is often more complex than that. Just because someone has helped you, there is no necessary reason why you should help back (the freeloader problem); and if you help someone else, how do you know that they will help you?

There are other biologically based explanations of altruism aside from kin selection and reciprocal altruism. Zahavi (1975, 1995) argues that altruism arises as a form of **sexual selection**, analogous to the peacock's tail. Peahens' preference for a larger plumage leads to peacocks with larger tails having a reproductive advantage. This sets up an evolutionary arms race in which tails become bigger and bigger despite not having a function (aside from reproductive success), and if anything they act as a handicap to the peacock. Turning the analogy back to altruism, being generous and helpful may serve the function of enhancing one's reputation (and mating success) by demonstrating wealth and the ability to provide. According to Zahavi (1995), this can occur without expectation of reciprocity: both the costs and benefits lie with the altruist.

Does altruism arise out of the same kind of evolutionary pressures as the peacock's tail? Altruism could be regarded as a display of wealth, or other positive attributes that are regarded as attractive, hence increasing the reproductive success of altruists.

A related argument, based on mathematical modeling of costs and benefits, was introduced by Nowak and Sigmund (2005). Rather than helping the same individual that helped you, they suggest that one only needs to offer help to those individuals who are likely to help you (including people who you may never meet again). They term this **indirect reciprocity**. As with Zahavi's theory, the concept of reputation is central to the idea of indirect reciprocity but, unlike Zahavi's theory, sexual selection is not the driving force of indirect reciprocity. In their simulations, Nowak and Sigmund (2005) assume that helping others increases one's reputation (termed 'image score') as well as incurring a cost (the cost to oneself in terms of foregone time, effort, etc.). Importantly, they assume that altruists know the image score of the person that they are deciding whether to help or not. Mathematically, the model is similar to that proposed by Hamilton (1964) for kin selection but in which the variable 'degree of relatedness' is replaced by 'image score' (i.e. reputation for helping). Over repeated iterations, cooperation is a stable outcome provided that people are discriminating in who they help: the benefits only tend to outweigh the costs when help is focused on those with good reputations for helping.

Altruism in humans

To many reading this, the examples of vampire bats sharing blood and insects sacrificing their fertility may sound like a far cry from human altruism. However, the evolutionary biological approaches described above are essentially mathematical models of those situations in which cooperation predominates over pure competition (based on costs and benefits accrued over generations). In that sense, the models do apply to humans as much as to any other species. The models make no strong claims about whether the actual mechanisms in the brain are conserved across such diverse species. For instance, Nowak (2006) argues that only humans are fully capable of *indirect* reciprocity because the decision to help (or not) involves consideration of the reputations of the individuals involved, and this knowledge may be obtained via language and social norms rather than personal history of exchanges with that person. Reputation also involves thinking about what others think of us, which is by no means simple (see Chapter 6).

Where humans do appear to differ from most other species is that they can reflect on their helping behavior and they appear to have some degree of control over their decision to help or not. However, even this point is up for debate. Whilst we have a sense that our decision to help or not is a rational free choice, much of our actual decision-making process could still be based on automatic and largely unconscious biases. For example, one study measured contributions of money towards milk for tea and coffee available in a Psychology Department common room (Bateson,

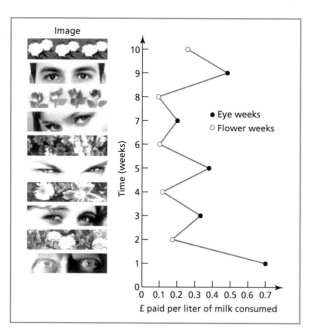

Contributions to a public good are increased when eyes are displayed on a donation tin rather than flowers. From Bateson, M., Nettle, D., & Roberts, G. (2006). Cues of being watched enhance cooperation in a real-world setting. *Biology Letters, 2,* 412–414. Copyright © 2006 The Royal Society. Reproduced with permission.

Nettle, & Roberts, 2006). The contributions were based on an honesty system where people were expected to leave money in a pot when they got a drink but there were no sanctions against free loading. More money was left when the pot happened to display a picture of eyes than when it displayed a picture of flowers. In another study, people were likely to donate more to charity after their mannerisms were subtly imitated (van Baaren et al., 2004), but these effects depend on participants not noticing that they are being deliberately mimicked (Lakin & Chartrand, 2003). Thus, it is not true to argue that all human altruistic behavior is an outcome of conscious decision making.

Whereas evolutionary theories provide a framework to understand the mechanisms by which a *population* engages in helping behavior, at the *individual* level there may be inbuilt motivational mechanisms that drive pro-social behavior. Whatever our true motivations, there is a deep-rooted sense that we help others because we care about their welfare rather than to spread our genes or receive favors in the future. As such, many theories of altruism suggest that empathy is a key component (e.g. Batson, 1991; de Waal, 2008; Piliavin, Dovidio, Gaertner, & Clark, 1981). In the model of Piliavin et al. (1981), witnessing a distressing situation leads to physiological arousal (e.g. increased heart rate) and emotional contagion (i.e. the observer also experiences high personal distress). In this model, the motivation to help is egoistic (to reduce one's own distress) rather than altruistic. However, whether or not a person helps is assumed to be determined by a cognitive appraisal (a cost–benefit analysis) rather than level of empathy per se. Helping someone may have a low cost to oneself (e.g. helping an elderly person

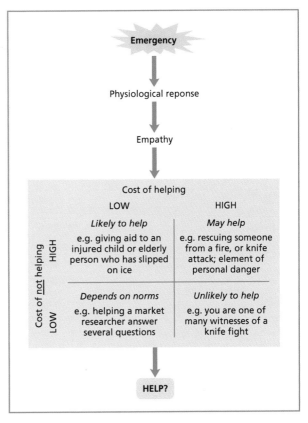

The model of altruism by Piliavin et al. (1981) includes several stages. The first stages are a physiological arousal response and an empathic (distress) response. The final stage is a cognitive appraisal in which the costs of helping and the costs of not helping are weighed. Although empathy provides the motivation for altruism, whether or not somebody acts depends on the perceived costs.

who has fallen over) or a higher cost (e.g. intervening in a knife fight). The costs of not helping also need to be considered and these may include negative emotions (e.g. guilt).

The **empathy-altruism model** of Batson (1991) extends the earlier model of Piliavin et al. (1981) but with an important difference – the motivation to help is considered to be primarily other-oriented (based on empathic concern for others) rather than self-oriented (to relieve one's own personal distress). In a series of studies, Batson and colleagues sought to distinguish between egoistic motives (self-oriented) and 'true' altruism (other-oriented) by describing in detail a particular scenario (e.g. relating to an eviction) and asking participants to state the level of help they would offer. In addition, the researchers would take self-report measures of their personal feelings (e.g. distress, fear of social disapproval) and their feelings towards the person in need of help (e.g. empathic concern). Empathic concern was found to predict helping behavior even when taking into account selfish motives

KEY TERM

Empathy-altruism model
The theory that the motivation to help is based on empathic concern for others.

such as a desire to escape aversive feelings (Batson, Duncan, Ackerman, Buckley, & Birch, 1981), social disapproval (Fultz, Batson, Fortenbach, McCarthy, & Varney, 1986), guilt (Batson et al., 1988), shame (Batson et al., 1988), and sadness (Batson et al., 1989).

The empathy-altruism model has been criticized on a number of grounds. Methodologically, the evidence is based on the assumption that people can accurately reflect on their own feelings and actions (or predict them) in imagined scenarios. Theoretically, the model hinges on the assumption that there are clear distinctions between self and other. Once that distinction becomes blurred, then so does the distinction between egoistic and altruistic motives. Batson and Shaw (1991) doubted that a merging of self and other genuinely occurs 'except perhaps in some mystical states' (p. 161). However, this claim has been challenged by research in both social psychology and, later, in social neuroscience. Cialdini, Brown, Lewis, Luce, and Neuberg (1997) used similar experimental designs to Batson but added a questionnaire measure of 'including others in the self' (Aron, Aron, & Smollan, 1992). When this measure was included it predicted helping behavior, but empathic concern did not. This has been extended by social neuroscience investigations showing shared neural substrates for processing self and other and its relevance to empathy (e.g. Decety & Chaminade, 2003). Specific examples of this include mirror systems (Rizzolatti & Craighero, 2004), but also regions such as the medial prefrontal cortex that responds to other people only when they are considered similar to one's self (e.g. Krueger et al., 2009). These studies do not rule out the role of empathy in altruism – quite the opposite – but they do suggest that the ideas of 'true' altruism and 'true' empathy (in which 'true' = selfless) are not very meaningful.

Several social neuroscience studies have investigated altruism directly. Moll et al. (2006) conducted an fMRI study of donating to charity. Participants were shown charity names and mission statements (e.g. 'Death with dignity: Allowing euthanasia for the terminally ill') and then given donation options that either incurred a cost to the participants (e.g. You = –\$2, Charity = +\$5) or no cost (e.g. You = \$0, Charity = +\$5). They could either choose to donate or not. In a separate condition, they were given a pure reward (e.g. You = +\$5, Charity = \$0). Choosing to donate to a charity activated some of the same neural circuitry (including the ventral and dorsal striatum) as receiving a pure reward – even when giving incurs a cost to oneself. The authors interpret this as a 'joy of giving' effect. However, other regions did differentiate these factors. A region in the ventromedial prefrontal cortex responded when participants decided to donate (but not to pure reward) and a region in the lateral orbitofrontal cortex was activated by decisions *not* to donate.

Tankersley, Stowe, and Huettel (2007) used fMRI to investigate the role of the right temporal-parietal junction (rTPJ) in altruism. Recall that this region responds when taking the perspective of another person (Ruby & Decety, 2004) and has been implicated in attributing mental states to others (e.g. Saxe & Wexler, 2005). In one condition, participants played a simple game (making a response as quickly as possible) in which they won money both for themselves and a chosen charity. In another condition, they watched the computer playing the game. After scanning, participants completed a self-assessment of their altruistic tendencies. Several regions, including the rTPJ, were activated more when watching someone else play the game as opposed to when playing it oneself. Importantly, activity in this region (and only this region) was modulated by participants' altruism ratings – the greater the altruism, the greater the difference between other versus self. This finding appears to be more consistent with Batson's (1991) idea of altruism as a selfless, other-oriented deed

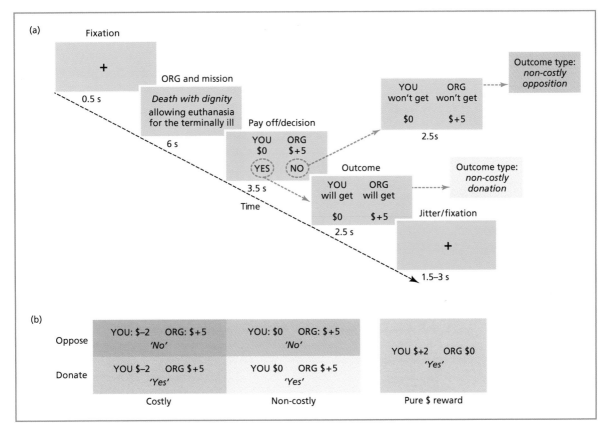

(a) An example trial; (b) a payoff matrix. In the experiment of Moll et al. (2006) participants have the option of donating to a charity or not, and that donation may come at a personal cost to them or not. In an additional condition, the participant receives a reward (without making any choice). Giving $5 to charity is similar – in neural terms – to receiving a gift of $5, which has been interpreted as a 'joy of giving' effect. From Moll et al. (2006). Copyright © 2006 National Academy of Sciences, USA. Reproduced with permission.

rather than a result of shared representations between self and other (Cialdini et al., 1997). However, Tankersley et al. did obtain an empathy measure but claim that the rTPJ response is related to altruism rather than empathy. This suggests a possible role for 'mentalizing' in altruism over and above emotional empathic responses.

In the studies of Moll et al. (2006) and Tankersley et al. (2007) the costs of helping and the costs of not helping were both relatively low. The situation was quite different in the study of Hein, Silani, Preuschoff, Batson, and Singer (2010), in which participants could opt to receive a painful stimulus in order to reduce pain delivered to another person. The other person was either a member of an ingroup or outgroup, defined in terms of allegiance to a soccer team. In this scenario both the cost of helping and the cost of not helping are relatively high. The decision to help ingroup members was linked to activity in the anterior insula, a region of the brain that is involved in emotional experience and pain perception. It was also influenced by subjective reports of empathic concern, as predicted by the empathy-altruism model. In contrast the decision not to help outgroup members was linked to activity in the ventral striatum, a region normally linked to reward processing, and in the study of Moll et al. (2006) it was linked to helping. The fact that it was associated with *not helping* in this study was interpreted in line with previous studies showing activity in this region associated with pleasure in other's punishment/misfortune (Singer et al., 2006). It suggests that different motivations to help (and their different associated costs and benefits) may have very different neural substrates.

Evaluation

The examples given in the section above, concerning charitable donations or helping people in distress, are often regarded as the clearest examples of altruism insofar as being 'selfless' acts, that is, there is no realistic chance of reciprocity. However, even in these scenarios there is a possibility for such behavior to be self-enhancing by reducing or preventing negative emotions (distress, guilt, and shame), increasing positive feelings (such as pride), and adding to one's public reputation. Although the debate about whether altruistic acts are egoist or selfless is interesting, the dichotomy itself is rather contrived. Humans are social creatures and our cognitive and neural apparatus cannot be easily divided into social and non-social, or self and other. Both the sexual selection theory and indirect reciprocity theory rely on the notion of reputation, and this is likely to be an important factor in motivating pro-social behavior in humans, whereas kin selection or direct reciprocity may tend to be the dominant mechanisms in other species.

OBEDIENCE AND COMPLIANCE – THE CLASSIC STUDIES OF MILGRAM AND ASCH

Obedience refers to the following of explicit directions from others who are regarded as having authority or perceived as taking responsibility. **Compliance**, on the other hand, involves a change in behavior arising in response to the attitudes of others, but without explicit orders to change. The classic example of obedience in the social psychology literature is Milgram's (1963, 1974) studies involving participants administering electric shocks to another person. The classic example of compliance is the work of Asch (1951, 1956) in which participants give a deliberate incorrect response to conform to the other (incorrect) responses of others in the group.

Asch's (1951, 1956) studies on conformity involved an apparent test of visual perception with groups of seven to nine people, only one of whom was naïve as to the true purpose of the experiment. On 12 out of 18 trials the 'participants' gave an agreed incorrect response (either too short or too long) before the naïve participant had his/her turn to give their response. Around half of all participants gave six or more incorrect responses that conformed to the group response. Participants initially reported feelings of uncertainty about their judgment, which evolved into a fear of disapproval. When the naïve participant could give their answer in private (writing it down) but everyone else publicly announced it, the conformity rate dropped to about a third of its previous level.

The aim of Milgram's (1963, 1974) obedience experiments was to find out how far people would go in obeying orders to inflict harm in a laboratory

KEY TERMS

Obedience
The following of explicit directions from others who are regarded as having authority or perceived as taking responsibility.

Compliance
A change in behavior arising in response to the attitudes/behavior of others, but without explicit orders to change.

set-up, under a variety of conditions. In the post-war decades, there was significant interest in how acts of atrocity can be performed by sane, apparently well-mannered, people with a 'normal' upbringing. Milgram had been particularly interested in the recent trial of Adolf Eichmann who had a role in implementing the Holocaust. Eichmann's defense was that he was simply obeying orders. Milgram's participants were males recruited from the community surrounding Yale University (although one study involving women found similar results). Participants were assigned as a 'learner' or 'teacher'. The learner had to remember pairs of words. The teacher had to administer an electric shock to the learner when they gave the wrong answer. The size of the electric shock was increased on each trial although, in reality, no shock was given – the 'learner' was part of the experiment and produced simulated cries of pain with increasing intensity. The 'teacher' would often attempt to stop the experiment but was then given a series of prompts to continue by the experimenter ('Please continue', 'The experiment requires that you continue', 'You have no choice, you *must* go on'). At 300 V the learner would fall silent and the teacher was told to treat this as a 'wrong answer' and continue.

The main findings of these experiments can be summarized as follows. The dependent measures are the point at which shocks cease to be administered and/or the proportion of participants who administer all shocks.

- Less intense shocks are given when the teacher and learner are in the same room; more are given when the learner is in another room but can be heard; and more still are given when the learner cannot be seen or heard.
- Obedience is reduced when the experimenter is not present in the room but relays instructions via telephone.
- The presence of two other obedient people increased shocks (relative to a lone teacher) but the presence of two disobedient people reduced the shocks administered.
- Obedience is lower when the experiment is conducted in a run-down office building for a private research firm, relative to when it was conducted at the University.

These studies suggest a variety of mechanisms, including empathy (enhanced when the learner is seen/heard), conformity (when other teachers are present), and the teachers' beliefs about the study (e.g. in the legitimacy of the study) and the experimenter (e.g. the extent to which he is directly aware of the learners distress). Although not specifically tested by Milgram, one would also predict greater obedience in situations in which the teacher believed the learner deserved a shock as a form of punishment. In such situations, people disengage from their natural empathic tendencies (Singer et al., 2006).

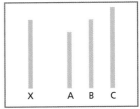

What is the length of line X? Is it A, B, or C? How would you respond if eight other 'participants' in the experiment had declared C to be the correct answer?

Soldiers are required to display obedience and, if ordered, to kill. The group processes of conformity, diffusion of responsibility, and negative attitudes towards an outgroup may enable this to occur in people who are otherwise empathic.

GAME THEORY AND SOCIAL DECISION MAKING

Many of the decisions we make affect the lives of others as well as ourselves, and the decisions that other people make also affect us. For example, if you decide not to help me, then how will that affect my decision to help you? Decisions to cooperate or compete typically involve complex trade-offs between costs and benefits: choosing not to cooperate may result in short-term benefits, but may reduce one's reputation and the longer term benefits of reciprocity. Social decision making relies, to some extent, on the same mechanisms as decision making in non-social situations (e.g. Lee, 2008). For example, lesions of the orbitofrontal cortex in mice result in them behaving impulsively – they favor small short-term gains over larger benefits accrued in the longer term (Rudebeck, Walton et al., 2006). In this instance, the reward is food. However, humans with lesions in this region also tend to make decisions for short-term personal gain, leading to difficulties in forming stable long-term social relationships and socially non-cooperative behavior (e.g. Eslinger & Damasio, 1985).

Game theory is a type of mathematical model that captures how an individual's success in making decisions is influenced by the decisions of others. In *economics*, it has been typically used to find the optimal decision for an individual (i.e. the one that has the greatest benefits for the least costs), taking into account the decisions of others. This optimal decision is termed the Nash equilibrium, named after the mathematician John Nash whose life was depicted by Hollywood in the film *A Beautiful Mind*. Stated simply, David and Nick are in a Nash equilibrium if David is making the best decision he can (taking into account Nick's decision) and Nick is making the best decision he can (taking into account David's decision). In *biology*, game theory has been used to model evolution based on the concept of fitness rather than decision (e.g. Maynard Smith, 1982). Hamilton's (1964) mathematical formulation of kin selection to explain altruism would be one example. In *psychology*, game theory is applied to 'real life' social decision making. Real decisions do not necessarily conform to the best decision for that individual (as predicted mathematically by the Nash equilibrium). Whilst such decisions could be classed as 'irrational' or errorful, they suggest instead that humans take into account factors other than their own personal gain when making social decisions. Examples of this come from the Prisoner's Dilemma (Axelrod & Hamilton, 1981) in which cooperation is often favored over self-interest. Similarly, in the Ultimatum Game (Guth, Schmittberger, & Schwarze, 1982) decisions are made on the basis of fairness (or altruistic punishment) rather than maximum personal gain. These findings from psychology have been augmented by research using neuroscience methods, giving rise to the new field of **neuro-economics**. Neuro-economics provides brain-based explanations of how individuals and groups make economic decisions such as assigning value to competing choices, exchange and reciprocity, and making best use of limited resources. For humans, the word 'value' has two meanings: it can refer to how much one is willing to give to obtain something (as

KEY TERMS

Game theory
A type of mathematical model that captures how an individual's success in making decisions is influenced by the decisions of others.

Neuro-economics
Provides brain-based explanations of how individuals and groups make economic decisions such as assigning value to competing choices, exchange and reciprocity, and making best use of limited resources.

in monetary value), but it is also a principle that one holds and defends (as in family values, or the value of fairness). Many of the results in neuro-economics can be construed as a trade-off between these different notions of value.

The Prisoner's Dilemma

There is a problem with reciprocal altruism in that it is not always the optimal strategy. The optimal strategy in many situations is to receive help from you, but for me to *not* return the favor back. This is exemplified by the **Prisoner's Dilemma** proposed by Axelrod and Hamilton (1981). A common version of the dilemma is as follows:

Two suspects are arrested by the police. The police have insufficient evidence for a conviction, and, having separated both prisoners, visit each of them to offer the same deal. If one testifies for the prosecution against the other and the other remains silent, the betrayer goes free and the silent accomplice receives the full 10-year sentence. If both remain silent, both prisoners are sentenced to only six months in jail for a minor charge. If each betrays the other, each receives a five-year sentence. Each prisoner must choose to betray the other (termed 'defect') or to remain silent (termed 'cooperate'). Each one is assured that the other would not know about the betrayal before the end of the investigation. How should the prisoners act?

(Adapted from Wikipedia.)

The different options can be represented by a **payoff matrix,** which lists the costs and benefits to each player based on the different independent decision options. For an *individual* the best solution (i.e. the Nash equilibrium), irrespective of what the other decides, is to defect (i.e. betray the other player). This is because defection leads to either 5 or 0 years, but cooperation leads to either 10 years or 6 months. The 'dilemma' arises because the best *collective* decision (i.e. the one that reduces the total amount of time spent in prison by both players) is for both to cooperate. The basic game can be altered by replacing the length of time in prison

Complex financial institutions are able to exist because of evolved cognitive mechanisms that permit social interactions between individuals – these include trust, norms of fairness, and responsibility.

KEY TERMS

Prisoner's Dilemma
A two-player game in which the best individual strategy is non-cooperation but the best collective strategy is cooperation.

Payoff matrix
In game theory, a matrix that lists the costs and benefits to each player based on the different independent decision options.

with other rewards or punishments, such as money. In general, participants often fail to choose the optimal solution based on maximized self-interest (i.e. always defect), suggesting they are taking into account the interests of the other player.

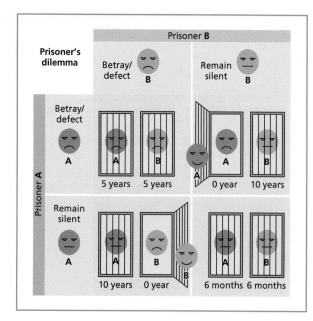

A payoff matrix for a typical Prisoner's Dilemma. If a player chooses to betray the other player they will get either 5 years or 0 years (depending on the other player's decision). If a player chooses to remain silent they will get either 10 years or 6 months in prison (depending on the other player's decision). As such the best decision is to betray. The game can be adapted and played over repeated trials by swapping years in prison with, say, monetary rewards. In this scenario, the best decision is less obvious as a betray decision may lead to retaliation.

In the version of the Prisoner's Dilemma outlined above the decision making between players is effectively simultaneous. Participants have no opportunities to retaliate, reward cooperation, or develop an understanding of the other player's strategy or motives. The iterated Prisoner's Dilemma has multiple rounds with the same players, but rather than a prison scenario a monetary scenario (or other reward/punishment) is set up. On any given trial, the greatest reward is obtained for a player who defects whilst their opponent cooperates. Thus, those who indiscriminately cooperate are easy to exploit. Those who always defect cannot be exploited but if both players choose to defect then they get less money than if both choose to cooperate. The dilemma here is that cooperation could lead to exploitation, but always defecting would lead to lower rewards if the other player does likewise. Computer simulations suggest that, in a multi-round game, the best solution is selective cooperation or **tit-for-tat** – cooperate initially and then copy the other player's last move by cooperating when someone has just cooperated with you and defect when someone has just defected (Axelrod & Hamilton, 1981). To put a social spin on this, tit-for-tat is fair (it always cooperates until someone else refuses to cooperate) and it is forgiving (if the other player switches back to cooperation there is no retaliation). This solution favors the benefits of cooperation but without running a high risk of exploitation.

Rilling et al. (2002) conducted an fMRI study of an iterative version of the Prisoner's Dilemma with monetary payoffs running for about 20 rounds. Mutual cooperation tended to be the most common outcome (followed by mutual defection). The mutual cooperation condition had the highest activity, relative to other conditions, in regions such as the ventral striatum and the orbitofrontal/ventromedial prefrontal cortex. However, note that the mutual cooperation condition is not associated with the highest monetary reward. The highest reward condition is when the other player cooperates and you betray. Rilling et al. (2002) suggest that the reason that activity in these reward-related regions is reduced in this betrayal condition is due to negative social emotions of guilt and of fear of retaliation on future trials. They suggest that activity in these regions could have as much to do with trust and camaraderie as winning. In support of this, the striatal activity associated with the mutual cooperation condition was abolished in a separate condition in which participants played an interactive computer game (playing tit-for-tat) rather than against a human player, even though the monetary payoffs were the same. In a subsequent study, Rilling et al. (2008) studied in more detail the neural response of unreciprocated cooperation (i.e. in which a participant opts to cooperate but their partner defects). This is associated with reports of

KEY TERM

Tit-for-tat
A strategy in which cooperation leads to cooperation and non-cooperation leads to non-cooperation on a trial-by-trial basis.

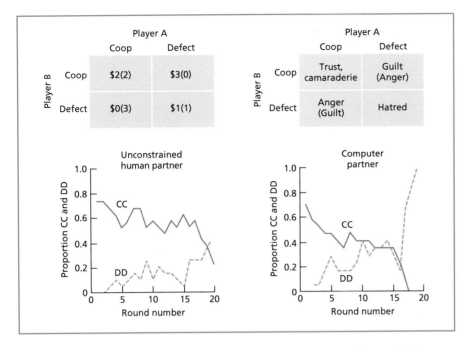

The payoff matrix for each player in the monetary version of the Prisoner's Dilemma used by Rilling et al. (2002). Player B's payoff is shown in brackets. The adjacent matrix shows the typical emotional feelings generated during the game. The bottom panels show the average pattern of results across 20 trials (CC = mutual cooperation; DD = mutual defection). When playing against another human, mutual cooperation prevails (but becomes increasingly exploitative towards the end of the game). When playing against a computer (programmed to respond using tit-for-tat) there is far less mutual cooperation and more exploitation. Adapted from Rilling et al. (2002). Copyright © 2002 Elsevier. Reproduced with permission.

anger, irritation, and disappointment. In terms of neural correlates, unreciprocated cooperation (as opposed to mutual cooperation) engenders activity in regions such as the insula (bilaterally) and the amygdala (on the left), which are associated with emotional processing.

Singer, Kiebel, Winston, Dolan, and Frith (2004a) had participants perform iterated Prisoner's Dilemma games with many 'players' (depicted in photographs) prior to fMRI scanning. The players were believed by the participants to be actual human players but were constructed by the experimenter to fall into one of three types: someone who always cheats, someone who always cooperates, and a neutral condition (where no interaction takes place). In addition, participants were told that some players were free to make their own decisions (intentional agents) whereas other players were just following orders as to how to play. Unlike Rilling et al. (2002), the scanning did not take place during the game but in a separate task later that consisted of viewing the previous set of faces and making a gender judgment. Viewing faces of people who previously cooperated (relative to neutral and/or people who defected) activated regions involved in face perception (e.g. the fusiform gyrus), emotion and reward processing (including the insula and ventral striatum), and 'theory of mind' (posterior STS, bordering rTPJ). These regions tended to show

Brain activity in the right posterior STS region (highlighted in top circle) and right fusiform (highlighted in bottom circle). These regions show greater activity when viewing faces of people previously encountered who had intentionally cooperated (i.e. free to choose their response) relative to those who were instructed how to respond. From Singer et al. (2004a). Copyright © 2004 Elsevier. Reproduced with permission.

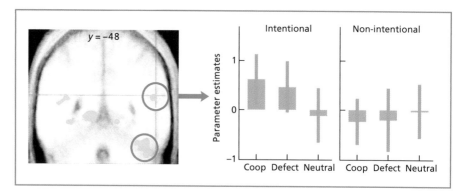

a greater response when people had also previously behaved intentionally, suggesting that these neural signals relating to trust do not reflect simple behavior or reward but also reflect attributions of responsibility. With regard to the reward-related activity (e.g. in the striatum), Singer et al. (2004a) propose that 'social fairness is experienced as rewarding per se' (p. 658). This accounts for the greater activity for intentional versus unintentional interactions even when level of monetary reward is equated.

The Ultimatum Game

In the **Ultimatum Game**, one player – the proposer – is given a pot of money and is instructed to decide what proportion of that money to give to a second player – the responder (Guth et al., 1982). The responder can then make one of two decisions. They either accept the offer, in which case it is split between both players according to the agreed amount, or they reject the offer, in which case neither player receives anything. The Nash equilibrium for a non-iterative (i.e. single exchange) Ultimatum Game is for the responder to accept any offer that is given. However, responders often reject offers of money. Offers below 20% of the pot of money are reliably rejected by most participants – that is, the responder often does not make decisions in a way that maximizes their own pot of money, but why? The standard answer to this is that the responder considers some offers as unfair and will reject offers of money in order to punish the proposer. To put it another way, one could say that the responder values the notion of fairness and their decision reflects a trade-off between a financial value (monetary gain) and a social one (fairness). On iterative versions of the Ultimatum Game the players engage in multiple rounds in which the same players act as responder and proposer. In this instance, a rejection of a low offer has a clearer benefit to the responder because it signals to the proposer that he/she needs to increase his/her offer. In this instance, it is not necessarily the best strategy to accept any offer because doing so would invite further low offers.

Sanfey et al. (2003) conducted an fMRI study of the Ultimatum Game in which participants acted as responders to offers from many different proposers (so that there was no chance of reciprocity). Although not known to participants, the offers from proposers were determined in advance to be either fair (i.e. a 50/50 split of $10) or unfair (e.g. an 80/20 split of $10 in the proposer's favor). There were two other conditions: one in which participants were told that the offer came from a computer, and one in which the participant received equivalent monetary rewards without any decision making or interaction. Participants were more likely to accept unfair offers when offered by a computer, which is consistent with the notion that rejection of unfair offers is motivated by a desire to punish. For human trials, a comparison of unfair versus fair offers showed activity in the bilateral insula, the anterior cingulate cortex, and the dorsolateral prefrontal cortex.

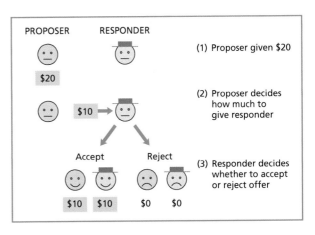

The insula activity was interpreted as providing an emotional signal related to unfairness (e.g. disgust or anger) that guides the behavior. Indeed, greater insula activity was found for unfair offers that were rejected relative to unfair offers that were accepted. Inducing feelings of disgust increases the rejection rates for unfair offers from a human player but not from a computer (Moretti & Di Pellegrino, 2010). Van't Wout, Kahn, Sanfey, and Aleman (2006) took a measure of emotional responsiveness (skin conductance response, SCR) during the Ultimatum Game and found a higher SCR when responders were given unfair offers relative to fair offers or relative to unfair offers from a computer. This suggests a role of emotion in the decision-making process. However, another study casts doubt on this. Civai, Corradi-Dell'Acqua, Gamer, and Rumiati (2010) used a similar design but asked participants to play (as a responder) either for themselves or for a third party. The SCR to unfair offers was found only when playing for themselves, even though the behavior was equivalent in both situations. This does not necessarily disprove the importance of emotion, but it does show that SCR may not be a reliable marker of it. It is important to establish what kind of mechanisms enable a person to play from another person's perspective, whether it matters whether the third party is known or liked, and so on, to clarify this important finding.

In addition to the insula, the study of Sanfey et al. (2003) highlights the importance of the dorsolateral prefrontal cortex (DLPFC). The DLPFC has a wider role in decision making and 'executive functions' and one general account of its function is in providing a biasing signal that makes some responses more likely and other responses less likely. However, in the context of the Ultimatum Game it is not clear whether this function lies primarily in biasing towards the more self-interested response or biasing towards responses based on fairness. To investigate this, Knoch, Pascual-Leone, Meyer, Treyer, and Fehr (2006) applied TMS over this brain region (either on the left or right) whilst the participant was acting as a responder to many different proposers (i.e. to avoid reciprocity). TMS applied over the DLPFC in the right hemisphere increases the tendency to accept unfair offers (Knoch et al., 2006) – that is, the responder behaves in a way that is more-selfish/less-social when this region of the brain is disrupted. It suggests that the normal function of this region is to bias towards responses based on fairness, possibly by inhibiting a more potent urge to act in terms of self-interest. Interestingly, the participant's judgments of fairness per se were unaffected by the TMS. This suggests that this region is indeed concerned with the response itself rather than in the computation of fairness (which is possibly a function of the insula). A similar finding is found for TMS over the right DLPFC in a different game (the Trust Game, described next). TMS over the right (but not left) DLPFC

The Ultimatum Game consists of two decisions between two players. The proposer must decide how much money to offer (between $1 and $20). The responder must decide whether to accept or reject the offer. It can either be played as a one-shot game (i.e. the proposer and/or responder changes every trial, or is terminated after a single trial) or iteratively (i.e. the same proposer and responder play repeated trials).

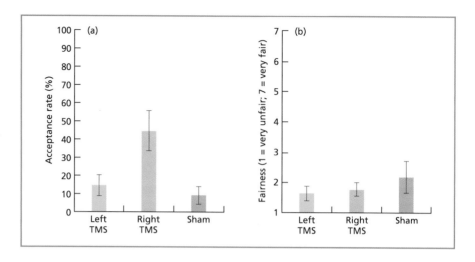

TMS over the right dorsolateral prefrontal cortex leads to people accepting more unfair offers in the Ultimatum Game (a), although their ability to decide what is fair/unfair is not compromised (b). From Knoch et al. (2006). Copyright © 2006 American Association for the Advancement of Science. Reproduced with permission.

reduced the amount of money returned to investors (Knoch, Schneider, Schunk, Hohmann, & Fehr, 2009). These participants behaved with more self-interest and were less concerned with building a good reputation.

The Public Goods Game and Trust Game

In the **Public Goods Game**, people may choose to contribute different amounts into a common pot of money but – crucially – everybody receives the same benefits irrespective of what they put in (see Fehr & Fischbacher, 2004). The real-life analogy here is that everyone pays different amounts of taxes, but the public benefits (e.g. clean streets, national security, education) are distributed evenly. Unlike the Ultimatum Game and Prisoner's Dilemma, many players can be simultaneously involved. A typical example of such a game is that four players are initially given $20 and they have to decide how much money to donate into a public pot (between $0 and $20). The amount donated then gets multiplied (e.g. doubled) and then returned equally to all four players. Let us consider three different scenarios:

1. All players contribute nothing. In this case, all players retain their initial $20 stake.
2. All players contribute everything. In this case the public pot of $80 ($20 × 4) gets doubled. All players receive $40 back.
3. Three players contribute everything and one contributes nothing. In this instance, the public pot of $60 ($20 × 3) gets doubled to $120 and then divided four ways. The three players who cooperated end up with $30 in total but the person who contributed nothing gets $30 on top of their $20 stake that they held onto, making $50 in total.

The dilemma here is that the best option, collectively, is for everyone to contribute. However, the best option for an *individual* decision maker is not to contribute (i.e. free load) but to hope that everyone else cooperates (in this regard the Public Goods Game and Prisoner's Dilemma are variants of the same game). When such games are played, most players contribute their stake in proportion to the stake of other players – that is, if other players contribute more to the pot then these players follow the social

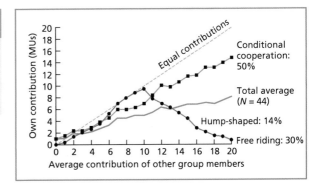

Typical results from a Public Goods Game. If other
players choose to contribute more, most people
decide to increase the amount that they will contribute
(conformity). However, a significant proportion of people
never opt to contribute anything (free riding). From Fehr
and Fischbacher (2004). Copyright © 2004 Elsevier.
Reproduced with permission.

norm and contribute more. However, as many as 30%
of players will consistently free load and contribute
nothing (e.g. Fischbacher, Gachter, & Fehr, 2001).

The **Trust Game** involves an exchange between
two players – an investor and a trustee (Berg, Dickhaut,
& McCabe, 1995). For example, the investor may be
given $20 and can choose how much of this stake to
give to the trustee. The money invested is then multiplied (e.g. doubled or tripled) and the trustee decides
how much to give back to the investor. In a single
exchange, the best option for the trustee (financially
speaking) is to keep all the money for himself/herself.
In iterative games, the trustee needs to return a fair
amount of money to the investee. An unfair return
would jeopardize future investment. Berg et al. (1995)
report that investors send about 50% on average,
although the amount sent back has a greater variance.

Both the Public Goods Game and the Trust
Game depend heavily on the notion of *trust* and in
both games there is a risk of exploitation. In the Trust
Game, the investor trusts the trustee to return a fair
sum and not steal the money. In the Public Goods
Game, an investor trusts that others will contribute
to the pot. Trust may depend on social norms as to
how to behave. These social norms may derive from
a basic sense of right/wrong that stems from our ability to understand how others feel (e.g. 'you will not
cheat me because you won't be able to live with the
guilt'). There is evidence that unconditional trust (i.e.
in which players always behave fairly) is associated
with greater activity in the medial prefrontal cortex
than conditional trust (Krueger et al., 2007; see also
McCabe, Houser, Ryan, Smith, & Trouard, 2001b).
Kreuger et al. (2007) interpret this in terms of 'mentalizing', such that players who adopt an unconditional
trust strategy are more likely to focus on the trusting
intentions of others. Trust is also likely to depend
on reputation for cooperation, and merely observing
cooperative individuals can evoke an extended network of brain regions involved in emotion processing
and theory of mind (Singer et al., 2004a).

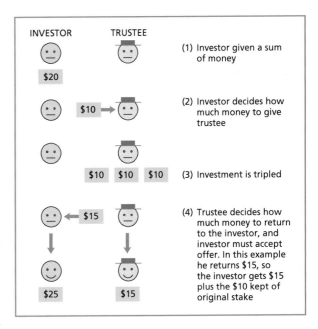

The Trust Game consists of two sequential decisions.
The investor must decide how much money to invest.
The sum invested then earns interest. The trustee
must decide how much to return to the investor. Will
the investor trust the trustee? Will the trustee keep
all the money for himself/herself or return a fair
amount?

Social norms may also be enforced by punishment. In one variety of the Public Goods Game, after each round the players have an opportunity to punish any other player that they wish to (Fehr & Gachter, 2000). For example, they could spend $1 in order to remove $2 from another player's balance. In these instances, free loaders can often expect small punishments from multiple players. Cooperation then tends to prevail on future trials. One might wonder what the motives are for punishing free loaders. Is it motivated by social norms of fairness, or is it motivated by self-interest? Although punishment comes at a cost to the punisher, it reaps rewards in the future because more people contribute to the pot after being punished. Falk, Fehr, and Fischbacher (2005) argue that punishment is motivated by social norms rather than self-interest because players still choose to pay for punishment even on one-shot interactions in which players never re-encounter each other. Functional imaging studies show that regions relating to reward are activated when an unfair player is punished (de Quervain et al., 2004). This was interpreted as satisfaction for punishing norm violations. In contrast to this idea, Rand, Dreber, Ellingsen, Fudenberg, and Nowak (2009) argue that cooperation is more likely to be sustained via acts of selective altruism than from punishment. They set up an alternative version of the Public Goods Game in which players could give other players rewards after each round (e.g. spending $1 to give another player $2). This was found to be more beneficial than punishment in eliciting cooperation.

To examine the neural substrates of trust, King-Casas et al. (2005) conducted an iterative version of the Trust Game using two linked fMRI scanners such that the neural responses of both an investor and trustee interacting with each other could be obtained simultaneously. Behaviorally, there was strong evidence for reciprocity. If the trustee increased the returns to the investor, then the investor would increase the amount invested with the trustee in the subsequent exchange. They divided the trials into three kinds depending on whether the reciprocity was fair (i.e. like for like), malevolent (i.e. trust was betrayed), or benevolent (i.e. trust was disproportionately rewarded). An unexpected (i.e. benevolent) reward was associated with activity in the dorsal striatum (the caudate nucleus) relative to the other conditions. This fMRI signal in the trustee's brain predicted the fMRI signal in another region of the investor's brain as he/she prepared his/her response. The timing of

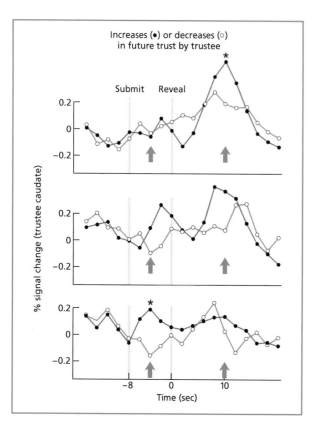

The graphs show the time course of BOLD activity in a region of the striatum of the trustee playing an iterative version of the Trust Game with another human player. The purple lines show activity in the brain that precedes a decision to increase trust (i.e. repay more) and the green lines show activity in the brain that precedes a decision to decrease trust (i.e. repay less). In the early phases of the experiment (top), the decision trust-related activity (difference between green and purple) occurs when the decision of the investor is revealed. Over time (middle and bottom) the peak of this activity shifts forward in time and occurs after the investor has made a decision, but before that decision is revealed to the trustee. Thus, activity in this region shows a transition between responding to the receipt of money and responding to the expectation of money on the basis of trust. From King-Casas et al. (2005). Copyright © 2005 American Association for the Advancement of Science. Reproduced with permission.

activity in the dorsal striatum changed during the course of the experiment: in early rounds it was found after the trustee learned about a favorable outcome; in later rounds it was found in the period in which the trustee was anticipating a favorable outcome. Their explanation of this finding was that the players had learned to trust each other (i.e. by developing a reputation) and anticipated each other's responses.

Reputation may be learned via communication (e.g. gossip) as well as via direct interactions with people. Delgado, Frank, and Phelps (2005) manipulated reputation in the Trust Game by using descriptions of people's moral character, which were overall positive, neutral, or negative. In terms of their subsequent actual interactions, the three types of character were equated in terms of their level of trust (i.e. returns on investment). They found that activity in the caudate nucleus was associated with reward learning, but only for interactions with neutral characters. Thus, their reputation really did precede them (as the saying goes) and it did so by blunting the brain's mechanism for learning about rewards.

CROSS-CULTURAL PERSPECTIVES ON SOCIAL NORMS OF FAIRNESS

Most of the experimental research on game theory is based on observations of university students within the Western world. Henrich et al. (2005, 2006, 2010) have attempted to extend this in 15 other cultures organized according to different principles (e.g. small-scale sedentary, semi-nomadic). The games used include the Ultimatum Game and the Public Goods Game. One of their key results was that across all cultures studied there is a tendency to not simply behave according to self-interest but to impose fairness norms in social interactions by engaging either in altruistic sharing or in punishing those who violate this norm (both of which incur a cost to the individual). The second key finding was that there is substantial variation across cultures in the extent to which these norms are enforced. In other words, all cultures are qualitatively similar but quantitatively different.

The larger the community of a human society, the more stringent are the notions of fairness. In the Ultimatum Game, people living in large communities expect a 50/50 division of the pot whereas people living in smaller communities may accept offers lower than 90/10 against them. Social norms of fairness may have a greater role when many interactions occur between strangers who are unlikely to meet again. From Henrich et al. (2010). Copyright © 2010 American Association for the Advancement of Science. Reproduced with permission.

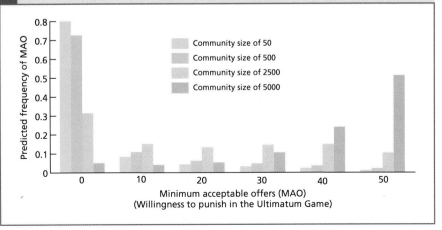

How are these cross-cultural differences to be accounted for? Individual-level variables such as formal education level or relative wealth compared to the rest of their group did not account for the data (Henrich et al., 2005). However, some group-level variables did. The extent to which societies engage in trade with strangers and the size of the communities lived in were both positively correlated with imposition of fairness norms and punishment for unfair acts (Henrich et al., 2005, 2006, 2010). This suggests that social norms for fairness may have a greater importance when interactions within a culture depend more on indirect reciprocity (with strangers who we may not meet again) than direct reciprocity (in which favors are returned to the same individual). Henrich et al. (2010) argue that the fairness found in large-scale societies is not just a product of our *biological* past but reflects a more recent *cultural* transition from small-group to large-group living.

Evaluation

The evidence reviewed above suggests that when an individual makes a decision in a social setting (i.e. that takes into account the actual or potential decisions of others) there is a tendency not to make decisions solely on the basis of self-interest but to take into account the interests of others.

In evaluating the contribution of game theory to the social neuroscience of interactions, it is worthwhile returning to the four possible reasons why people cooperate that were introduced at the beginning of the chapter. First, the suggestion that social interactions are rewarding in their own right receives a great deal of support. For example, many studies demonstrate greater activity in reward-related regions when interacting with another person relative to a computer or an agent who is unable to exercise choice over their decisions, even though these situations are otherwise equivalent (e.g. in terms of monetary gain, cognitive demands). Of course, this hinges on the assumption that activity in a given region such as the ventral striatum and orbitofrontal cortex does indeed reflect reward prediction and/or reward fulfillment. At present these assumptions appear to be plausible but more research is needed to characterize the actual cognitive operations that are performed by these regions. The second idea is that we cooperate because of the benefits of reciprocity. Although reciprocity may be the mechanism that drives cooperation from an evolutionary standpoint, it need not be the main motivational influence at the level of individual minds and brains. Are people motivated to interact because of expected reciprocal benefits or because interactions are rewarding in their own right (and the benefits of reciprocity are something of a by-product)? Evidence from game theory suggests that both punishment for non-cooperation and a desire to conform tend to increase the levels of cooperation.

Much of the evidence cited in this section has come from fMRI. Converging evidence from neuropsychology or developmental disorders (such as autism) has not been extensively brought to bear on cooperative games in the same way as it has on, say, theory of mind (but see Koenigs & Tranel, 2007; Sally & Hill, 2006). This will be an important enterprise in the future for better understanding which regions are necessary and what precise functions they perform in social and non-social contexts.

SUMMARY AND KEY POINTS OF THE CHAPTER

- Altruism refers to helping others at some cost to one's self, and it may come about in a population by helping others who are likely to help us. This can be achieved via kin selection (if you are a helper then your kin may be likely to carry a helping trait) or by reciprocity based on who has helped you in the past (direct reciprocity) or who has a good reputation for helping (indirect reciprocity).
- In humans, altruism may be motivated by a concern for the welfare of others (empathy), which is facilitated by shared neural circuits for self and other.
- Experiments based on game theory suggest that humans tend not to maximize their own gain, but to act towards the collective good. Decisions take into account social norms (e.g. of fairness) and unfair decisions tend to be punished, even if the cost of punishment is personally disadvantageous.
- Social decision making involves a network of brain regions, including those involved in emotions (e.g. the insula), cognitive control (e.g. lateral prefrontal cortex), and reward (e.g. the ventral striatum). Abstract social principles, such as trust, are starting to be understood in terms of the functioning of this network.
- Social interactions tend to be rewarding in their own right. For example, interacting with an unseen human tends to activate the brain's reward network more than interactions with a computer, even when the behavioral outcomes of those interactions are the same.

EXAMPLE ESSAY QUESTIONS

- How is human altruism similar to, and different from, that in other species?
- What is the role of empathy in altruism?
- What is 'trust' from a social neuroscience perspective?
- Do people matter more than money in human decision making? Discuss using evidence from game theory and social neuroscience.

RECOMMENDED FURTHER READING

- Glimcher, P. W., Camerer, C., Poldrack, R. A., & Fehr, E. (2009). *Neuroeconomics: Decision making and the brain*. San Diego: Academic Press. An excellent, although advanced, collection of chapters.
- Tomasello, M. (2009). *Why we cooperate*. Cambridge, MA: Bradford Books. An accessible book that presents evidence primarily from children and primates.

CHAPTER 8

CONTENTS

All you need is love   180

Attachment   184

Separation, rejection, and loneliness   194

Summary and key points of the chapter   199

Example essay questions   199

Recommended further reading   199

Relationships

Human relationships primarily consist of friends, family, and a romantic partner. We invest a huge amount of time and effort into cultivating and maintaining these relationships. Even though most of us no longer live in close-knit communities of extended families, we find new ways of staying in touch with our inner circle. For example, undergraduates spend an average of 30 minutes a day keeping in touch with friends via Facebook alone (Pempek, Yermolayeva, & Calvert, 2009). Why? We affiliate with others because we like it, and we like it because it is good for us. It is good for us not only for the material benefits that accrue from cooperation, but also because it has protective effects on our health. Uchino, Cacioppo, and Kiecolt-Glaser (1996) reviewed 81 studies investigating the effects of social support on cardiovascular-, immune-, and endocrine-related health. Social support in terms of supportive family interactions and the presence of an intimate and confiding relationship has a protective effect against these conditions. In contrast, loneliness and lack of intimacy may have the opposite effect: for instance, being associated with greater cognitive decline in old age (Wilson et al., 2007).

At what point does a series of interactions between two people become a relationship? This is not a straightforward question to answer, but the answer almost certainly lies in their psychological state rather than the interactions themselves. For example, one might go into the same shop every day to buy groceries and, as part of this routine, chat to the shopkeeper. This does not necessarily mean that you are having a relationship with the shopkeeper! Mutually beneficial interactions (e.g. buying–selling) are not sufficient for a relationship. The appropriate test might be this: imagine you go in one day and the shopkeeper is not there because someone else is standing in. You carry out the same interactions as before (including the chatting) but there is something missing and your needs are not fully met. In this example, it would be reasonable to conclude that there was a relationship (of sorts) going on here – a **social bond**, to use another term.

Poets and artists have tried to capture love for millennia. Are neuroscientists going to fare any better?

Social bonds are characterized by a sense of happiness or well-being in the presence of the bonded other, and a sense of wanting or longing (perhaps even distress) in their absence. These bonds may develop from contact and prosocial interactions, but ultimately they need not depend upon them (e.g. the same prosocial interactions are less rewarding if the other person is substituted). In Chapter 7, we saw evidence of how trust development involves a shift, over time, from reward mechanisms being triggered by a positive interaction towards reward mechanisms being triggered by the interaction itself, even before the positive/negative outcome is known (King-Casas et al., 2005). This is likely to be one important neural mechanism for establishing a social bond.

An attachment is a powerful type of social bond that tends to be limited to particular kinds of relationships. Historically it was used primarily to describe infant–mother relationships (or other caregivers) but over the years it has been applied to describe romantic relationships. The emotion that is associated with being in an attachment relationship is **love**. The severing of an attachment relationship (e.g. death, divorce) can be associated with strong negative feelings of grief or distress. This chapter will start by considering love: whether it is a unitary concept and how it may be represented in the brain. This theme is continued by considering attachment more generally, focusing on infant–mother bonds and romantic bonds. Finally, the chapter will consider separation, social exclusion, and loneliness.

ALL YOU NEED IS LOVE

Love is generally not classified as a basic emotion although some researchers question this (e.g. Shaver, Morgan, & Wu, 1996). Recall that Ekman's (1992) criteria for a basic emotion include: having a universal expression; having evolved for a specific purpose; and having a distinct neurological substrate. It is true that love is not associated with an expression. Love is not straightforward to define and appears to be rather diverse in form: for instance, the love we have for our family feels quite different to that for a new boyfriend or girlfriend. Nevertheless, in other respects it is 'basic': it can be ascribed an evolutionary function (to form and maintain attachments, e.g. to ensure parental care) and it does appear to have its own neurological signature (discussed below).

Types of love

Social psychologists typically define love in terms of several underlying factors. One of the most influential models along these lines is Sternberg's (1986, 1988) **triangular theory of love**, in which different types of love arise from three factors: passion (broadly, sexual attraction), intimacy (feelings of warmth, closeness, and sharing), and commitment (resolve to maintain the relationship through difficulty). For example: passion without commitment or intimacy is associated with infatuation (e.g. teenagers in love with pop stars); commitment and intimacy without passion are associated with companionate love; a combination of all three factors gives rise to consummate love; and so on.

Of course, the type of love experienced may change over time. Anthropological studies suggest that the passionate component of love only lasts for the first

6 months to 3 years of a relationship, which is normally enough time to ensure conception (Jankoviak & Fischer, 1992). Many relationships continue to endure for decades in which other aspects of love (e.g. intimacy) may grow. There may also be cultural differences. For example, arranged marriages may begin only with a sense of commitment but the other components (intimacy and passion) may arise through mutual respect and shared experiences (Gupta & Singh, 1982). Beyond the romantic/passionate phase of a relationship, Adams and Jones (1997) identified three factors that maintain commitment to a relationship:

1. Personal dedication: due to ongoing positive feelings towards the partner.
2. Moral obligation: due to social norms or religious/cultural beliefs.
3. Costs versus benefits of leaving: financial and emotional costs; availability of an alternative relationship.

	Passion	Commitment	Intimacy
No love	✗	✗	✗
Infatuation	✓	✗	✗
Empty love	✗	✓	✗
Liking	✗	✗	✓
Fatuous love	✓	✓	✗
Romantic love	✓	✗	✓
Companionate love	✗	✓	✓
Consummate love	✓	✓	✓

The triangular theory of love aims to explain different kinds of love in terms of the presence or absence of three underlying factors: passion, commitment, and intimacy. Based on Sternberg (1988).

WHAT FACTORS DETERMINE WHO WE FALL IN LOVE WITH?

The 'rules' of attraction have been extensively studied by psychologists. Many of these factors apply to friendships as well as romantic relationships.

- *Proximity*. This provides an opportunity to contact. The famous example of proximity is Festinger's study of friendship formation in a housing complex. People were more likely to form friends with people on the same floor relative to other floors or other buildings (Festinger, Schacter, & Back, 1950).
- *Familiarity*. People are more liked when they are seen more often, in both experimental (Jorgensen & Cervone, 1978) and naturalistic settings (Moreland & Beach, 1992).
- *Physical attractiveness*. This was considered previously in Chapter 5. However, one factor not previously considered is attractiveness of a potential partner relative to one's own perceived attractiveness. The **matching hypothesis** states that people are more likely to form long-standing relationships with those who are as equally physically attractive as they are (Walster, Aronson, Abrahams, & Rottman, 1966). In this study, male and female students were randomly paired together at a college dance (although they believed they were

KEY TERM

Matching hypothesis
States that people are more likely to form long-standing relationships with those who are as equally physically attractive as they are.

being paired on personality). Women judged to be attractive (by independent raters) were more likely to be asked on a date, but it was similarity in relative attractiveness between male and female partners that predicted whether they would still be dating after 6 months.

- *Similarity of attitudes.* Newcomb (1961) asked students to fill in a questionnaire about their attitudes and values before moving into shared accommodation. Over the course of the semester they then noted down people who they were attracted too. Initially attraction was related to proximity but then shifted to similarity of attitudes.
- *Reciprocal liking.* If you are told that someone likes you then you are more likely to like them (Dittes & Kelley, 1956).

Neuroscience and love

Research in social neuroscience has not been based on social psychology models such as Sternberg's, but it has attempted to look at different kinds of love (for a critique see Fusar-Poli & Broome, 2007). Bartels and Zeki (2000) compared brain activity, using fMRI, in response to viewing one's romantic partner relative to a long-term friend. In a subsequent study, Bartels and Zeki (2004) contrasted the viewing of images of one's own child with another acquainted child (i.e. maternal love). In both cases, there was activity in a number of regions that the authors claim are rich in the neuro-hormones vasopressin and oxytocin (of which more later) and 'reward centers' such as the striatum. (There were also differences between maternal and romantic love.) Intriguingly, there were a number of deactivations, including the amygdala and regions that have traditionally been linked to 'mentalizing', such as temporal poles, temporo-parietal junction, and medial prefrontal cortex. A deactivation, in this context, means that the activity for the familiar acquaintance was greater than for the loved one. They argue that love involves a pull–push mechanism in the brain by which regions involved in critical social assessment (e.g. of others' intentions) are deactivated whilst regions involved in reward and attachment are activated. This may provide a basis for the unconditional nature of love that distinguishes it from context-sensitive emotional responses.

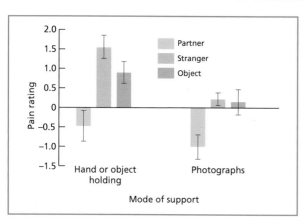

In this study, the thermal pain threshold of a group of women was determined. They were then given a number of painful stimuli during which they held their partner's hand, held a male stranger's hand, or squeezed a ball, or during which pictures of these stimuli were presented on a computer screen. Their pain threshold was reduced in the partner conditions. From Master et al. (2009). Copyright © 2009 Association for Psychological Science. Reprinted by permission of SAGE Publications.

As well as being rewarding, the mere presence of a loved one may act as a buffer against stress or pain. Showing female participants pictures of a male loved one whilst the participant received a painful (but non-harmful) stimulus reduced the reported pain intensity (Master et al., 2009) and this analgesia correlates with activity in a number of regions, including the reward-related nucleus accumbens (Younger, Aron, Parke, Chatterjee, & Mackey, 2010). Similar behavioral results are found when the participant held their partner's hand (but could not see him) relative to holding the hand of a stranger or squeezing a ball (Master et al., 2009).

The term 'falling in love' implies a lack of control and some researchers have likened the feeling to those of certain clinical disorders (e.g. Marazziti, 2009; Tallis, 2005). For example, mood elation with excess energy and sleeplessness has been likened to the manic phase of bipolar depression, and the craving and intrusive thoughts about the loved one have been likened to obsessive–compulsive disorder (OCD; Marazziti, 2009). Of course this need not reflect clinical disturbance but may be an important biological mechanism that enables us to put aside inhibitions and abandon our comfort zones (e.g. neglecting friends and family to spend time with a new partner). There is evidence for hormonal and neurochemical changes in people who are in the stages of romantic love and, interestingly, these appear to be confined to the early 'passionate' phase of a relationship rather than the later 'companionate' stages. For example, the concentration of a particular serotonin transporter was lower in participants in the passionate phase of love and comparable to patients with OCD relative to controls (either single or in longer term relationships) and different from the

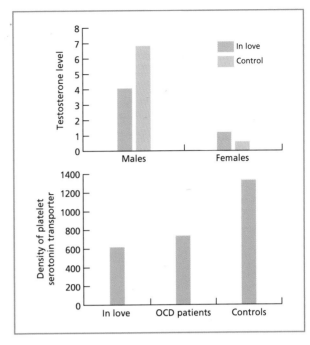

Top: The testosterone levels of males in love are lower than male controls, and those of females in love are higher than female controls. Data from Marazziti and Canale (2004). Bottom: The level of a serotonin transporter (measured in terms of a binding parameter) is lower in people in love relative to controls, and is comparable to patients with OCD. Data from Marazziti et al. (1999).

same participants tested 12–18 months later (Marazziti, Akiskal, Rossi, & Cassano, 1999). Similarly, testosterone levels are lower in males and higher in females during the early stages of a romantic relationship but return to normal 12–18 months later, even when they are maintaining the same relationship (Marazziti & Canale, 2004). The specific details of these hormonal/chemical changes are not especially relevant here, but several general points are worth highlighting. It is interesting to note the similarity between being in love and a clinical 'disorder' such as OCD. It is also interesting to note that the changes had disappeared as the relationship matured out of the early passionate phase.

Evaluation

Love is an emotion that is elicited in the presence of (or by the thought of) another person with whom there is an attachment. There appear to be different facets to love (for instance, as articulated by Sternberg's theory) but this need not imply that love is culturally created or unworkable within a scientific framework. There is evidence for a common (and partly different) circuit for different kinds of love (e.g. maternal vs. romantic), and for major biological changes associated with the act of falling in love.

ATTACHMENT

An **attachment** is a long-enduring, emotionally meaningful tie to a particular individual (Schaffer, 1996). Attachments in young children provide them with comfort and security and are characterized, behaviorally, in terms of proximity seeking (i.e. maintaining close physical proximity) and separation distress when proximity is not maintained. Attachment is found in all animal species in which the young are initially in need of care and protection, as is most vividly illustrated by the case of **imprinting**. This phenomenon is found in many birds, in which newly hatched chicks follow their presumptive parent around. Research reveals that there is a narrow window of opportunity, between 15 hours and 3 days, for a gosling to imprint and the movement of a stimulus is deemed critical (Tinbergen, 1951). Once imprinted, the bird is virtually unable to learn to follow a new foster parent (although see Bolhuis, 1990).

In humans, the system is rather more flexible in terms of who attachment relationships are formed with, the number formed, and when they may occur. Nevertheless, there are strong trends. The mother is normally the first (and strongest) attachment relationship and this tends to emerge at around 7–8 months of age, as assessed in naturalistic settings (e.g. distress when an infant is left with a babysitter; Schaffer & Emerson, 1964), in laboratory environments (e.g. Kotelchuck, Zelazo, Akgan, & Spelke, 1975), and across a wide range of cultures (Van Ijzendoorn & Kroonenberg, 1988). Attachment is assessed objectively in terms of some measure of separation distress (e.g. crying) or proximity-seeking behavior (e.g. attempting to follow or reach out to the adult). Shortly after a specific attachment is formed, infants tend to experience a fear of strangers suggesting that this is a consequence of having an attachment relationship (Schaffer, 1996).

Central to biologically based accounts of attachment is the notion that interactions with attachment figures (e.g. mother) activate the reward-based mechanisms

Research into imprinting was studied extensively by the biologist Konrad Lorenz, shown here followed by a line of goslings who, upon hatching, have apparently mistaken him for their mother. Copyright © Science Photo Library.

of the brain. In the behaviorist tradition, the reward was considered to be a learned association between a stimulus (e.g. mother) and the set of rewards that she provides (e.g. food, warmth). This notion was challenged, most notably by Harlow (1958) in 'The Nature of Love', by showing that newborn rhesus monkeys were selective about the objects they would seek comfort from. For example, if given a choice between an artificial wire 'monkey' that provides milk (from a bottle) and an artificial cuddly 'monkey' (who provides no milk) they would spend more time with the latter, going over to the wire monkey only when hungry. Rather than thinking of attachment as a learned response, others such as Bowlby (1969) and Harlow (1958) argued that it is an innate response (or a primary reinforcer) that has been shaped by the evolutionary need for care and protection. In addition to triggering reward-based mechanisms, the brain's other significant reaction to attachment formation is a reduction in its stress response (e.g. Spangler & Grossmann, 1993). This may contribute to the sense of security associated with attachment relationships.

Individual differences in mother–infant attachment

The presence of a stable caregiver (or caregivers) during a sensitive period in the first year may be necessary for an attachment to be formed, but the quality of that attachment may depend on the quality of the interactions. The classic laboratory paradigm for assessing different types of attachment in infancy is the **Strange Situation Test** developed by Ainsworth and colleagues (Ainsworth, Blehar, Waters, & Wahl, 1978). In this test, the infant is placed through a series of seven episodes involving separation from the parent, interactions with a stranger, and finally reunion with the mother. Each episode lasts around 3 minutes but can be curtailed if the infant is distressed.

PERCENTAGE OF INFANTS FORMING ATTACHMENTS TO PARTICULAR INDIVIDUALS		
Individual	At onset of attachment	At 18 months
Mother	95	81
Father	30	75
Grandparent	11	45
Other relative	8	44
Sibling	2	24
Other child	3	14

Bowlby (1969) originally suggested that humans tend to form a single attachment relationship (to their mother), but empirical evidence suggests otherwise. By the age of 18 months, most infants have more than one attachment relationship. This may act as a 'biological insurance policy' in the event of loss of the attachment figure. Adapted from Schaffer and Emerson (1964).

Harlow performed a variety of studies on attachment in primates, for example, showing that infant monkeys prefer a cuddly artificial 'mother' over a wire 'mother' even if the latter provides milk. This evidence ran against the behaviorist tradition of the time, which believed that maternal love was a learned response to having needs met. Reproduced with kind permission of Harlow Primate Laboratory, University of Wisconsin.

KEY TERM

Strange Situation Test
A measure of attachment during infancy in which the infant experiences separations and reunions with a stranger and with an attachment figure.

AINSWORTH'S STRANGE SITUATION TEST		
Episode	People present	Events
1	Mother, infant	Infant explores; mother watches/plays
2	Mother, infant, stranger	Stranger enters and is silent at first, then talks to mother, then interacts with infant
3	Infant, stranger	Mother leaves; stranger interacts with infant
4	Mother, infant	Mother returns and settles infant; stranger departs
5	Infant	Mother leaves; infant is alone
6	Infant, stranger	Stranger enters and interacts with infant
7	Mother, infant	Mother returns and settles infant; stranger departs

The infant's behavior is scored according to how it deals with stress and the way that it responds to the mother, particularly during the reunion episodes (4 and 7). Using this paradigm, Ainsworth et al. (1978) identified three types of attachment style. **Securely attached** infants get moderately upset by the departure of the mother and greet her positively in reunion. Insecurely attached infants fell into two types: **insecure/avoidant** types who avoid contact with the mother, especially at reunion; and **insecure/anxious** types who show high levels of stress on separation and are difficult to console at reunion. Most infants are classed as securely attached and this finding holds cross-culturally, although the balance between the two insecure categories varies more across cultures (Van Ijzendoorn & Kroonenberg, 1988).

Ainsworth et al. (1978) attributed the different attachment styles to the different quality of interactions between mother and infant, and to mothering styles in particular. Mothers of securely attached infants were assumed to be able to pick up on the infant's signals and respond to them appropriately and consistently. Mothers of insecurely attached

The Strange Situation Test has become the standard measure of infant attachment. The infant's behavior is scored according to how it deals with stress and the way it responds to the mother, particularly during the reunion episodes (4 and 7).

KEY TERMS

Securely attached
A type of attachment characterized by proximity seeking with the attachment figure and distress at separation.

Insecure/avoidant
A type of attachment characterized by ambivalence towards the attachment figure and avoidance of contact at reunion.

Insecure/anxious
A type of attachment characterized by extreme distress at separation.

infants tended to be either neglectful in their care (which was linked to insecure/ avoidant infants) or inconsistent in their care (which was linked to insecure/anxious infants). Others have noted that the infant's temperament may be important for eliciting particular kinds of attachment (e.g. Belsky & Ravine, 1987).

The classification has good reliability over time and appears to have external validity in that it predicts the quality of other kinds of social behavior. Grossman (1988) found that the classification at 1 year was 87% successful at predicting the classification (based on teacher/parent ratings) at 6 years, and that securely attached 6-year-olds had better concentration and independent playing. Moreover, 4-year-olds who had previously been classified as securely attached during infancy are rated as more popular by their peers and they develop more secure friendships, particularly with friends who were also previously classified as securely attached (Park & Waters, 1989).

There are a small number of developmental social neuroscience investigations of different attachment styles in human mothers and infants. Swingler, Sweet, and Carver (2010) recorded ERPs from infants viewing either their mother's face or a stranger's face and related this to signs of distress and visual search for the mother in the Strange Situation Test. Greater distress and visual search (i.e. signs of secure attachment) during the separation phase of the Strange Situation were linked to greater amplitudes and longer latencies of ERP components elicited by the mother's face for an attention-related component (termed Nc) and a P400 component that has been linked to face recognition (and may be an infant analogue of the N170 in adults). Fraedrich, Lakatos, and Spangler (2010) performed a complementary study by examining the ERP response of mothers to images of infants' faces (in various emotional expressions) and related this to attachment style (note that: they did not compare images of their own infant with a stranger infant). Mothers with insecure attachments were found to have a greater amplitude to the N170 component (which is related to perceptual processing of faces), but mothers with secure attachments showed a greater P300 (which is related to more semantic aspects of face processing). The findings of these two studies reveal that attachment style manifests itself in face perception and attention-related components. It is, however, likely that a far wider network is implicated in attachment and this has been more clearly demonstrated in human adults using fMRI and also in animal models of attachment. These are considered below.

Attachment in adult relationships

It has been argued that attachment style carries over to adult romantic relationships (e.g. Hazan & Shaver, 1987; Simpson, 1990). In adults, these tend to be assessed using questionnaires that ask about attitudes towards relationships. For example, the questionnaire of Hazan and Shaver (1987) was designed to classify adults into the same three categories as those proposed by Ainsworth et al. (1978). For example, it contains statements such as 'I find it difficult to trust others completely' (insecure/ avoidant) and 'I often worry that my partner doesn't really love me' (insecure/anxious). Those with an anxious adult attachment style may worry about rejection and tend to show heightened vigilance to cues of support or criticism, whereas those with an avoidant attachment style may report less need for close relationships and may tend to be distrustful of support/criticism from others. More recent questionnaire-based measures of adult attachment do not attempt to divide into categories but instead describe them according to two dimensions: avoidance (high to low) and anxiety (high to low) (e.g. Fraley, Waller, & Brennan, 2000) In this scheme, a secure attachment style would be associated with low scores on the avoidance dimension and low scores on the anxiety dimension.

There is some evidence to suggest that adult attachment styles are *partially* conserved from infancy. Main, Kaplan and Cassidy (1985) administered the Strange Situation Test to mothers of 1-year-old infants, and then asked the mothers 5 years later about their own attachment experiences to their mothers. Mothers who reported their own early attachment experiences as secure tended to have infants classed as secure on the Strange Situation Test. A 20-year longitudinal study that compared Strange Situation performance at 12 months with adult attachment style shows a 72% correspondence over this period between secure and insecure attachment styles (Waters, Merrick, Treboux, Crowell, & Albersheim, 2000). Negative life events (e.g. parental divorce, life-threatening illness) were related to changes in attachment style. Both of these studies, however, examined attachment within families rather than romantic attachments.

Studies of adult romantic attachment styles reveal a detailed picture of the underlying neural systems. Gillath, Bunge, Shaver, Wendelken, and Mikulincer (2005) presented women with various relationship scenarios (e.g. break-up) during fMRI and correlated brain activity with individual attachment styles measured in terms of level of relationship anxiety and avoidance. Low levels of attachment anxiety were associated with high levels of activity in orbitofrontal cortex (given negative scenarios), whereas high levels of attachment avoidance were linked with high activity in lateral prefrontal regions (for negative and positive scenarios). This suggests different styles of control mechanism in the maintenance of relationships.

Other studies provide evidence of the role of emotion and reward processing regions in attachment styles. Lemche et al. (2006) presented participants with brief unconscious statements relating to negative attachment outcomes (e.g. 'my Mum rejects me') and compared these to a neutral condition. After these statements were presented, a probe sentence was presented that the participants had to respond to (e.g. 'other people like me', 'I trust my friends'). The degree of interference from the unconscious primes has been shown to relate to attachment insecurity and, in this study, this measure was related to activity in the amygdala as well as a skin conductance response during the task. The results were interpreted as providing evidence of the role of the amygdala in attachment *insecurity* rather than attachment per se.

Vritcka, Anderson, Grandjean, Sander, and Vuilleumier (2008) investigated adult attachment style measured in terms of avoidance and anxiety dimensions. They used a difficult perceptual task in which they received feedback on their own performance in terms of a happy or angry facial expression. Activation of striatum and ventral tegmental area was enhanced to positive feedback signaled by a smiling face, but this was reduced in participants with avoidant attachment. This indicates relative impassiveness to social reward in the avoidant group. Conversely, a left amygdala response was evoked by angry faces associated with negative feedback. This correlated positively with anxious attachment, suggesting an increased sensitivity to social punishment. This study differs from the others in that attachment style is entirely incidental to the task. The results suggest that attachment style affects how different individuals respond to, and interpret, social rewards (e.g. smiles) and punishments (e.g. anger).

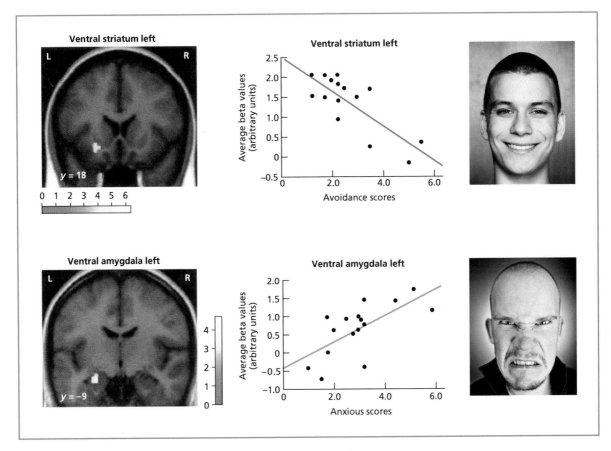

Self-reported relationship avoidance scores correlate negatively with activity in the ventral striatum when participants get feedback about a win with a smiling face (i.e. those who have an avoidant approach to relationships may find smiles less rewarding). Self-reported relationship anxiety scores are correlated with activity in the amygdala when participants get feedback about a loss with an angry face (i.e. those who are anxious about the quality of attachment may react with fear to social disapproval signaled by anger). Note that low avoidance and low anxiety scores are indicative of secure attachment. From Vritcka et al. (2008). Copyright © 2008 Vritcka et al. Photos on right are not part of the original figure.

Role of key hormones and neurotransmitters in attachment

Consistent with the notion that early attachment relationships provide a model for all future long-term relationships (e.g. romantic partners) is neurobiological evidence for shared substrates between parent–infant bonding and partner–partner bonding in sexual relationships (e.g. Gonzalez, Atkinson, & Fleming, 2009; Wommack, Liu, & Zuoxin, 2009). These studies emphasize the fact that attachment is bidirectional (not just from infant to caregiver) and by no means limited to infancy. Much of the evidence in this area has come from animal models of two mammal species: Prairie voles and Montane voles (Carter et al., 1995). These two animals are similar in many respects but differ from each other in one crucial respect. **Prairie voles** form a long-enduring attachment relationship (termed **pair-bonding** in this literature) with

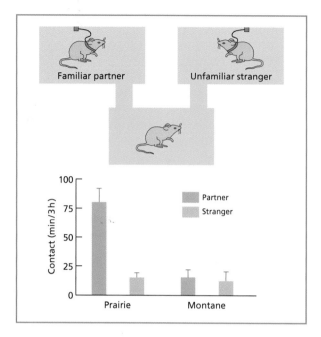

The standard experimental set-up for measuring partner preference in voles involves placing a vole in a chamber connected to two other chambers containing both a familiar vole and an unfamiliar stranger. The amount of time spent in each chamber is measured. If a Prairie vole spends at least 24 hours with another vole and they are allowed to mate, then this induces a durable partner preference (an attachment). The Montane vole shows no such loyalty. It actively avoids both its previous partner and the stranger. From Wommack et al. (2009).

an opposite sex partner whereas **Montane voles** are promiscuous. Prairie voles share a variety of responsibilities (nest building, nest guarding, some aspects of parental care), they remain together not just in the breeding season, and on the death of a partner they rarely find a replacement (Carter et al., 1995). Montane voles do not engage in joint parental care and show little evidence of forming attachment relationships.

Pair-bonding is assessed using a partner preference paradigm in which a vole is free to choose between two compartments: one containing a familiar vole and one an unfamiliar vole, both of the opposite sex. Prairie voles reliably affiliate with a familiar vole that they have lived with for 24 hours and mated with, and they avoid unfamiliar voles, whereas Montane voles show no such preference (Williams, Catania, & Carter, 1992).

Two hormones that have been extensively studied in attachment formation are **oxytocin** and **vasopressin**. Both are peptides that are unique to mammals but have similarities with other molecules found in other species. Oxytocin is synthesized by nuclei in the hypothalamus, and these nuclei connect to various structures (including the amygdala) and to the pituitary gland, from where they are released into the blood stream (e.g. Carter, 1998). In mothers, oxytocin can result in uterine contraction (during labor) and milk production (during breastfeeding). After birth, it appears to be involved in attachment formation but not the maintenance of attachments. Virgin rats given an injection of oxytocin display maternal behavior (Pedersen, Ascher, Monroe, & Prange, 1982) and maternal rats given a chemical lesion that disrupts its action show reduced maternal behavior after birth (i.e. during attachment formation) but not when given several days after birth (i.e. during attachment maintenance) (Insel & Harbaugh, 1989). The monogamous Prairie vole and the promiscuous Montane vole have similar oxytocin cell populations but differ substantially in the distribution of oxytocin *receptor types* in regions such as the nucleus accumbens, amygdala, and certain hypothalamic nuclei (Insel & Shapiro, 1992). Oxytocin's action in the central amygdala is associated with stress-reducing, anxiolytic, effects (Bale, Davis, Auger, Dorsa, & McCarthy, 2001). In humans, oxytocin reduces the

The sociable and faithful Prairie vole (left) and its 'love them and leave them' cousin, the Montane vole (right).

amygdala's response to fear-inducing stimuli measured using fMRI (Kirsch et al., 2005) and oxytocin may have a general social affiliative role, as one recent study with humans illustrates. Kosfeld, Heinrichs, Zak, Fischbacher, and Fehr (2005) administered oxytocin, using a nasal spray, to participants playing a social exchange game involving trust (The Trust Game, described in Chapter 7). Participants given oxytocin displayed higher levels of trust (in terms of the amount of money they were willing to invest) but the effects were specific to social interactions. When the same experiment was performed with a computer making random decisions (but with the same risk of winning/losing) there was no increase in trust.

The vasopressin molecule is structurally similar to oxytocin and has grossly similar behavioral effects insofar as it promotes attachment. For example, disruption of both oxytocin and vasopressin affects the ability of Prairie voles to form partner preferences under the normal conditions of 24 hours of mating contact. However, it differs from oxytocin's action in a number of ways. In Prairie voles, males are more responsive to vasopressin and females to oxytocin (Cho, DeVries, Williams, & Carter, 1999). This may enhance gender-specific behavior linked to social relationships, such as the male tendency to act aggressively towards an intruder, which is associated with vasopressin (Gobrogge, Liu, Jia, & Wang, 2007). Whilst oxytocin tends to reduce anxiety, vasopressin has the opposite effect (Huber, Veinante, & Stoop, 2005).

Both oxytocin and vasopressin may exert additional influences on the dopaminergic reward-based system, which originates in the ventral tegmental area and projects through the ventral striatum (including the nucleus accumbens) into the frontal cortex. Increased release of dopamine in this region is linked to sexual activity (Pfaus et al., 1990), although this is true of both monogamous and promiscuous species. The role of dopamine in attachment may be related to its interaction with hormones at key sites. For example, activation of both dopamine and oxytocin receptors within the nucleus accumbens is required for partner preference formation in Prairie voles (Liu & Wang, 2003). Also, dopaminergic disruption in the medial prefrontal cortex can lead to partner preference formation in the absence of mating, suggesting that its role is not limited to sex-based reward (Smeltzer, Curtis, Aragona, & Wang, 2006).

The **hypothalamic-pituitary-adrenal (HPA) axis** is activated during stress and produces changes in stress-related hormones such as cortisol (Erickson, Drevets, & Schulkin, 2003). Securely attached babies secrete less cortisol than insecurely attached babies (Spangler & Grossmann, 1993), consistent with the view that secure attachment is less stress inducing. In rats, removal of the adrenal gland in late pregnancy reduces maternal behavior but hormone injections reverse the effect (Graham, Rees, Steiner, & Fleming, 2006). In Prairie voles, a stressful event has different

KEY TERM

Hypothalamic-pituitary-adrenal (HPA) axis
A neural pathway that is activated during stress and produces changes in stress-related hormones such as cortisol.

effects on affiliative behavior in males and females (DeVries, DeVries, Taymans, & Carter, 1996). Males are more likely to affiliate with a female during stress, whereas females are less likely to affiliate with a male and more likely to choose the company of another female. Oxytocin is known to modulate the HPA by reducing the release of stress-related hormones in rodents (Neumann, Toschi, Ohl, Torner, & Kromer, 2001). Heinrichs, Baumgartner, Kirschbaum, and Ehlert (2003) administered either oxytocin or placebo (via nasal sprays) to a group of human participants undergoing a stressful event – the Trier Social Stress Test (TSST). This involves a number of phases, including preparation and delivery of a speech to an unresponsive audience. In addition to the administration of oxytocin/placebo the participants either did or did not have social support from a best friend during the preparation phase. Both oxytocin and social support reduced the levels of cortisol associated with public speaking, and the two factors interacted with each other – that is, there were greater benefits when both occurred together relative to that expected from the effects of each alone. This may explain the known health-related advantages of having social support for a range of physical and mental health problems that tend to increase with stress (e.g. Uchino et al., 1996).

This seemingly straightforward link between oxytocin and stress is complicated by other evidence. Resting levels of oxytocin (in women) and vasopressin (in men) in the blood plasma are *higher* in people who report relationship dissatisfaction (Taylor, Saphire-Bernstein, & Seeman, 2010) and this has been linked to greater,

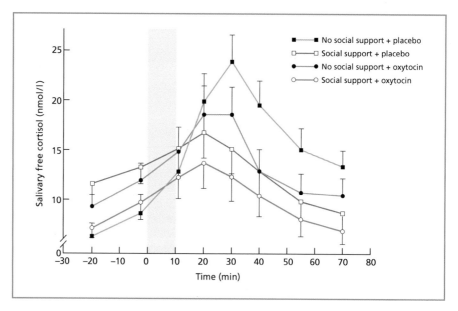

The Trier Social Stress Test (TSST) involves a number of phases, including preparation and delivery of a speech to an unresponsive audience. Levels of the stress-related hormone cortisol (measured in saliva) are lower when participants have received oxytocin and when they have received social support from a best friend in the preparation phase (and these two factors interact – the effects are significantly greater when combined). The shaded bar represents the period of public speaking. This is followed by mental arithmetic in front of a panel of evaluators. From Heinrichs et al. (2003). Copyright © 2003 Elsevier. Reproduced with permission.

not less, HPA activity (Taylor et al., 2006). How can we explain these seemingly contradictory findings relating to the administration of oxytocin (in which increased oxytocin promotes affiliation and reduces stress) and resting levels of oxytocin (in which high levels are associated with poor relationships and high stress)? Taylor and colleagues suggest that oxytocin (and vasopressin) signals the need to affiliate closely with others (and is therefore higher in relationship distress), in addition to responding to actual rewarding affiliative behavior. These may involve different pathways in the brain (that are not yet fully understood). There are other possibilities too. For instance, it is unclear how resting levels of a hormone (in the blood stream) may relate to their levels (and function) in the brain. It is possible, in theory at least, that higher levels of oxytocin in the blood stream are linked to less oxytocin-related activity in the brain, leaving no paradox to explain.

Evaluation

Attachment manifests itself behaviorally as distress at separation from, and comfort in the presence of, the attached other. The presence of an attached other can activate the reward-based system (e.g. the nucleus accumbens) and deactivate the fear system (e.g. amygdala) and stress system (the HPA), but the initial development of an attachment may further depend on the effects of hormones such as oxytocin and vasopressin, which act on certain receptors in the brain. Mammals that form monogamous relationships (e.g. Prairie voles) tend to have far more of these receptors than mammals that do not (e.g. Montane voles). Aside from attachment formation, oxytocin and vasopressin may act generally to enhance affiliation (e.g. trust) and reduce stress. In infants, different attachment styles (secure, anxious, avoidant) may reflect the quality of mother–infant interactions, but may generalize to some extent to different relationship behaviors and attitudes in childhood (peer–peer interaction) and adulthood (in terms of romantic partners). Different adult attachment styles modulate activity within the social/emotional brain both when thinking about relationships (e.g. a break-up) and in social interactions more generally (e.g. receiving positive/negative feedback from another person).

EFFECTS OF EARLY INSTITUTIONALIZATION ON ATTACHMENT

In 1990, after the collapse of the dictatorship in Romania, a series of orphanages were discovered in which abandoned children and infants had minimal care and lacked physical and mental stimulation. These children were subsequently adopted into 'good' homes and their progress has been followed in a series of longitudinal studies. One of the most general findings of this research is that whilst their intellectual and cognitive abilities have, over years, approached the expected levels for their age, their social behavior still shows disturbances (Beckett, Castle, Rutter, & Sonuga-Barke, 2010; Kumsta et al., 2010). This includes a lack of pretend play (O'Connor, Bredenkamp, & Rutter, 1999) and social disinhibition characterized by a lack of awareness of social boundaries

Many of the Romanian orphans who were discovered in 1990 in poor conditions were adopted into stable families. Their subsequent cognitive and social development has been followed into adulthood by researchers in Europe and North America. © Getty Images.

and an indiscriminate friendliness towards strangers (O'Connor & Rutter, 2000). The latter behavior is also found in high-quality institutions in which there is rich mental and physical stimulation but in which there is a high turn-around of staff providing care (Tizard & Rees, 1975). As such, it suggests that this aspect of social behavior – indiscriminate friendliness – is related to unavailability of a stable attachment figure rather than deprivation per se (Roy, Rutter, & Pickles, 2004). There may be a critical window for this to occur. Normal attachments begin within 6–9 months (as does a fear of strangers). Chisholm (1998) assessed attachment patterns (at the age of 4.5 years) in adopted-away orphans and found that: infants adopted before 4 months of age tended to have normal attachment patterns; those adopted between 4 and 8 months had formed attachments that tended to be insecure but they were not over-friendly to strangers; and those adopted after 8 months of age showed both insecure attachments and over-friendliness. Children adopted after the first 8 months of life also showed higher levels of stress-related salivary cortisol than those who had been institutionalized for less than 4 months in their first year of life (Gunnar, Morison, Chisholm, & Schuder, 2001). This may be a consequence of a failure to form an attachment during the most sensitive period. Children who had previously been adopted away at 1–7 years of age were found to have reduced resting levels of brain activity (measured by PET) in regions including the orbitofrontal cortex, amygdala, and hippocampus (Chugani et al., 2001).

SEPARATION, REJECTION, AND LONELINESS

The previous section on attachment considered how social bonds may be formed. This section concentrates on the effects of disruption of social bonds, such as the failure to form social bonds (social exclusion), separation distress and grief (where a previous attachment is temporarily or permanently lost), and perceived isolation and lack of intimacy (loneliness). Although these are emotionally negative and stress-inducing events, they are also motivating events. We are motivated to reconnect with others (i.e. establish new social bonds or better social bonds) in order to avoid these punishing consequences.

The pain of separation and social exclusion

Separation or rejection is often described as painful (e.g. 'hurt feelings'). One suggestion is that these social aspects of pain have piggybacked onto evolutionary older mechanisms that represent physical pain (MacDonald & Leary, 2005). Regions in the anterior cingulate cortex respond to the affective and sensory components of physical pain and, to some extent, watching someone else in pain activates the same regions, particularly those that correspond to the affective component (Singer et al., 2004b). But is it associated with the 'pain' of social exclusion? Eisenberger, Lieberman, and Williams (2003) conducted an fMRI study of a cyberball game involving three players, including

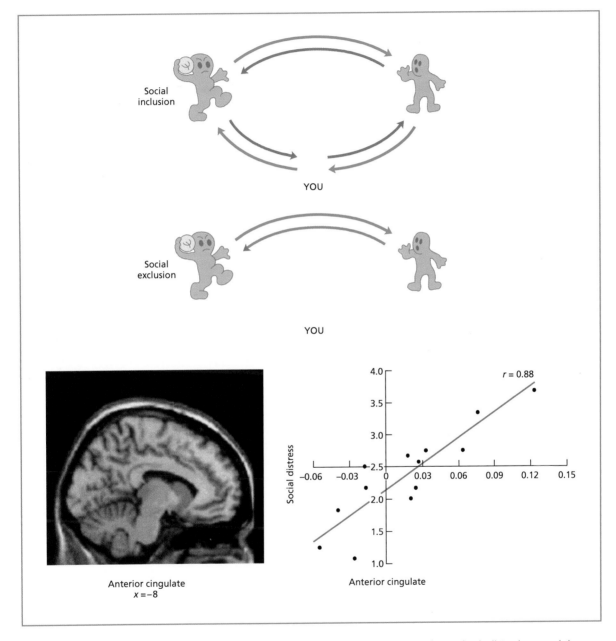

In the cyberball game a participant must decide which of two other players to throw the ball to. In a social exclusion condition, two of the players always send the ball to each other and never to the participant. In a social inclusion condition, all players get to play. Social exclusion tends to activate the anterior cingulate and this correlates with subjective levels of distress. Bottom figures from Eisenberger et al. (2003). Copyright © 2003 American Association for the Advancement of Science. Reproduced with permission.

the one person being scanned. Players could opt to throw the ball to one of the two other players. However, after a while the game was fixed such that two players consistently threw to each other, excluding the person in the scanner. There were two other conditions: one in which the player was included, and one in which they were excluded but given the cover story of 'due to technical difficulties'. The anterior cingulate and the anterior insula were activated more by social exclusion (both real and due to technical difficulties) than social inclusion. Activity in the anterior cingulate correlated with self-reported distress during social exclusion. The anterior insula is also known to be involved in pain perception and bodily experiences in general. A region in the prefrontal cortex (right ventrolateral PFC) was linked to social exclusion, but not exclusion due to 'technical difficulties', which they interpret as playing a controlling role in limiting the distress of social exclusion.

Other research shows that sensitivity to pain (measured objectively) predicts sensitivity to social rejection (measured by self-report) on the social exclusion condition of the cyberball game (Eisenberger, Jarcho, Lieberman, & Naliboff, 2006) – that is, those who have a lower tolerance to thermal pain tend to report higher levels of unpleasant feelings following exclusion. Opioids have pain-killing (analgesic) effects both for physical pain and for separation distress (Kalin, Shelton, & Barksdale, 1988), and it has been found that individual differences in the mu-opioid receptor gene in humans are related both to self-reported distress to social exclusion and to changes in brain activity in the anterior cingulate in the cyberball task (Way, Taylor, & Eisenberger, 2009). A meta-analysis of findings in social psychology revealed that although social exclusion was associated with reported feelings of negative emotions (e.g. distress, sadness) there was a general trend for it to be accompanied by feelings of emotional blunting – for both positive and negative emotions (Blackhart, Nelson, Knowles, & Baumeister, 2009). This is again consistent with the notion that there may be an analgesic response to social exclusion, which results in a general emotional numbness.

The research of Panksepp (2005; Panksepp et al., 1980), primarily on non-human animals, has highlighted the role of opioids in separation distress: for instance, when pups and chicks are separated from their mothers. Unlike the functional imaging studies conducted on humans, Panksepp (2005) emphasizes the role of subcortical regions, which may be too small to reliably detect using fMRI but have been established as crucial in animal models. For example, the periaqueductal gray (or central gray) region is rich in opioid receptors and electrical stimulation of this region has analgesic effects (Hosobuchi, Adams, & Linchitz, 1977). However, it may also respond to 'social pain', such as when infant animals are separated from mothers (Rupniak et al., 2000), and it is also implicated in reactive aggression (Siegel et al., 1999), which often takes the form of a social threat to resources, such as threats to one's infants (Panksepp, 2005). There is some evidence for cross-species commonalities. For example, when human participants are asked to recollect experiences during fMRI that have made them sad (Damasio et al., 2000), a similar brain network is activated to those regions found to be activated during separation distress in the guinea pig (Panksepp, 1998). This includes the anterior cingulate, dorsomedial thalamus, and periaqueductal gray (Panksepp, 2005).

Grief

Grief is an intense feeling of loss that occurs as a result of permanent separation (normally death) from a loved one. O'Connor et al. (2008) conducted an fMRI study of women who had lost a mother or sister due to breast cancer in the last 5 years. The grieving women consisted of two groups. One group were diagnosed with 'complicated grief' in which there was no sign of abatement of the sense of loss, yearning for the loved one, or preoccupation with thoughts of them. The other group were classified as having non-complicated grief. Both groups were shown photographs of either the loved one or a stranger with grief-related words (e.g. 'dying') or neutral words superimposed on them. Both groups showed activity in a number of regions relating to pain (dorsal anterior cingulate, insula, periaqueductal gray) when presented with images of the deceased relative to the stranger. This is consistent with an overlap between 'social pain' and physical pain found in the more artificial scenario of social exclusion in a game. However, there was some evidence that those with 'complicated grief' activated the nucleus accumbens more than those with non-complicated grief (although it was found for grief-related words and not images of the deceased). This activity was correlated with subjective reports of yearning, and was interpreted as evidence of an ongoing attachment to the deceased and an inability to fully adapt to their loss.

> **KEY TERM**
>
> **Loneliness**
> A perceived social isolation and/or lack of intimacy.

Loneliness

Loneliness is a related concept to social exclusion but differs from it in a number of ways. It is more akin to a *trait* (a relatively stable disposition) than a *state* (induced by a particular situation). It can also differ from social exclusion in that it need not entail an element of rejection by others. It could be an outcome of shyness or social anxiety, for instance. It could also reflect *perceived* isolation (e.g. relating to lack of intimacy) rather than actual isolation. However, loneliness and social exclusion do tend to go together because lonely people often withdraw from social situations, the net result being that others do not include them in social activities or actively reject them. Lonely people may also look for rejection cues in other people and interpret ambiguous cues threateningly. This, of course, reinforces the feeling of loneliness (e.g. Cacioppo & Hawkley, 2009).

Cacioppo et al. (2009) showed images of social scenes or non-social scenes, matched for ratings of emotionality, to lonely and non-lonely people during an fMRI study. For pleasant stimuli, non-lonely people showed greater activity in the ventral striatum for social images than non-social images, but for lonely people the reverse was true. This is broadly consistent with behavioral ratings, showing that lonely people judge images of social situations as less pleasant. For unpleasant stimuli, non-lonely people activated the temporo-parietal junction region (part of the mentalizing network) in response to social relative to non-social images. This pattern was absent in the lonely people, although they tended to activate their visual cortex more than non-lonely people. Perhaps

Queen Victoria experienced what clinical psychologists now refer to as 'complicated grief' following the early death of her husband, Prince Albert. For several years after his death she insisted that his personal effects should be laid out (e.g. hot water for shaving) and she appeared to be in a deep depression. She wore black for her remaining 40 years of reign. This photograph, taken 2 years after her husband's death in 1863, shows her in full mourning regalia.

Loneliness can be construed as a vicious circle in which perceived social isolation leads to hypervigilance to social threats (e.g. of not being liked), which leads to further isolation. From Cacioppo & Hawkley (2009). Copyright © 2009 Elsevier. Reproduced with permission.

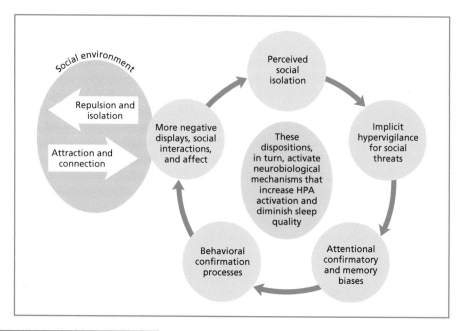

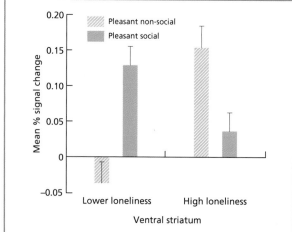

Lonely people show a different brain response to social scenes relative to non-social images. The graph shows activity in the ventral striatum and is consistent with the view that social situations are less rewarding to this group. From Cacioppo et al. (2009). Copyright © 2009 by the Massachusetts Institute of Technology. Reproduced with permission.

lonely people are biased to attend to the scene itself, whereas non-lonely people focus on other people.

Perceived loneliness has important health consequences. For example, perceived loneliness predicts elevated blood pressure even after statistically removing effects of objective social support and isolation (Hawkley, Masi, Berry, & Cacioppo, 2006). Cognitive decline in the elderly and risk of Alzheimer's disease are predicted by perceived loneliness even after amount of social contact and size of their social network is factored out (Wilson et al., 2007). Experimentally inducing participants to think of a lonely future has detrimental effects on subsequent performance on intelligence measures, even though their current objective levels of social support are not altered by that manipulation (Baumeister, Twenge, & Nuss, 2002). The biological mechanism by which these effects are exerted is unknown, but it is known that social isolation is associated with increased activity of the stress-related HPA axis (Adam, Hawkley, Kudielka, & Cacioppo, 2006) and affects gene expression relating to the body's defense system from infection (Cole et al., 2007).

Evaluation

Social exclusion may tend to rely on some of the same neural substrates as physical pain. This may result in temporarily blunted emotions as a counter (analgesic) response. An extreme form of this may be grief resulting from a bereavement. Lonely people have a tendency to process social stimuli differently, even when there is no element of social exclusion implied.

SUMMARY AND KEY POINTS OF THE CHAPTER

- An attachment is a long-enduring, emotionally meaningful tie to a particular individual, such as between mother and infant or between romantic partners. Love is the emotion that is elicited in the presence of (or by the thought of) another person with whom there is an attachment. Love is associated with activity in the reward circuitry of the brain (e.g. the nucleus accumbens).
- Individual differences in attachment can manifest themselves in perceptual/attention-based measures of face processing (in ERPs). In adults, attachment anxiety appears linked to the amygdala (given a negative appraisal) whereas attachment avoidance has been linked to lateral prefrontal activity and reduced reward-related responsiveness to positive feedback from others.
- Animal models based primarily on the monogamous Prairie vole have been important for exploring the role of neuro-hormones such as oxytocin and vasopressin on attachment formation. This may be achieved by interfacing with the dopaminergic reward system and by reducing the activity of the stress response (e.g. in the amygdala and the HPA axis).
- Social exclusion, separation, and grief may tend to rely on some of the same neural substrates as physical pain. Lonely people have a tendency to process social stimuli differently, even when there is no element of social exclusion implied.

EXAMPLE ESSAY QUESTIONS

- Is love too subjective a concept to study scientifically?
- What have animal models contributed to our understanding of attachment?
- Do mother–infant bonding and romantic bonding share common mechanisms?
- Is 'social pain' arising from separation or rejection related to physical pain?

RECOMMENDED FURTHER READING

- Cozolino, L. (2006). *The neuroscience of human relationships: Attachment and the developing social brain*. New York: WW Norton. Basic but a good starting place.
- De Haan, M., & Gunnar, M. R. (2009). *Handbook of developmental social neuroscience*. New York: Guilford Press. Contains good up-to-date chapters on relationships and attachment (particularly animal models).

CHAPTER 9

CONTENTS

Identity and the self-concept   202

Ingroups, outgroups, and prejudice   213

Herds, crowds, and religion   220

Summary and key points of the chapter   223

Example essay questions   224

Recommended further reading   224

Groups and identity

In 1971, a group of 21 students at Stanford University were randomly assigned to be either guards or prisoners in a mock prison set-up (Zimbardo, 1972; Zimbardo, Maslach, & Haney, 1999). Being assigned to be a prisoner or a guard is to become part of a group that has certain norms and expectations. Despite receiving no detailed instructions as to how to behave, over the course of 6 days the guards became increasingly brutal in their behavior (waking the prisoners at night; enforced physical exercise; stripping naked) and the prisoners began to show signs of psychological disturbance, leading to the experiment being terminated early by the watching psychologists. The results of the Stanford Prison Experiment illustrate how easily people are able to adjust their behavior, and identity, to fit with the norms and expectations of a group – even when assignment to a group is arbitrary. A more recent replication of the study in the UK also emphasized how individuals come to identify with an arbitrarily imposed group and adjust their behavior to fit their role, although the outcome was rather different (Reicher & Haslam, 2006). In this study, there was less evidence of brutality from the guards and they were overcome by the prisoners, who had more readily come to identify with their roles.

One might expect that social neuroscience is ill suited for addressing questions at the group level. For instance, the Stanford Prison Experiment would not lend itself well to social neuroscience investigations. However, group influences exert their pressure on individual minds (and brains) and social neuroscience, like much of social psychology, can make a distinction between individuals acting as group members (women, men, straights, gays, Blacks, Whites, etc.) and individuals acting as individuals. This chapter considers several issues relating to groups and identity.

What elements comprise our sense of identity? Race? Nationality? Gender? Beliefs? Our social identity is defined in terms of the various groups that we belong to.

KEY TERM

Prejudice
Negative orientation
towards an outgroup.

Firstly, the chapter considers the various components that are typically considered to comprise 'the self'. These components consist of those that operate primarily at the level of the individual (e.g. our own personality, our sense of being in control of our actions) and those that operate primarily at the level of the group (e.g. our social identity, cultural beliefs, and traditions). The second part of the chapter considers the way in which groups are assigned and evaluated, giving particular attention to the issue of **prejudice**. The final section considers various aspects of collective behavior and beliefs, including religion.

WHAT IS A GROUP?

It would be rather too simple to describe a group as a collection of individuals, although at some level this is exactly what it is. Not all collections would be considered as a group – for example, are blue-eyed people a group or are the collection of people who happen to be with you in the shopping mall or on an airplane a group? A *psychological* group would be connected by virtue of perceived relatedness, common goals, or by the way that individuals influence each other. Johnson and Johnson (1987) listed seven ways in which a group may be defined in psychology:

1. A collection of individuals who are interacting with one another.
2. A social unit consisting of two or more persons who perceive themselves as belonging to a group.
3. A collection of individuals who are interdependent.
4. A collection of individuals who join together to achieve a goal.
5. A collection of individuals who are trying to satisfy some need through their joint association.
6. A collection of individuals whose interactions are structured by a set of roles and norms.
7. A collection of individuals who influence each other.

IDENTITY AND THE SELF-CONCEPT

What is it that makes me who I am? One approach to answering this question is to think about all the things that apply to me that do not apply to other people. At the most basic level, one could answer the question by saying that I occupy my own body. I cannot occupy other people's bodies and they cannot occupy mine. Similarly, I can move and control my own body (and I can create my own thoughts and ideas) but I cannot control other people, and other people cannot control me. Another thing that applies to me that distinguishes me from all others is that I have my own personal history in terms of what I have done, the places I have been, and so on. In neural terms, there is a unique pattern of synaptic connections in everyone's brain that reflects their unique history. Another approach to answering the question 'what is it that makes me who I am?' is in terms of a social identity. Instead of answering the question by listing all the things that make someone unique, one could answer the question by listing all the features that connect us with the people around us. Humans are social creatures and it would be odd if we did not define ourselves in social terms. Our social or cultural identity is determined by our membership of various groups (e.g. ethnic, religious, national, political, socio-economic) and the shared traditions (skills, beliefs, rituals) that bind groups together.

Different facets of the self

In terms of our experience, the self feels unitary and indivisible. For some, the idea of 'the self' seems to imply a non-materialistic soul and therefore should not be taken seriously by the neurosciences. Others regard the notion of the self as an illusion that is constructed by the brain (e.g. Dennett, 2003). Under this account, one could study the self scientifically (just as one can study visual illusions scientifically) but the research agenda becomes one of understanding how the brain creates self-awareness rather than trying to identify the self with a sub-system(s) in the brain. However, most contemporary views of the self based on cognitive processes regard the self as a collection of systems that operate according to different principles and draw on different kinds of knowledge. Under these accounts, the self is not equated with **self-awareness** (our conscious feeling of a unitary, ongoing self) but with a variety of mechanisms, many of which are not available for conscious report. A divide-and-conquer approach, in which clearly identified sub-components are identified, is much more amenable to neuroscientific study than an ill-defined notion of the 'self'. For example, Neisser (1988) distinguished between five possible kinds of self and Gallagher (2000) distinguished between two (these are listed below). However, in both cases these are likely to consist of several underlying mechanisms, many of which may not be unique to the self. Gillihan and Farah (2005) surveyed a range of potential self-mechanisms (e.g. face and body ownership, agency, personality traits, autobiographical memory) and concluded that the mechanisms that underpin the self-concept are generally not unique to the self, but also involved in other circumstances.

> KEY TERM
>
> **Self-awareness**
> Our conscious feeling of a unitary, ongoing self.

Neisser (1988) made a distinction between five kinds of self-knowledge:

1. The *ecological self* consists of one's sense of being located within the body.
2. The *interpersonal self* consists of the sense of oneself as a locus of emotion and social interaction.
3. The *extended self* consists of one's sense of existing over time.
4. The *private self* is linked more closely with self-awareness and the feeling that 'I' can reflect upon conscious experiences and feel ownership of them.
5. The *conceptual self* is linked to semantically relevant knowledge about oneself in terms of both social role (and group identities) and personal knowledge.

Gallagher (2000) broadly divides the self into two levels:

1. The *minimal self* consists of the feeling that we own our body and control our actions.
2. The *narrative self* consists of our social identity and autobiographical memory that extend through time and enable one to construct a conscious sense of unity.

The self-concept can be construed in terms of a variety of different mechanisms that draw on different kinds of information. Some self-related processes (e.g. on the right) are independent of culture, whereas others (e.g. on the left) are defined entirely by our cultural setting. Other mechanisms (in the middle) can best be construed as a mixture of culturally determined and culturally independent information.

'The self-concept'

The cultural/collective self	The ongoing self	The sensorimotor self
• *Group membership* (e.g. ethnic, national, linguistic, or ideological identity) • *Shared beliefs, skills, and rituals* (e.g. religions, sports, eating/drinking practices, materialistic ideals)	• *Personal memories* (autobiographical memories and personal semantic memories) • *Personality traits* (stable patterns of behavior that distinguish individuals in a group) • *Motivation* (maintain self-esteem)	• *Sense of agency* ('free will', sense of being in control of one's own actions) • *Sense of embodiment* (sense of existing within the space of one's own body)

The figure above provides an outline of some of the different mechanisms that contribute to the self-concept that are drawn from prominent reviews (Boyer, Robbins, & Jack, 2005; Gallagher, 2000; Gillihan & Farah, 2005; Neisser, 1988). For ease of explanation, they are divided into three different kinds of mechanism but other divisions and groupings are possible. At one end there is the sensorimotor self, which consists of the sense that the self is located within the body (**embodiment**) and that we control our own actions and thoughts (**agency**). Although these mechanisms may be separate from each other, they can be conveniently grouped together on the principle that these self-related mechanisms are assumed to be independent of culture and shared by all – as in Gallagher's (2000) notion of a minimal self. At the other end there are culturally determined aspects of self, such as the groups that one identifies with (e.g. religious, ethnic, political) and the set of skills, beliefs, and rituals that are culturally acquired (e.g. wearing make-up, not eating pork, Protestant work ethic). Another set of mechanisms can be identified that give rise to the sense of the self as continuing over time. Key components here are our own set of personal memories, and also our personality traits, which are regarded as relatively stable over time and determine individual differences in behavior (e.g. wall-flower vs. party animal). Both our unique past experiences and our personalities distinguish us from other individuals within our culture, although they are nevertheless strongly influenced by our culture. For example, culture dictates the kinds of things that will be learned and encountered, and it also constrains how we behave and think.

The key component mechanisms are considered below in terms of possible neuroscientific mechanisms.

Sense of agency

There is a sense in which we feel in control of our own actions. We can readily distinguish situations in which we move our arm versus situations in which our arm is moved by someone else. The basic assumption is that when we ourselves generate an action, then we can use the motor program to predict what the action will feel like – for instance, the velocity, trajectory, and force on contact (Wolpert, Ghahramani, & Jordan, 1995). This is termed a **forward model**. If there is a match between

KEY TERMS

Embodiment
The sense that the self is located within the body.

Agency
The sense that we control our own actions and thoughts.

Forward model
The use of motor programs to infer the sensory consequences of actions.

predicted effects of the action and the actual action, then the action is attributed to the self, otherwise it may be attributed externally (e.g. Sato & Yasuda, 2005).

The event that generates the initial motor plan to act, and the prediction of what the action will feel like, may typically occur unconsciously. People report a conscious intention to act that is later in time than the unconscious preparation to act that can be measured in the brain. When EEG electrodes are placed over the motor cortex and participants are instructed to make a button press whenever they want to, then the brain starts to generate an electrical potential several hundred milliseconds before the person reports any intention to press the button (Libet, 1985; Libet, Gleason, Wright, & Pearl, 1983). Thus, both the awareness of one's intentions to act and the action themselves arise out of largely unconscious processes. We *infer* that our intention to act causes the action, but this is not necessarily so (Wegner, 2002). The feeling that we are in control of our actions may nevertheless be important for creating a sense of moral and social responsibility for our behavior.

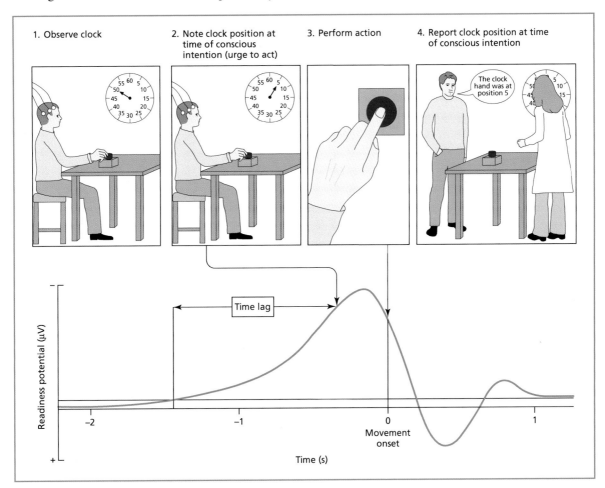

The feeling that we are in control of our actions is related to our sense of agency. Does the fact that the brain shows a change in motor cortex activity (measured by EEG) long before the participants report having any intention to act (in terms of reporting the time of the intention relative to the moving clock hand) mean that the sense of agency is an illusion with no real function? From Haggard, 2008. Copyright © 2008 Nature Publishing Group. Reproduced with permission.

Patients with schizophrenia who report **delusions of control** may have a disrupted sense of agency and, hence, a disrupted sense of (one aspect of) self (Frith, Blakemore, & Wolpert, 2000). Many patients with schizophrenia report that they feel that their actions and their thoughts are under external control – for example, 'They inserted a computer in my brain. It makes me turn to the left or right' or 'Thoughts come into my mind from outer space'. One suggestion is that there is a failure to monitor action intentions with action outcomes, thus giving rise to a feeling of external control. Frith and Done (1989) report that schizophrenic patients reporting alien control had problems in correcting errors made on a joystick game when the corrections had to be made internally (but they could correct it when given external visual feedback). This was initially explained in terms of a failure to compare conscious intentions with action outcomes, but is now interpreted in terms of the forward model comparison between an (unconscious) prediction of action outcomes and the actual action outcome (Frith et al., 2000). This monitoring explanation has also been extended to account for thoughts as well as actions. Many patients with schizophrenia report auditory hallucinations, and it has been suggested that these hallucinations reflect their own inner thoughts that are misattributed externally as heard speech (Frith, 1992).

Sense of embodiment

There is a sense in which our self is located within the space occupied by our own bodies. Researchers such as Damasio (1999, 2003) argue that bodily awareness lies at the core of self-awareness. Bodily senses not only include the sense of touch but also the sense of where our limbs are located in space (**proprioception**), which is given by stretch receptors in the muscles and ligaments, and also internal senses (**interoception**) that convey the internal state of the body and may include pain, core temperature, hunger, heart rate, and breathlessness. According to Damasio (1999) the reinstatement of bodily sensations (in the brain, rather than in the body itself) is a key aspect of emotion representation and decision making. As such, in his view, the sense of embodiment accounts for self-motivated behavior in addition to the feeling of the self being located within the body. Simulation theory also places embodiment and agency at the heart of self-awareness and social cognition by assuming that we understand other people's actions, emotions, and sensations by mapping them onto our own sensory and motor mechanisms (e.g. Gallese, 2003).

A disruption of the sense of embodiment can accompany various brain lesions. **Out-of-body experiences,** in which the participant reports being in a location different from their actual physical location, can arise following damage to the right temporo-parietal junction (Blanke, Landis, Spinelli, & Seeck, 2004). This region has been implicated in taking the perspective of others both in terms of both bodily perspective (e.g. Blanke et al., 2005) and sharing their mental states (e.g. Decety & Lamm, 2007). Certain illusions can also produce something akin to an out-of-body experience in neurologically normal participants (Lenggenhager, Tadi, Metzinger, & Blanke, 2007). If they see, using virtual reality, an image of someone placed in front of them and if the virtual person and the participants are stroked on the back in synchrony then the participant may report feelings such as 'I felt as if the virtual character was my body'. The participant also has difficulty in locating him/herself in space when displaced by the experimenter, and he/she gravitates towards the virtual body. The same is not found when an object rather than a body is stroked.

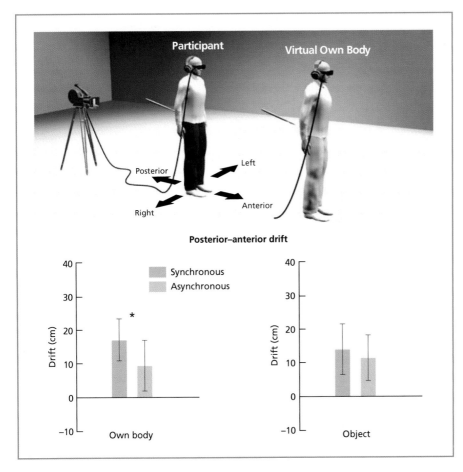

Posterior–anterior drift

Is it possible to create an experience similar to an out-of-body experience in an experimental setting? If you see an image of yourself or another person projected in front of you (using virtual reality) and if that person's back is seen to be stroked at the same time as your own back is stroked, then you tend to lose your sense of location in space and report feeling that the virtual character is your own body. The effect is not found if the stroking is out of synchrony, or if an object is seen instead of a body. Adapted from Lenggenhager et al. (2007).

Memory and the self

Conway and Pleydell-Pearce (2000; also Conway, 2005) highlight the importance of memory in creating a sense of an ongoing self. Their so-called self-memory system consists of two components. The first component is autobiographical knowledge, which consists of both fact-based knowledge about one's life (e.g. our occupations and the places that we have lived) and event-based knowledge about episodes in a particular time and place. This autobiographical knowledge is assumed to be hierarchically organized. The second component, which they term the 'working self', comprises the goals and motivational state of the individual. These two components interact together such that retrieving information about the past cannot be fully separated from one's current goals and aspirations. As Conway (2005) puts it, 'memory is motivated'. According to this model, one is always viewing the past through the prism of our current goals, knowledge, and beliefs and this may be sufficient to create a feeling of unity over time. There is evidence that people tend to remember the past in terms of their current knowledge and beliefs. Marcus (1986) asked participants to rate their attitudes on various political issues in 1973 and again in 1982. In the 1982 session they were also asked to recall the attitudes that they held in 1973. A systematic bias was found such that their previous attitudes were judged to be closer to their presently held ones. People also have a tendency to remember

According to the model of Conway (2005), our autobiographical memory consists of knowledge of specific episodes (episodic memory) organized around particular life-defining themes, such as one's occupations and friends. According to this model, memories are not retrieved as carbon copies, but rather are reconstructed based upon one's current motivations and beliefs. Copyright © 2005 Elsevier. Reproduced with permission.

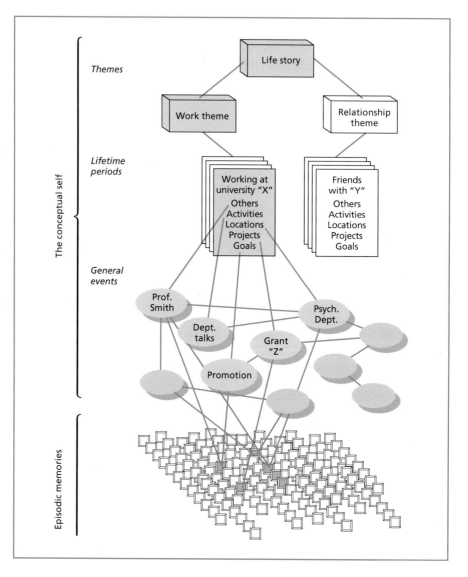

the past in a self-enhancing manner. Those who identify strongly with an ingroup are worse at remembering acts of historical violence by their ingroup (Sahdra & Ross, 2007) and self-enhancing memories are judged to feel more recent than (matched) memories that reflect badly on oneself (Ross & Wilson, 2002).

Le Doux (2002) in 'Synaptic Self' also argues that memory, encoded through synaptic connectivity, is at the core of what constitutes the self. Unlike Conway and Pleydell-Pearce (2000), Le Doux (2002) does not give prominence to one particular memory system (e.g. autobiographical knowledge) but incorporates all memory systems. Le Doux's (2002) approach is inherently reductionist – it reduces psychological mechanisms (e.g. social identity, memory, personality) to a single neurobiological mechanism. (Non-reductionist approaches attempt to link together psychological and neuroscience levels of explanation but do not attempt to replace the former with the latter.) Le Doux's account enables social and cultural influences to be taken into account, as they affect synaptic learning in the same way as any other learned information. Le Doux (2002) also enables more hardwired aspects of the self (potentially including our sense of agency and embodiment, personality traits) to be incorporated in his model of the self, by arguing that synaptic connectivity is a result of genetics as well as learning. Whilst Le Doux (2002) offers one of the few unitary accounts of the self, the account fails to explain the detailed findings outlined in this section. Why do out-of-body experiences occur from disruption of those particular brain regions? Why do collectivist cultures differ from individualistic ones in the particular brain region that they do? And so on.

DO AMNESICS HAVE A DISRUPTED SENSE OF SELF?

Before beginning to answer this question it is important to stress that patients with amnesia do not have global memory impairments. Amnesia typically arises from damage to the medial temporal lobes (the hippocampus and surrounding structures). These structures appear to be particularly important for certain types of memory, but not all types of memory. Specifically, they are important for what has been termed explicit memory or declarative memory (Squire, 1992; Tulving, 1983). These are consciously accessible memories and can be further divided into episodic/autobiographical memory (i.e. memory of events that are specific in time and place) and semantic memory (i.e. memory of factual and conceptual knowledge). Moreover, although amnesia normally affects the ability to form new declarative memories (called anterograde memory) as well as the ability to recall memories from the past (called retrograde memory), older declarative memories from childhood and early adulthood are often relatively spared.

One of the most striking and well-documented cases of amnesia is HM. HM began to experience epileptic seizures at the age of 10 and, by the time of leaving high school, his quality of life had deteriorated to a point where surgeons and family decided to intervene surgically. The procedure involved removing the medial temporal lobes, including the hippocampus, bilaterally (Scoville & Milner, 1957). What the surgeons did not foresee was that HM would develop one of the most profound amnesias on record. Despite acquiring few episodic memories over the intervening decades (he died in 2008), HM maintained consistent beliefs, desires, and personality traits (Corkin, 2002). He was described as altruistic, and exhibited courteous social behavior. Although he dreamed of being a neurosurgeon he attributed his failure to do so to the fact that he wore glasses (the blood would spurt on them) rather than to his memory failure. He was unsure of his age or whether he had gray hair, and he did not always recognize himself in photos. However, he did not seem shocked to look in a mirror, suggesting that he has some up-to-date knowledge (perhaps implicit) of his appearance.

Several studies have shown that amnesic patients have good insight into their own personalities when asked to decide how much a personality trait applies to them (e.g. Klein, Loftus, & Kihlstrom, 1996; Klein, Rozendal, & Cosmides, 2002). For example, Klein et al.'s (2002) amnesic patient tended to rate the same personality traits over time (correlation of .69) with similar reliability to non-amnesic controls (correlation of .74), and his scores agreed with a family member's rating of his personality (correlation of .62) in the same way as was found for controls (correlation of .64). This suggests that amnesic patients retain certain core self-knowledge in the face of severe problems in recalling specific episodes.

Knowledge of personality traits

Contemporary theories of personality describe it in terms of a limited number of traits (e.g. extraversion–introversion; conscientious–spontaneous), and each individual can be described in terms of their position on these trait dimensions (e.g. high in extraversion, low in conscientiousness). These traits are found to be relatively stable across time (i.e. within individuals) and also relatively stable across cultures.

Making trait personality judgments about oneself relative to a familiar other (George W. Bush) activates a region in the ventromedial prefrontal cortex (Kelley et al., 2002). However, making judgments of personality per se, relative to judging whether the word is in UPPER/lower case, activates a region implicated in semantic memory retrieval in the left lateral prefrontal cortex (Kelley et al., 2002). This suggests a possible specialized role for the medial prefrontal cortex in thinking about the self, rather than in personality judgment per se. Other studies have clarified this finding. Mitchell, Banaji, and Macrae (2005a) have shown that activity in the medial prefrontal region is not specific to self as it is also found for other people when they are judged to be similar to oneself. This may also be related to cultural differences in self-concept, as discussed in the next section. A recent conceptualization of the functional role of the medial prefrontal cortex is in terms of binding together the different information that makes up social events – actions, agents, objects, goals, beliefs (Krueger et al., 2009). More dorsal regions are implicated when making inferences about the actions of others (i.e. in short stories, sentences, or single words) whereas inferences about the self concerning one's own memories, traits, and goals are associated with activity in a more ventral region that is likely to have greater connectivity to emotion-related regions. As such, there is evidence of a self–other continuum in this region of the brain but the distinction is not absolute and its function may be rather general rather than acting as a 'self-module'. This will be returned to again in the next section.

The medial prefrontal cortex (mPFC) is a crucial part of the social brain but its function is not well understood. It is activated when we are required to make judgments about other people in a wide variety of tasks, but there is some relative specialization between judgments about ourself (more ventral regions; in orange) versus other people (more dorsal regions; in blue). People who are close to us (e.g. family) or who are perceived to be similar to us tend to activate the self-related regions (in yellow). From Krueger et al. (2009). Copyright © 2009 Elsevier. Reproduced with permission.

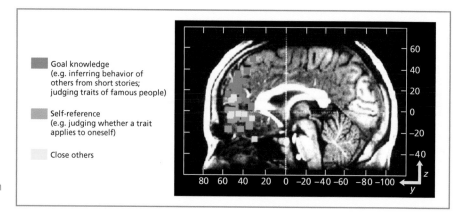

Goal knowledge (e.g. inferring behavior of others from short stories; judging traits of famous people)

Self-reference (e.g. judging whether a trait applies to oneself)

Close others

Cultural and social identity

Several prominent theories in social psychology make a distinction between one's personal identity (e.g. relating to one's own personality and experiences) and social identity related to membership of various groups (e.g. Tajfel & Turner, 1986). The extent to which one's behavior is influenced by these different identities depends on what information is salient at a given point in time and it may, to some degree, vary across cultures.

In East Asian countries (such as China, Japan, Korea) people tend to regard themselves as having an **inter-dependent self** in which their own identity, beliefs, and goals are strongly connected with the people around them (Markus & Kitayama, 1991). The emphasis is on fitting in and attending to others. In contrast, in Western cultures (e.g. North America, Europe) people tend to perceive their own beliefs and goals as largely independent – as having an **independent self**. The emphasis is on self-discovery and personal achievement. In Japan, the success of Olympic athletes tends to be explained both in terms of personal attributes (e.g. skill, hard work) and their social background, whereas in the USA the personal attributes are given the main focus (Markus, Uchida, Omoregie, Townsend, & Kitayama, 2006). This distinction also lies at the heart of the difference between a so-called **collectivist culture** and an **individualist culture**.

Although these social differences are environmental in origin (dependent on cultural immersion), it has been claimed that they lead to more widespread changes in cognitive style outside of the social domain, and to wider differences in the way that the brain is organized and utilized (Han & Northoff, 2008). For example, Westerners tend to pay more attention to objects than background context, whereas East Asians tend to attend more to relations and contexts than objects. This manifests itself both on tests of visual perception and attention (e.g. detecting changes to objects versus backgrounds; Masuda & Nisbett, 2006) and in terms of BOLD activity in areas involved in object perception (Gutchess, Welsh, Boduroglu, & Park, 2006). Lin and Han (2009) attempted to manipulate self-construal within the same group of Chinese participants by priming them with essays containing either the inter-dependent pronoun 'we' or the independent pronoun 'I'. Priming with 'I' increases local processing over global processing, whereas priming with 'we' produces the opposite profile. ERP studies show that such changes occur in early (100 ms) components of perception (Lin, Lin, & Han, 2008).

KEY TERMS

Inter-dependent self
A form of social identity in which individual beliefs and goals are strongly connected with the people around them.

Independent self
A form of social identity in which individual beliefs and goals are viewed as independent from the people around them.

Collectivist culture
A culture in which the goals of the social group are emphasized over individual goals.

Individualist culture
A culture in which the goals of the individual are emphasized over that of the social group.

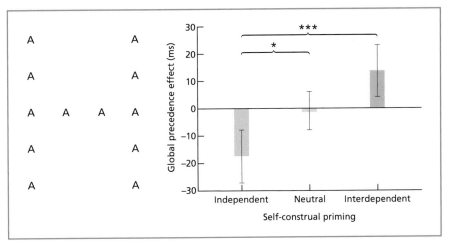

Participants were asked to detect the presence of a letter (H or S) that could appear at either the local or global level (in this example, the H appears at the global level, made up of local As). Priming with the inter-dependent pronoun 'we' increases detection of global letters over local ones, whereas the independent pronoun 'I' has the opposite effect. Graph on right taken from Lin and Han (2009).

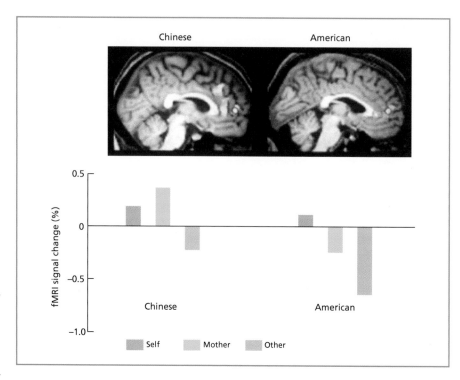

Zhu, Zhang, Fan, and Han (2007) compared the neural substrates of self-concept in Chinese and Western participants using fMRI. In their task, they presented participants with trait adjectives (e.g. brave, childish) and asked them to judge their appropriateness to themselves, to their mother, or to a famous person. In both cultures, a region in the ventromedial prefrontal cortex was activated when making self-referential judgments but was not activated when making trait judgments about a famous person (see also Kelley et al., 2002, discussed above). However, the two cultures differed when making judgments relating to their mothers. Whereas the Chinese showed a similar neural response to self and mother judgments in this region, the Westerners response to their mother was similar to that of the famous person. This is consistent with different self-concepts in which the self can be construed as inter-dependent with kin and close acquaintances in some cultures.

Evaluation

Broadly speaking, the self-concept entails two different kinds of processes: *differentiation,* which is the notion that the self is a unique entity with a distinct personal history and dispositions, occupying a particular body, and in control of his/her own thoughts and actions; and *assimilation,* where one's identity is embedded in a particular social and cultural context and we identify ourselves with one or more groups and share the values and practices of those groups. The latter process may differ cross-culturally (e.g. the notion of independent versus inter-dependent selves). As such, 'the self' should not be construed as a discrete entity or linked to one brain mechanism.

Does social neuroscience lead to genuinely new insights into self and identity beyond answering the question of *where* the relevant mechanisms might be located in the brain (Willingham & Dunn, 2003)? The fact that thinking about the self activates the medial prefrontal cortex is interesting (this is a 'where' answer), but what is more interesting is the range of conditions that also do so: others perceived to be similar; cross-cultural differences; people who are more humanized; and so on. From this, we gain clues as to *how* the self is represented. For instance, the results point to the conclusion that the self is indeed represented socially (in terms of associations with particular others). Of course, similar conclusions have been drawn from questionnaire studies (e.g. Aron et al., 1992). However, questionnaire studies measure how people *think* the self may be represented, which may not always reflect the true underlying mechanism (although measuring what people think is certainly of interest in its own right). To give another example, measuring when people think they created an intention to act is of limited use in itself but when coupled with brain measures of when activity in the motor cortex changes (Libet, 1985) or TMS manipulations that disrupt the perceived action–intention link (Haggard, Clark, & Kalogeras, 2002) we learn a lot about the possible underlying mechanisms of agency that go beyond a description of where the mechanisms are in the brain.

KEY TERMS

Stereotyping
Perceiving members of a given category as possessing various common attributes.

Prejudice
Negative orientation towards an outgroup.

INGROUPS, OUTGROUPS, AND PREJUDICE

Our social identity can be construed in terms of a collection of different group memberships, for instance relating to one's nationality, race, religion, political allegiances, and so on. Inter-group biases vary as a function of the perspective from which the groups are judged. An Asian woman may view her maths ability more favorably when her ethnic identity is highlighted relative to when her gender is highlighted (Shih, Pittinsky, & Ambady, 1999). Similarly White women show evidence of a more negative attitude towards Black women when race is highlighted than when gender is highlighted (Mitchell, Nosek, & Banaji, 2003).

Although group membership operates across many dimensions, these dimensions are not always independent (or are independent but are believed to be otherwise). For example, most people who are British are also White, have English as their first language, and so on. These correlated attributes can give rise to **stereotyping**, in which a collection of attributes become linked together. Rather than considering people in terms of their unique constellation of features, there is a tendency to consider them in terms of social categories (e.g. race, gender, age). This enables a wide range of information from long-term memory to be brought to the fore (Macrae & Bodenhausen, 2000). Related to stereotyping, there is also a tendency to divide continuous dimensions into discrete categories. Racial categories are the most common example (e.g. Black vs. White), although skin color is not a reliable marker of ethnicity. Stereotyping and categorization are not always bad things. They enable us to assimilate a large amount of information and make generalizations and predictions. However, they can lead to negative outcomes such as **prejudice**.

Stereotypes enable us to make quick inferences about people, but should we 'judge a book by its cover'?

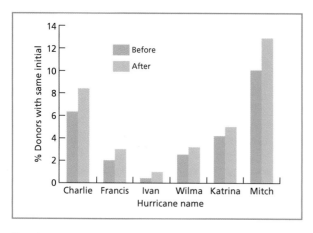

People are more likely to donate to hurricane Charlie if their name begins with the letter C (and similarly for other hurricanes). The graph compares donations before and after a hurricane. Different baseline percentages reflect different name letter frequencies (e.g. not many names begin with the letter I). This is often regarded as an implicit form of egotism or positive self-regard. Data from Chandler et al. (2008).

Liking of the self as an origin of ingroup preference

Ingroup favoritism over outgroups can be demonstrated even when people are randomly assigned to arbitrary groups (e.g. Rabbie & Horwitz, 1969). Even having graphic icons randomly assigned to oneself, relative to another, is sufficient to increase ratings of aesthetic attractiveness of those icons (Feys, 1991). One possible reason for a preference for ingroups over outgroups is that self-related attributes are more likely to be found in one's ingroup than outgroup, and there is a general tendency for self-related attributes to be regarded favorably (Greenwald et al., 2002).

There is evidence that people have an unconscious preference for self-related material. Nuttin (1985) found that when shown pairs of letters, people tend to prefer letters in their own name – the **name letter effect (NLE)**. Participants are typically not aware of the origin of this bias (Nuttin, 1985) and the bias remains when one takes into consideration the fact that letters from one's own name will be more frequently encountered (Hoorens & Nuttin, 1993). Evidence from other stimuli is consistent with this. Pelham, Mirenberg, and Jones (2002) found that people are statistically more likely to choose occupations and cities to live in whose names share letters with their own first or last name (e.g. Denise the dentist from Denver). People are also more likely to donate to a hurricane relief fund if their own name shares the same initial as the hurricane name (Chandler, Griffin, & Sorensen, 2008).

Outgroup prejudice and racism

Prejudice refers to negative attitudes, emotions, or behaviors to members of a group on the basis of their membership of that group (Brown, 1995). Racism, sexism, and ageism can be regarded as specific variants of prejudice. This definition presupposes a tendency to treat individuals categorically and, unsurprisingly, stereotyping has been viewed as a key cognitive mechanism in prejudice research (Allport, 1954). The origins of the negative orientation may be attributable to historical, political, or economic reasons (e.g. that enable one group to prosper at the expense of another). However, it may also reflect a mere consequence of comparing outgroups against one's ingroups (in which the latter may be 'naturally' favored).

One problem in conducting research on prejudice is how to measure it. Questionnaires, such as the Modern Racism Scale (McConahay, 1986), are based on self-reported racist attitudes. However, whilst it was common in the 1950s and 1960s for people in Western countries to openly report racist attitudes, our society norms have changed such that it is no longer considered acceptable to express such views. This raises the possibility that racist beliefs may still be common but not openly endorsed. For example, scores on the Modern Racism Scale change according to the ethnicity of the person who is administering it (Fazio, Jackson,

Dunton, & Williams, 1995). This social desirability problem has been addressed through the use of more implicit measures of racism, in which either the participant is unaware that his/her beliefs are relevant to the task or in which the measure that one takes is considered to be relatively automatic and immune to participants' conscious attempts to hide their beliefs. For example, electromyography (EMG) can measure muscle activity associated with smiles or frowns that are not easily visible to the naked eye. Vanman, Paul, Ito, and Miller (1997) found that frown-related (corrugator muscles), but not smile-related (zygomatic muscles), EMG activity increased when White Americans were asked to imagine working in a cooperative task with a Black American, despite self-reporting potential Black partners more favorably. Implicit measures have featured prominently in the social neuroscience literature on prejudice, but remain controversial (Fiedler, Messner, & Bluemke, 2006).

One of the most commonly used implicit measures of racism is the **Implicit Association Test (IAT)** (Greenwald, McGhee, & Schwartz, 1998). This records participants' response times to categorizing words and names based on speed-button presses to two keys. For example, people may be instructed to press one button when they see names that are more likely to be given to Black people (e.g. Leroy, Aisha) and another button when they see names likely to denote White people (e.g. Amanda, Matthew). At the same time, they are instructed to also press one button when they see pleasant words (e.g. friend, sunrise) and another button when they see unpleasant words (e.g. murder, vomit). The Black/White names are in upper case and the pleasant/unpleasant words are in lower case. The crux of the measure lies in how these two discriminations

Skin color and race vary along a continuum but we tend to make categorical judgments about them. For example, Barrack Obama, 'America's First Black President' has one 'white' parent and one 'black' parent (according to conventional labeling) but Obama is nevertheless generally categorized as 'black'.

(Black/White vs. pleasant/unpleasant) are paired together. In the congruent condition, the response buttons are assigned according to expected prejudices – pleasant and White are responded to on the same button, and unpleasant and Black are responded to on the other button. In the incongruent condition, the response mappings are reversed – pleasant and Black are responded to on the same button, with unpleasant and White allocated to the other button. Greenwald et al. (1998) reported a difference of around 200 ms between these conditions from White participants. For example, 'murder' and 'LEROY' are responded to faster when they are paired on the same button.

The most obvious advantage of this method over self-report is that it is regarded as less susceptible to social desirability effects and may therefore be a more reliable index of true prejudicial beliefs. Another advantage is that the method can easily be adapted to explore other issues relating to stereotyping and ingroup/outgroup biases such as national identity (Devos & Banaji, 2005), political allegiances (Knutson, Wood, Spampinato, & Grafman, 2006), and stereotyped gender attitudes (Nosek,

KEY TERM

Implicit Association Test (IAT)
Gives a measure of differential association of two target concepts with an attribute, by pairing different concepts and attributes to the same or different response key.

In the Implicit Association Test (IAT) people are asked to categorize names according to whether or not they denote Black or White people (e.g. Matthew, Leroy) and they are asked to categorize nouns according to whether they are pleasant or not. White participants are faster to make these judgments when the same response is required to 'Black' and 'unpleasant'. Can we use this to infer implicit racism?

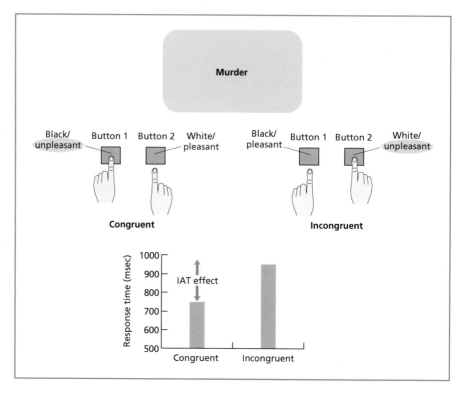

Banaji, & Greenwald, 2002). The biggest problem with this method is ambiguity over what it is really measuring. Does it really reflect deeply held beliefs or does it reflect cultural knowledge of stereotypes? For example, White Americans are familiar with the stereotype of a Black American (e.g. violent, athletic, good dancers) even if they, apparently, do not personally endorse them (Devine & Elliot, 1995). Another issue is how implicit measures relate to self-report measures of racist attitudes and actual discriminatory behavior? Hofmann, Gawronski, Gschwendner, Le, and Schmitt (2005) report a weak correlation of only +.20 between the IAT and various self-report measures, suggesting that they are measuring quite different things. However, the acid test is to determine which measure predicts actual discriminatory behavior. The IAT has been shown to predict various behavioral signs (e.g., speaking time, smiling, speech errors, and hesitations) in encounters with an outgroup race, more so than self-report measures (McConnell & Leibold, 2001). But does this reflect prejudice (i.e. negative evaluation of an outgroup) rather than discomfort due to lack of outgroup familiarity? Is the IAT really sensitive to long-term attitudes or to the current situational demands? For instance, the IAT score can be reduced after presenting White participants with images of famous and well-liked Black people such as Martin Luther King and Denzel Washington (Dasgupta & Greenwald, 2001).

Most social neuroscience studies relating to prejudice fall within the implicit measures approach. The methods of cognitive neuroscience (e.g. EEG, fMRI) fit well within this approach because it is possible to measure brain-related responses independently from participants' behavior and verbal reports. As such, these studies are susceptible to many of the problems of interpretation leveled at implicit behavioral measures such as the IAT. However, some genuinely new insights have emerged from these studies, including the suggestion that implicit associations may be divided between qualitatively different systems (affective vs. semantic) rather than belonging to a single network (as in Greenwald et al. 2002), and that control of prejudice need not imply *deliberate* control by the participant (for a review see Amodio, 2008).

Hart et al. (2000) showed Black and White American participants both Black and White faces during fMRI. Participants were asked to classify the gender of the face (i.e. race processing was incidental). They found that there was a slower decline in amygdala activation over the scanning session for the outgroup relative to the ingroup. This was discussed in terms of the ingroup being more familiar (i.e. unrelated to prejudiced beliefs). However, an alternative interpretation might relate amygdala activation to negative evaluation of the outgroup (i.e. relating to prejudiced beliefs) based on the assumption that the amygdala is crucial for the emotion of fear and the detection of threat (e.g. Le Doux, 2000). In support of this, Phelps et al. (2000) conducted a similar study in which White participants viewed pictures of unfamiliar Black and White faces during fMRI. To maintain attention to their face, they were asked if each face was the same as the previous one. Outside of the scanner, participants completed both an implicit measure of racism (the IAT) and a self-report measure (the Modern Racism Scale). Whilst activity in the amygdala did not differ between the viewing of Black and White faces, differences in amygdala activation (Black–White) across participants correlated with the implicit, but not explicit, measure of racism.

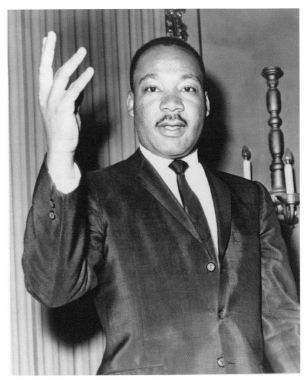

Measures of implicit racism such as the IAT (a behavioral measure) and amygdala activity in fMRI (a neurophysiological measure) disappear when positive Black images, such as of Martin Luther King, are displayed. This suggests that these measures are not driven by the *perception* of an outgroup member per se.

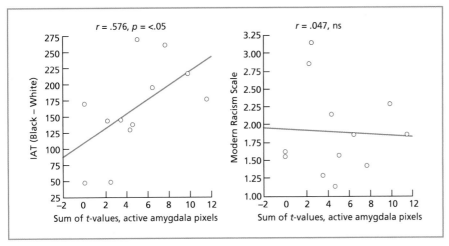

Activation of the amygdala in response to White participants viewing Black faces correlates with performance on an implicit (IAT) but not explicit measure of racism. Is the amygdala activity related to fear? From Phelps et al. (2000). Copyright © 2000 by the Massachusetts Institute of Technology. Reproduced with permission.

Subsequent research has confirmed amygdala activation by White participants in response to Black faces. However, the results are somewhat inconsistent and appear to depend on subtle aspects of the task. In their second study, Phelps et al. (2000) reported that the correlation between amygdala activation and the IAT score was not found when well-liked famous Black people (e.g. Martin Luther King) were used as stimuli. Lieberman, Hariri, Jarcho, Eisenberger, and Bookheimer (2005) report that amygdala activation is greater when participants must match faces according to race using pictures (e.g. matching a Black face to a different Black face or a White face) rather than verbal labels (e.g. matching a Black face to either 'Caucasian' or 'African-American'). This study is noteworthy in that it was administered to separate groups of White and Black participants, enabling a direct comparison between them. Lieberman et al. (2005) found that the amygdala response was greater for Black faces, relative to White faces, in *both* Black and White participants. (Note that the earlier study of Hart et al., 2000, used participants of both races but only analyzed the data in terms of ingroup vs. outgroup by pooling across the two races.) They suggest that this reflects negative cultural stereotypes towards African-Americans rather than prejudice per se. In a similar vein, Amodio, Harmon-Jones, and Devine (2003) looked at the startle response of White participants shown White, Black, and Asian faces. A defensive eye blink (measured by EMG) is part of the normal startle response to, say, hearing a sudden loud noise. This startle response is amplified if the person feels threatened, and this amplification is modulated by the amygdala (Davis, 1992). The White participants showed a heightened startle response to a loud noise when accompanied with Black faces relative to White and Asian faces. This is consistent with a negative evaluation of Blacks rather than relating to outgroup perception per se, given that both Black and Asian groups are outgroups to the White participants. The results of these three studies (Amodio et al., 2003; Lieberman et al., 2005; Phelps et al., 2000, experiment 2) converge on the conclusion that amygdala activation in response to different races does not reflect a general negative bias towards all members of racial outgroups but, rather, a more complex evaluation that takes into account familiarity with particular members of the outgroup together with sociocultural stereotypes (that are not necessarily endorsed).

Amodio (2008) proposes that knowledge relevant to the IAT is represented in at least two different memory systems – a semantic memory system and an affective system of associations (based on classically conditioned associations). A person may know about group stereotypes (i.e. they represent them in their semantic memory) but they do not personally hold negative views about that group (i.e. they are not represented as a threat association). Such a person would not normally be considered prejudiced but could conceivably show racial bias on the IAT. In order to separate these factors, Amodio and Devine (2006) devised an alternative version of the IAT in which one category of response was Black vs. White and the other category of response was changed from an affective one (pleasant vs. unpleasant) to a semantic one (physical vs. mental). According to Devine and Elliot (1995), most White Americans recognize the stereotype of Black Americans being good at physical activities (e.g. basketball, dancing) and White Americans being good at math, reading, etc. Importantly, both physical ability and mental aptitude are regarded in a positive light and so are matched for overall valence, unlike the original IAT. Amodio and Devine (2006) found that performance on the original, affective version of the IAT was uncorrelated with performance on the purely stereotypic version of the IAT.

Another way in which performance on implicit measures of race processing (including the IAT and various neuroscience measures) may be fractionated is in terms of control processes. Whilst all members within a given culture are likely to know about common stereotypes and prejudices, people will differ in the extent to which they act upon them. Knowing about stereotypes and prejudices but not acting upon them may require greater online control. Related to this is the idea, mentioned at the outset, that people have different motives for not exhibiting prejudice. Plant and Devine (1998) attempted to classify people according to whether they had internal reasons for rejecting prejudice (e.g. endorsing statements such as 'I attempt to act in non-prejudiced ways towards Black people because it is personally important to me') versus those who report conforming to external norms (e.g. endorsing statements such as 'I attempt to appear non-prejudiced towards Black people in order to avoid disapproval from others'). It is a moot point as to whether people who report some external motivations for avoiding prejudice should be regarded as prejudiced or not. However, those who report *only* internal motivations could be reasonably described as non-prejudiced. Indeed, those reporting only internal motivations to avoid prejudice show reduced interference on the IAT test (Devine, Plant, Amodio, Harmon-Jones, & Vance, 2002) and also a reduced startle response to Black faces accompanied by a loud noise (Amodio et al., 2003).

Many of the functional imaging studies described above correlated performance on measures such as the IAT, administered outside the scanner, with performance on different tasks performed inside the scanner, typically involving presentation of Black and White faces. An alternative approach is to perform the IAT during fMRI scanning itself (e.g. Beer et al., 2008; Knutson, Mah, Manly, & Grafman, 2007). A comparison of brain activity on incongruent relative to congruent trials in these studies reveals a network of different regions, including those involved in executive functions (i.e. control of cognition). One important region within this network is the anterior cingulate cortex. This region has been implicated in detecting potential conflicts in responses, for instance in situations in which a single stimulus could potentially give rise to two responses (e.g. the Stroop test; see Chapter 4). The IAT is conceptually related to the Stroop task, in that the stereotypically incongruent condition (Black+good) leads to greater response conflict than stereotypically congruent condition (Black+bad).

Amodio et al. (2004) conducted an ERP study that was designed to explore the response conflict to negative stereotypes by the anterior cingulate cortex. Although ERP is not well suited to detecting *where* neural activity is occurring, there is a particular ERP deflection termed error-related negativity (or ERN) that has a known locus within the anterior cingulate cortex (Dehaene et al., 1994). This deflection occurs when people make an error and the peak of the deflection occurs within 100 ms of the error occurring (Gehring et al., 1993). As such it is an early and automatic signature of errorful behavior, and it occurs independently of whether the participant is aware of having made an error (O'Connell et al., 2007). Amodio et al. (2004) used a task different from the IAT in which Black or White faces are briefly presented, for 200 ms, followed by either a picture of a tool or a gun (based on Payne, 2001). The participants' task is simply to decide if a tool or gun was presented (i.e. ignoring the face). However, the combination of Black + gun is consistent with a negative Black stereotype and Black + tool is inconsistent with this stereotype. Thus, a Black face followed by a tool should elicit greater response

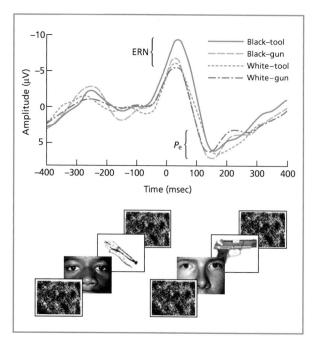

The event-related negativity (ERN), recorded using EEG, is greater when Black faces are paired with tools rather than guns because the correct response (tool) conflicts with the stereotype (i.e. association of violence with Blacks). Graph from Amodio et al. (2006). Copyright © 2006 Oxford University Press. Reproduced with permission. Stimuli from Payne (2001). Copyright © 2001 American Psychological Association. Reproduced with permission.

conflict and be more error prone (relative to White face followed by tool, as a control comparison). This is found both in terms of behavioral responses (Payne, 2001) and in terms of the error-related negativity almost certainly arising from the anterior cingulate (Amodio et al., 2004). This was found for both those who were internally versus externally motivated to avoid prejudice, although a slightly later EEG component (termed the Pe) did distinguish between these different motivations (Amodio, Kubota, Harmon-Jones, & Devine, 2006). This later component, unlike the earlier one, is sensitive to whether one is aware/unaware of having made an error (O'Connell et al., 2007).

Evaluation

Recent research in both social psychology and social neuroscience has focused on implicit measures of prejudice. These measures are almost certainly multi-faceted, that is, different cognitive mechanisms are likely to contribute to a single behavioral measure (such as the IAT). Current research in this area suggests that there are different mechanisms involved in representing the knowledge of stereotypes (e.g. semantic memory) versus one's affective valuation of them (e.g. in terms of potential threat), and that attempts to control prejudice may engage different mechanisms depending on the extent to which one is motivated only by internal beliefs (i.e. truly non-prejudiced) or also due to an external desire to conform (i.e. potentially prejudiced).

It is to be noted that, by far, most of the social neuroscience studies relating to prejudice are based on the model of a White American ingroup and a Black American outgroup. A small number of studies have looked at the reverse (i.e. Black ingroup, White outgroup) but very few have looked at other races or non-racial forms of prejudice (e.g. based on religion, sex, nationality). Race is a perceptually salient category. For example, ERP studies show that Black and White faces diverge as early as 120 ms (Ito & Urland, 2003). It would be interesting to contrast this with prejudice based on more arbitrary categories: for example, using faces that are ambiguous in terms of ethnicity.

HERDS, CROWDS, AND RELIGION

Chapter 7 considered how decisions made by an individual are influenced by the decisions and behavior of others, and by culturally prescribed norms (e.g. concerning fairness). This section extends these ideas to behavior and actions at the inter-group level rather than the inter-individual level. Two particular examples will be considered: crowds and religion.

Herds and crowds

In a recent review, Raafat, Chater, and Frith (2009) use the term **herding** to denote 'convergent social behavior that can be broadly defined as the alignment of the thoughts or behaviors of individuals in a group (herd) through local interaction and without centralized coordination'. The most interesting issue here is to establish the mechanisms by which this comes about. Specifically, does it happen through a form of unconscious spontaneous 'contagion' by which our actions and emotions may become primed by seeing the behavior of others, or is it the result of a more rational deliberation reflecting, say, a conscious attempt to conform or to understand the goals of the group? In terms of findings from social neuroscience, there are few if any concrete results although the familiar argument between mirror systems and mentalizing rears its head again here (Raafat et al., 2009).

There is, however, a longer tradition in the social psychology of crowds. A crowd appears to be organized, but is in fact self-organizing rather than externally coordinated. Le Bon (1903) and later Freud (1922) argued that the person in a crowd experiences a 'loss of self' in which unconscious emotional impulses take over. In support of this, there is evidence that people behave differently when they can hide behind anonymity, such as wearing a uniform that they would not normally wear (e.g. Diener, 1980). This has been termed **deindividuation**, and refers to the process by which people lose their individual identity and replace it with the norm of a group or a particular role in the group (e.g. acting as guards and prisoners in the Stanford Prison Experiment). Contrary to earlier explanations, contemporary theories such as that proposed by Reicher (1984) do not argue that the individual's identity is *lost* when he/she is in a crowd (or when characteristics of the group are emphasized) but rather that the individual's identity becomes synonymous with that of the group (i.e. it *changes* but is not lost). Such changes may be temporary or they may be long term, for example if living in collectivist cultures in which deindividuation may be the norm.

Religion

Religions are found in every human culture and consist not only in the belief of super-natural beings with human-like (e.g. desires, emotions) and non-human-like (e.g. omnipresent) abilities, but also of complex social practices and rituals. The multi-faceted nature of religion makes a single account of it, in social or cognitive terms, unlikely. However, several of the basic mechanisms discussed so far can be considered relevant. For example, the persistence of religious beliefs in the face of contradictory evidence resembles the conformity findings of Asch (1951). Others have commented on how wearing certain religious clothes (e.g. the full-length veil or chador) is a form of deindividuation in which the social norms of the group are emphasized (Jahoda, 1982).

KEY TERMS

Herding
Convergent social behavior in a group through local interaction and without centralized coordination.

Deindividuation
The process by which people lose their individual identity and replace it with the norm of a group or a particular role in the group.

There are many things that a scientific account of religion must explain. One issue is how and why behavior becomes ritualized and, hence, self-perpetuating.

Some aspects of religiosity may have their origins within the individual rather than at the social, group level. For example, Persinger (1983) has argued that the temporal lobes may support certain kinds of religious and mystical experiences based upon the unusual experiences reported during epileptic seizures emanating in this area. These include dream-like states, a feeling of presence, and intense meaningfulness. However, double-blind studies in which magnetic stimulation, using TMS, is applied to the temporal lobes of non-epileptics (i.e. in which neither the participant nor the experimenter knows when the stimulation occurs) have failed to provide convincing support for this idea (Granqvist et al., 2005). Moreover, explanations along these lines fail to account for the social nature of religion. We tend to acquire the religious beliefs of those around us as a result of interacting with others rather than via individual discovery.

Some account is needed of why religious beliefs tend to be so widespread. One common socio-psychological explanation of religion is in the promotion of pro-social behavior within groups (e.g. Bloch, 2008; Bulbulia, 2004). As Bulbulia (2004) put it: 'groups that pray together stay together, and so flourish against other groups' (p. 673). Other manifestations of religious beliefs, including rituals and monuments, would serve to identify one's own group membership and bring them together. Religions also tend to promote good deeds within the group and offer protection to the faithful against threats (real or imagined), which encourages compliance. Religious rituals may be inherently rewarding (and self-perpetuating) because they are believed to bring rewards or protect from harm. Boyer (2008) draws an analogy with non-religious rituals in people with obsessive–compulsive tendencies that are also perceived as cleansing or protecting in some way. Prayers offered by believers do indeed activate the reward circuits of the brain – that is, the belief in a reward can be rewarding in its own right (Schjødta, Stødkilde-Jørgensenb, Geertza, & Roepstorff, 2008).

A more cognitive explanation for religious beliefs views them as an extension of theory-of-mind mechanisms. Guthrie (1993), in his book 'Faces in the Clouds', argues that we have a natural tendency to project human-like mental states into the world. Similarly, Tomasello (1999) argues that the evolutionary origins of human culture lie in a tendency to infer 'hidden forces' in the world. In the social realm, the hidden forces consist of the mental states of others – people's thoughts, desires, and beliefs. But this default way of thinking may extend towards postulating mental states to inanimate objects (e.g. the Sun) and unseen or supernatural agents (as in God). We may also tend to explain unpredictable physical occurrences in terms of these hidden agents (e.g. an unexpected flood caused by God's anger). The pro-social behavior and the theory-of-mind explanations may explain different aspects of religiosity. The former may explain why religions spread and become ritualized, whereas the latter may explain the nature of some of the beliefs (e.g. projection of agency).

Kapogiannis et al. (2009) attempted to separate different aspects of religious belief into different dimensions by analyzing the extent to which participants, with varying degrees of religiosity, agree with various statements (e.g. 'God's will guides my acts', 'God is wrathful'). They identified three dimensions with potentially different cognitive and neural mechanisms (assessed using fMRI). These dimensions may be rooted in other mechanisms involved in social cognition, imagery, and logical inference. Their three dimensions are as follows:

1. *God's perceived level of involvement* (i.e. the extent to which a super-natural agent is believed to influence events). Kapogiannis et al. (2009) linked this component with a theory-of-mind mechanism, although statements reflecting *less* involvement of God activated these regions *more,* which is not straightforward to explain.
2. *God's perceived level of emotion* (e.g. anger, love, forgiveness). This tended to be linked to areas of the brain involved in emotional regulation and inference rather than traditional areas of emotional processing (such as amygdala, insula).
3. *Knowledge source* (both experiential and doctrinal sources of knowledge based on previous experience). Experiential aspects of religion were linked with activation in areas associated with visual imagery (see also Beauregard & Paquette, 2006).

In summary, although we do not yet have a good social or cognitive neuroscientific theory of religious beliefs, we are starting to see what such a theory might look like and what sort of elements it may contain.

SUMMARY AND KEY POINTS OF THE CHAPTER

- Our self-awareness is constructed out of different cognitive operations and is multi-faceted (e.g. awareness of our body, our actions, of existing over time). However, many self-related mechanisms occur without us being aware of them.
- Some aspects of 'the self' do not differ across cultures (e.g. that we occupy our own bodies) whereas other aspects are embedded within our cultural context (e.g. our social identity).
- Our social identity is determined by the groups to which we consider ourselves to belong and our perceived roles within these groups (e.g. independent vs. inter-dependent self). Different social identities

may activate different neural networks within the brain (including the medial prefrontal cortex), and have different consequences for cognition.

- Groups to which we belong (our ingroups) may tend to be viewed favorably because we tend to view ourselves favorably. This may also lead to negative attitudes towards outgroups.
- Social neuroscience studies of prejudice often reveal activity in the amygdala associated with perception of a racial outgroup. However, it is unclear whether this is related to negatively held cultural stereotypes or negatively held personal beliefs, but it does not seem to be an index of 'outgroup processing' per se.
- Neuroscientific and behavioral measures of cognitive control may provide an implicit measure of racist attitudes and may distinguish between different motivations for control (conformity to norms vs. personal beliefs).
- Religious activities and beliefs are an example of group behavior that may arise from a complex mix of cognitive (e.g. over-extension of agency), emotional (e.g. rewarding nature of rituals), and social functions (increase group cohesiveness and identity).

EXAMPLE ESSAY QUESTIONS

- How can 'the self' be understood from a social neuroscience perspective?
- How do cross-cultural differences in social identity affect the brain and cognition?
- What has evidence from social neuroscience contributed to our understanding of prejudice?
- How can factors arising at both the group and individual level account for the pervasiveness of religious beliefs?

RECOMMENDED FURTHER READING

- Amodio, D. M. (2008). The social neuroscience of intergroup relations. *European Review of Social Psychology, 19*, 1–54. A good review of the social neuroscience of prejudice.

- Krueger, F., & Grafman, J. (2012). *The cognitive neuroscience of beliefs*. New York: Psychology Press. Contains chapters on religion, politics, stereotypes, and prejudice.

- Le Doux, J., Debiec, J., & Moss, H. (2003). *The self: From soul to brain*. New York: New York Academy of Sciences. (Also published in journal form as Volume 1001 of *Annals of the New York Academy of Sciences*.) A varied collection of papers from a mainly cognitive neuroscience perspective that discuss the self.

CHAPTER 10

CONTENTS

The neuroscience of morality   229

Anger and aggression   239

Control and responsibility: 'It wasn't me. It was my brain'   252

Summary and key points of the chapter   254

Example essay questions   255

Recommended further reading   255

Morality and antisocial behavior

10

Antisocial behavior can be defined as any behavior that violates the social norms of a particular culture. Broadly speaking, we can discriminate between two different kinds of social norms: conventional norms and moral norms. Examples of **conventional norms** might include not swearing or vomiting in public, dressing neatly for a job interview, and shaking hands when being introduced. Breaking these norms might be considered rude or offensive, but it does not normally lead to physical or mental harm of another person. Examples of **moral norms** include not hitting other people or respecting the property rights of other people (e.g. not committing theft or deliberate damage). Whereas conventional norms exist via consensus, moral norms may stem from a natural concern for the welfare of others, including our ability to respond empathically towards others (Turiel, 1983). As such, conventional norms tend to vary more across cultures than moral norms. Breaking moral norms is also more likely to incur punishment than breaking conventional norms, thus moral norms are more likely to be enshrined in law. The law can therefore be regarded, in psychological terms, as defining those collectively agreed upon social norms that, if broken, require punishment to be metered out on the offender. As such, antisocial behavior and criminality are related but not the same thing: the latter is a sub-set of the former (e.g. swearing and offensive language can be construed as antisocial but it tends to be legal).

Moral judgments involve an evaluation of whether a particular behavior is 'right' or 'wrong' – or, perhaps more accurately, whether the intention behind the behavior is 'right' or 'wrong'. The terms right and wrong have at least three meanings:

1. Whether it feels right or wrong to me (e.g. emotional reactions of pride or guilt).
2. Whether society deems it to be right or wrong (e.g. in terms of the law).
3. Whether the consequence of an action is likely to be net positive or net negative (a rational cost–benefit analysis).

These different aspects of right and wrong may be linked to different kinds of cognitive and neural mechanisms within individual brains. Ideally, these different sources of information should converge on the same outcome – that is, what feels right to me is also deemed right by society and tends to result in positive outcomes (for both me and others). In other situations they may not converge. For instance, some antisocial behavior may be perceived as right in the eyes of the perpetrator but wrong in terms of the law (e.g. revenge for an insult) or may lead to a net negative outcome (e.g. if caught). Similarly, certain moral dilemmas, often devised by psychologists and philosophers, lead to competing sources of information about right and wrong. Consider the two dilemmas below (originally from Thomson, 1976, 1986):

> *The Trolley Dilemma: A trolley car (or tram) is out of control. If it continues along the current tracks it will crash into five hikers crossing*

KEY TERM

Utilitarian
The moral worth of an action is determined by its outcome – maximized positive outcomes for minimized negative ones.

the line. If it is switched to a side track it will kill only one hiker who happens to be crossing on this part of the line. You control the switch. What do you do?

The Footbridge Dilemma: A trolley car (or tram) is out of control. If it continues along the current tracks it will crash into five hikers crossing the line. There is a footbridge going over the track with a fat man standing on it. If you push the fat man off the footbridge on to the line it will kill him, but it will stop the train before it gets to the five hikers, thus saving them. You are standing next to the fat man. What do you do?

For these dilemmas, most people are not aware of a consensual answer so point 2 above is eliminated. For instance, we are not explicitly taught the correct answer in school. As such, it involves a competition between our intuitions and a reasoned judgment. One sort of reasoned judgment is a **utilitarian** decision about the costs and benefits of the different outcomes. The utilitarian judgment that leads to the greatest benefit for the least cost is to kill one and save five in both situations. However, most people, when given these scenarios, flip the switch in the Trolley Dilemma, thereby killing one, but do not push the fat man off the footbridge, thereby killing five (Petrinovich & O'Neill, 1996). These judgments normally occur intuitively and

Both the Trolley Dilemma and the Footbridge Dilemma involve a decision between saving five lives and killing one person. In the Trolley Dilemma, the decision concerns whether to flick a switch that sends the trolley onto a side track (killing a hiker crossing the track). In the Footbridge Dilemma, the decision is whether to push a large person off a footbridge to stop the runaway trolley from killing five people.

people find it hard to explain their reasoning, or to give a coherent explanation as to why they gave different answers in the two scenarios (Cushman, Young, & Hauser, 2006). Some people do give consistent answers to both questions (flip the switch, push the man) but they would perhaps draw the line at killing one to save five people in the Organ Donor Dilemma of Hauser (2006):

> *The Organ Donor Dilemma: Five people are rushed into an ER each requiring a different organ transplant if they are to stay alive. No organs are available. However, there is another patient in the waiting room with only minor injuries; all his/her vital organs are intact. Do you kill this person in order to save the five in need of organs?*

Moral dilemmas such as these have been extended into social neuroscience research and these particular examples will be returned to later. This chapter will consider, amongst other things, what brain regions support this kind of decision making and whether atypical groups (e.g. murderers, patients with certain brain lesions) perform normally or abnormally on such tasks. The first part of the chapter is concerned primarily with morality and the second part is concerned with anger, aggression, and antisocial behavior. Finally, the issue of responsibility and control over actions is discussed.

THE NEUROSCIENCE OF MORALITY

This section will consider two broad ideas: the nature of the mechanisms that support moral thinking, contrasting the role of emotional versus cognitive processes; and whether the line drawn between moral and conventional norms is universal or variable across both groups and individuals.

Judging right from wrong: 'Cognitive' versus emotional processes

Should a man steal a drug to pay for his wife's life? Is it okay for someone to lose their life if five other lives get saved? If someone calls you 'pig' is it okay to hit them? Examples such as these are often used in experiments in the psychology of morality, and have more recently been used in social neuroscience. How are such decisions reached? Broadly speaking, two kinds of processes have been postulated: a mechanism based on emotional evaluations (or gut instincts) to these questions, or a more deliberate attempt at reasoning through the problem (e.g. considering the basis for the judgment, and weighing the alternative answers). The latter is sometimes referred to as 'cognitive', although this presupposes a view that cognition and emotion are separate kinds. A better way of labeling this dichotomy might be between moral intuition (which tends to be more heavily emotion based, although it need not be) and moral reasoning (Haidt, 2007). Within this conceptual space, there are researchers who highlight the importance of emotions (e.g. Haidt, 2001), those that highlight the importance of reasoning (e.g. Kohlberg, Levine, & Hewer, 1983; Piaget, 1932), and those who argue for a mix of both (e.g. Greene, 2008). Some argue that the distinction itself is not valid (e.g. Moll & Schulkin, 2009), and still others propose an alternative to the dichotomy (e.g. Hauser's, 2006, notion of an intuitive moral grammar).

Haidt (2001, 2003) argues that most of our moral judgments are guided by our emotions, and he proposes the existence of a set of *moral emotions* that are linked to such decisions. Under this account the role of cognitive/reasoning mechanisms is to provide a post hoc justification of the decision rather than influence the decision itself. These moral emotions can be grouped into various sub-groups, or families, according to criteria such as whether the emotion is self-focused (e.g. guilt, pride) or other-focused (e.g. anger, pity) and whether it is critical (e.g. guilt) or praising (e.g. pride). The different kinds of emotion may be linked to (and motivate) different kinds of behavior. For instance, pity and compassion ('other-suffering' in this taxonomy) may elicit altruistic acts whereas anger and moral disgust may elicit aggressive or antisocial acts ('other-critical' in this taxonomy). Self-critical moral emotions (guilt, shame, embarrassment) may tend to protect against antisocial acts. Feelings of anger may have a particularly important role to play in eliciting antisocial behavior. **Anger** is the emotion felt when someone else is judged to have intentionally violated a social norm. Displays of anger (in the face, body, or voice) serve a dual function of notifying the other person that they have crossed a line, and also signaling that further retributive action may be taken against them (e.g. violence). Anger is an inter-personal stop signal. **Moral disgust** involves a judgment about the moral standing of another person relative to oneself in terms of their general disposition to engage in acts that are deemed to be wrong (Tybur et al., 2009). Unlike anger, it is not necessarily associated with a triggering incident but may be based on, say, outgroup membership alone. Disgust may tend to follow from a process of **dehumanization** (i.e. treating certain human groups as animal-like). Acts of genocide are often pre-empted by characterization of the outgroup as inhumane or animal-like in their behaviors (Bandura, Barbaranelli, Caprara, & Pastorelli, 1996).

According to many researchers, moral emotions differ from basic emotions (considered in Chapter 4) in that they consist of a blend of basic emotions plus

	CATEGORIES OF MORAL EMOTIONS (SUB-DIVISIONS)				
	Self-conscious		**Other-conscious**		
	Self-critical	Self-praising	Other-critical	Other-praising	Other-suffering
Guilt	√				
Shame	√				
Embarrassment	√				
Pride		√			
Indignation/anger			√		
Contempt/disgust			√		
Pity/compassion					√
Awe/elevation				√	
Gratitude				√	

Source: Adapted from Haidt (2003).

Moral emotions are emotions that are related to the behavior of oneself (in relation to others) or the behavior of others (in relation to oneself or others).

cognitive appraisals of the triggering event (e.g. Moll, de Oliveira-Souza, Zahn, & Grafman, 2008). A cognitive appraisal in this context refers to thoughts that accompany the emotion, such as what the other person is thinking, future outcomes, or the background context to the event (further discussion of this idea is given in Chapter 4). Moll et al. (2002) presented pictures of three kinds of emotional scenes to participants undergoing fMRI: images of moral violations (e.g. images of physical assaults, abandoned children), images of aversive scenes (e.g. dangerous animal), and pleasant images. These were matched for their self-reported arousal. The moral violation and aversive images were matched in terms of how negatively they were judged, but the moral violation images were judged as more morally unacceptable than the other affective stimuli. All affective stimuli (relative to a neutral set of images) tended to activate regions linked to emotional processing, such as the amygdala and insula, but moral emotions (relative to other affective stimuli) additionally activated regions such as the orbitofrontal gyrus, the medial pre-frontal cortex (PFC), and the right posterior superior temporal sulcus (STS). The medial PFC and right posterior STS have been linked to 'theory of mind' (Amodio & Frith, 2006; Saxe, 2006), whereas the orbitofrontal cortex is implicated in the regulation of social behavior (Rolls, 1996). Similar results were obtained for the moral emotions of embarrassment (Berthoz, Armony, Blair, & Dolan, 2002) and guilt (Takahashi et al., 2004) elicited by reading verbal narratives describing a norm violation relative to neutral narratives: for example, 'I left the restaurant without paying' (guilt) and 'I mistook a stranger for my friend' (embarrassment).

According to Moll and colleagues there is not a conflict between emotions and cognition that needs to be resolved by higher control, but rather moral emotions *are* an integration of emotion with cognitive appraisal (Moll, Zahn, de Oliveira-Souza, Krueger, & Grafman, 2005). However, there might be situations in which there is an extra stage of 'conflict resolution' in order to decide between different courses of action. If an action has both negative and positive outcomes (e.g. hurting others to save your own reputation) then this may require an additional control mechanism to overcome more typical thinking (i.e. hurting others = wrong). Also, the distinction between basic emotions and moral emotions is not likely to be as straightforward as implied by this research. For example, anger is listed both as a moral emotion (Moll et al., 2008) and a basic emotion (Ekman, 1992). It also assumes that basic emotions occur without any kind of cognitive appraisal, which may not be the case. There just might be more scope for cognitive appraisal when viewing an image of someone being hit rather than viewing an image of someone in pain. The distinction between basic emotions and moral emotions can be regarded as an extension of the broader debate as to what emotions are, when they are used, how they differ across species, and so on.

What regions of the brain are activated when viewing (or thinking about) scenes involving moral transgressions, such as child abandonment? Is it the same pattern found when viewing other emotional stimuli that do not involve a transgression?

Some of the studies cited above involve violations of social norms but they do not necessarily involve moral norms (i.e. an intentional act against a victim or victims) – or, at least, these two factors have not always been directly compared as different experimental variables (e.g. Berthoz et al., 2002; Takahashi et al. 2004). Mistaking a stranger for a friend may be embarrassing but it is perhaps not immoral. Finger, Marsh, Kamel, Mitchell, and Blair (2006) directly contrasted these scenarios using fMRI in a 3 × 2 design. One factor was the type of transgression (moral, conventional, neutral) and the other was whether onlookers saw the transgression or not. Participants had to silently read the narratives. For example, a moral transgression may involve a narrative describing a car crash in which you killed someone (either with onlookers or not). A conventional transgression may involve a narrative about vomiting in a public place (either with onlookers or not). The two transgression conditions, relative to the neutral condition, were associated with activity in lateral prefrontal regions. However, whereas this pattern was found for moral transgressions irrespective of whether onlookers were present, it was only found for social transgressions in the presence of onlookers. They suggest that the activation of cognitive control mechanisms may depend both on the nature of the transgression and the social context.

Greene and colleagues have taken a rather different approach in that they required participants to make decisions on moral dilemmas during functional imaging, rather than passively processing moral scenes or narratives (Greene, Nystrom, Engell, Darley, & Cohen, 2004; Greene et al., 2001). These studies were based on the Trolley and Footbridge Dilemmas (introduced above) together with other conceptually similar scenarios. Greene et al.'s (2001) explanation for the discrepancy between these scenarios is that flipping a switch in the Trolley Dilemma is *impersonal* (less emotionally based) whereas pushing the man off the footbridge in the Footbridge Dilemma is a *personal* act. Their imaging results were consistent with this. Contemplating personal scenarios such as in the Footbridge Dilemma activates regions linked to emotional processing (amygdala, posterior cingulate cortex) and social cognition (superior temporal sulcus, medial prefrontal cortex), whereas consideration of the impersonal moral dilemmas was associated with relatively greater neural activity in regions implicated in the control of behavior (dorsolateral prefrontal cortex, inferior parietal lobe). For those people who did opt to push the fat man off the footbridge (i.e. thereby consistently basing their judgment on cost–benefit analysis), their brain activity showed the more cognitive pattern rather than the more typical emotional pattern on these trials (Greene et al., 2004).

The theory proposed by Greene (2008) is that different senses of right/wrong are implicated in emotional versus reasoned moral decisions. Emotional judgments appeal to generic principles ('it is wrong for me to harm someone') whereas the reasoned/'cognitive' judgments are based on a cost–benefit analysis that focuses on the consequences of actions (it is better to kill one than to kill five). Other evidence is consistent with this. Increasing the cognitive demands by getting participants to perform an irrelevant task during the moral dilemma slows down decisions based on reasoning about consequences but not emotions (Greene, Morelli, Lowenberg, Nystrom, & Cohen, 2008).

Hauser (2006) offers a somewhat different account noting that, in the Footbridge Dilemma, people would be unlikely to push the man off the footbridge even if they had to flip a switch to do so (i.e. thereby making the dilemma impersonal). The key difference between the scenarios, according to Hauser, lies in the different status of the fat man versus the lone hiker: by pushing the man off the bridge he is being used

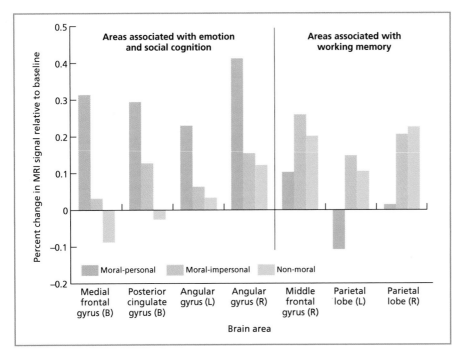

Greene et al. (2001) gave participants different kinds of moral dilemma during fMRI scanning (and a non-moral control condition). Dilemmas involving personal harm to someone else (e.g. the Footbridge Dilemma) were more likely to engender activity in areas of the brain processing emotions, whereas dilemmas involving impersonal harm (e.g. the Trolley Dilemma) were more likely to engender activity in areas associated with working memory. (B = bilateral; L = left; R = right.) From Greene et al. (2001). Copyright © 2001 American Association for the Advancement of Science. Reproduced with permission.

as a means-to-an-end (he is an innocent bystander) whereas in the Trolley Dilemma the lone hiker and the five passengers have an equivalent status in that one or other of their deaths is a foreseen consequence. Hauser attributes this to fast-acting intuitions (a 'moral grammar') but they need not be emotionally based. This does not invalidate the findings of Greene et al. (because the personal/impersonal distinction may still be important) but it suggests that there are other factors in need of further investigation.

WHY DOES LACK OF EMPATHY NOT LEAD TO ANTISOCIAL BEHAVIOR IN AUTISM?

People with autism are rarely knowingly cruel or hurtful. Their behavior could perhaps be considered as *asocial* rather than *antisocial*. However, if one accepts that a core deficit in autism is a lack of empathy (e.g. Baron-Cohen, 2009) then this observation might appear surprising, especially when one considers that a lack of empathy has been postulated to explain the violent behavior of psychopaths (Blair, 1995). De Vignemont and Frith (2008), however, note that the nature of the empathic difficulties could be very different between these two groups. Psychopaths show reduced responsiveness (measured autonomically) to emotional stimuli, particularly distress cues (Blair, Jones, Clark, & Smith, 1997), whereas people with autism are generally considered to have normal responses to basic emotional stimuli – rather, their difficulty may lie in being able to reflect upon their emotions or in using emotions to predict the behavior of others.

To date, the evidence suggests that people with autism have normal moral reasoning. Leslie, Mallon, and Di Corcia (2006) found that autistic children pass tests of moral reasoning. For example, they identify it as 'bad' to steal someone else's cookie when it makes them cry but that it isn't 'bad' to eat one's own cookie even if the other person (who greedily wants two cookies) starts to cry. Blair (1996) also found that autistic children understand the moral/conventional distinction. However, it would be important to carry out further tests that examine behavior on tests of moral intuition, such as the Footbridge Dilemma, in which there is a high emotional response that conflicts with a more rational answer. People with autism may tend to rely on rule-based logic when performing moral reasoning rather than on emotional judgments (Kennett, 2002). The personal account of autism written by Temple Grandin (1995, pp. 131–132) gives an illuminating description of her understanding of morals:

> Many people with autism are fans of the television show Star Trek. I have been a fan since the show started. When I was in college, it greatly influenced my thinking, as each episode of the original series had a moral point. The characters had a firm set of moral principles to follow, which came from the United Federation of Planets. I strongly identified with the logical Mr. Spock, since I completely related to his way of thinking.
>
> I vividly remember one episode because it portrayed a conflict between logic and emotion in a manner I could understand. A monster was attempting to smash the shuttle craft with rocks. A crew member had been killed. Logical Mr. Spock wanted to take off and escape before the monster wrecked the craft. The other crew members refused to leave until they had retrieved the body of the dead crew member. To Spock, it made no sense to rescue a dead body when the shuttle was being battered to pieces. But the feeling of attachment drove the others to retrieve the body so their fellow crew members could have a proper funeral. It may sound simplistic, but this episode really helped me to understand how I was different. I agreed with Spock, but I learned that emotions will often overpower logical decisions, even if these decisions prove hazardous.

Universal morals, moral variations, and absent morals

Are there any moral rules that do appear to be innate and leave little or no scope for cultural variation? The responses to moral dilemmas such as the Footbridge and Trolley Dilemmas show relatively small variability across gender, age, religion, and politics (Banerjee, Huebner, & Hauser, 2011). One of the best documented examples of a cross-cultural taboo is that against incest. For example, which of the following scenarios do you find the most reprehensible?

1. Licking ice cream off a toilet seat.
2. Having consensual, contraceptive sex with your brother or sister.

For most people, the second scenario is far more disgusting but people find it hard to justify their reasoning (Haidt, 2001). One could make the claim that is because society considers it the most reprehensible (and my norms reflect those of my society), but this would push the problem to another level: Why does society consider it reprehensible? It is hard to argue that the consequences of this second case are much worse than the first case. The fact that incest is wrong appears to be one example of a moral norm that is innate, rather than determined by society. Evidence for this comes from the **Westermarck effect** (Westermarck, 1891). This states that we tend not to be sexually attracted, as adults, to people who we knew in the earliest years of life, up to around 6 years. Given that we do not know (for sure) who is genetically related to us, we appear to have evolved a mechanism that applies to people who are *likely* to be our kin, namely those that we grow up with. One consequence of this is that we are less likely to develop sexual attractions towards non-kin who we spend time with in the early years. Evidence for this comes from children reared together in a Kibbutz (Shepher, 1971), and the success/failure of Taiwanese marriages arranged at different ages of childhood (Wolf, 1995). An fMRI study contrasting disgust elicited by thoughts of incest also revealed a partially different network compared to contamination-related disgust (Borg, Lieberman, & Kiehl, 2008). The incest taboo suggests that it is theoretically possible to evolve mechanisms that determine the nature of our moral code.

However, there is variability too. There are important differences in the extent to which particular situations are cast as conventional versus moral. Cultural or inter-individual variability does not necessarily disprove the notion of an innately disposed moral disposition because innateness may specify the dimensions on which it can vary, or pose certain constraints on what can vary. As in most domains, 'nature' and 'nurture' are more likely to be collaborators rather than competitors.

Three different examples are considered below in which there are within-culture individual differences between where the line is drawn between moral versus conventional norms. These consist of: liberal versus conservative moral attitudes; individual motivations for not endorsing racial stereotypes; and psychopathy.

1. *Liberal versus conservative moral attitudes.* Graham, Haidt, and Nosek (2009) asked participants from the USA to rate how relevant a variety of situations were to moral judgments. Across the board, people tended to rate issues relating

KEY TERM

Westermarck effect
The tendency not to be sexually attracted, as adults, to people who we knew in the earliest years of life.

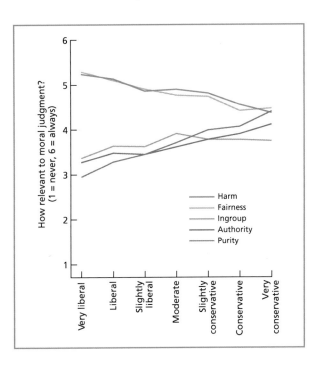

While most people regard issues relating to harm and fairness to be moral issues, people differ more in the extent to which they believe that issues relating to authority (e.g. respect for authority), group identity (e.g. group loyalty), and purity (e.g. control of desires) are morally relevant. Self-declared 'conservatives' from the USA regard authority and purity to be as morally relevant as the traditional moral areas of harm and fairness. From Haidt (2007). Copyright © 2007 American Association for the Advancement of Science. Reproduced with permission.

to harm of others and fairness as being morally relevant, as would be expected (given that this is how morals are normally defined). What is more surprising is that some people apply moral judgments to issues such as respect for authority and ability to control desires, as well as to norms relating to the rights and welfare of people. This was particularly true of individuals who self-declared themselves as having conservative rather than liberal moral foundations. Graham et al. (2009) argue that definitions of morality that emphasize the rights of the individual are too narrow and that a second factor of 'maintaining social order' is needed to account for the full spectrum of what many people consider to be moral.

2. *Moral versus conventional suppression of racist attitudes.* While most people do not openly express racist attitudes, they differ in their motivations for doing so. For some people the motivation comes from the fact that it is not considered socially acceptable to express racist stereotypes (e.g. due to political correctness), whereas for other people the motivation comes from the fact that they believe that it would be wrong to do so (Plant & Devine, 1998). This distinction can now be explained in terms of whether different individuals within a culture regard this as a conventional norm or a moral one. These individual differences in the moral/conventional status of not expressing racist beliefs manifest themselves on various measures (see Chapter 9 for more details).

3. *Moral versus conventional distinction in* **psychopathy**. Psychopathic individuals show a high level of **instrumental aggression** (goal-directed) and **reactive aggression** (threat-related) that is not tempered by any sense of guilt or empathy with the victim (Hare, 1980). Blair (1995) tested 10 people convicted of murder or manslaughter on a version of the moral/conventional norm test based on that used with children. Children as young as 3 years make a distinction between moral violations and conventional ones on the basis of, say, it being okay to talk in class if an authority figure gives permission but it not being okay to hit someone even if given permission (Smetana, 1981, 1985). The psychopathic group failed to make this distinction, justifying their answers by saying that moral norms arise out of consensus and authority rather than a concern for the welfare of others. The question of *why* they fail to understand the moral/conventional norm distinction is an interesting one. It may relate to other deficits (e.g. in empathy, processing distress cues), and this possibility will be returned to later.

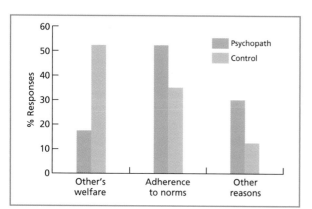

Psychopaths take a more flexible approach to moral norms. Whilst they are aware of the norms of right and wrong, they are more likely to endorse the view that these norms can justifiably be broken (e.g. for personal gain or if an authority figure is absent). In this graph, when asked to justify why a moral transgression is wrong they tend to appeal to social norms (e.g. 'it is illegal') rather than the welfare of others (e.g. 'it will hurt them'). Data from Blair (1995).

KEY TERMS

Psychopathy
High levels of aggression that are not tempered by any sense of guilt or empathy with the victim.

Instrumental aggression
Aggression that is self-initiated and goal-directed.

Reactive aggression
Aggression that occurs in response to threat.

These three examples suggest that notions about what constitutes a conventional or moral norm vary within a culture, and are likely to do so between cultures (within limits). But are there any instances in which a sense of morality is absent? Although psychopaths fail to make the moral/conventional distinction they do perform normally on moral dilemmas such as the Trolley and Footbridge Dilemmas (Cima, Tonnaer, & Hauser, 2010). However, there is one clinical group who do not perform normally on such tasks.

Patients with lesions to certain regions of the frontal lobes (particularly the orbitofrontal and ventromedial regions) often display inappropriate social behavior. They also have high levels of reactive aggression (Grafman et al., 1996) but not necessarily the high levels of instrumental aggression displayed by psychopaths (Blair, 2003). Earlier studies with this patient group relied on tests of moral reasoning (rather than moral intuition) that ask about general situations (e.g. 'is it wrong to do X?') rather than personal behavior (e.g. 'how would you respond in situation X?'). These studies tend to reveal intact moral reasoning – that is, the patients know how one *ought* to behave (Bird et al., 2004; Blair & Cipolotti, 2000; Saver & Damasio, 1991). One exception to this rule may be in orbitofrontal lesions acquired during childhood. This may result in a failure to develop adequate social knowledge in the first place (Anderson, Bechara, Damasio, Tranel, & Damasio, 1999). The two cases in this study acquired their lesions before the age of 2 years and, as adults, they had a string of convictions for petty crimes. Both responded abnormally on tests of moral reasoning.

More recent studies have examined other types of moral dilemma that assess the patients' own moral intuitions, rather than their knowledge of moral norms (Ciaramelli, Muccioli, Ladavas, & di Pellegrino, 2007; Koenigs et al., 2007). These studies reveal clear differences in this domain, but only on certain kinds of moral dilemma. For example, Koenigs et al. (2007) contrasted personal dilemmas (such as the Footbridge Dilemma) with impersonal dilemmas (such as the Trolley Dilemma) and noticed that the patients performed differently on the former but not the latter. In these dilemmas, they tend to always choose to save five and kill one, irrespective of context. However, their responses cannot be construed as 'errors' but rather as a systematic bias to respond rationally rather than on the basis of emotion. Other evidence suggests they have emotional disturbances: these patients are judged by family members to exhibit low levels of empathy, embarrassment, and guilt, and on an objective assessment they

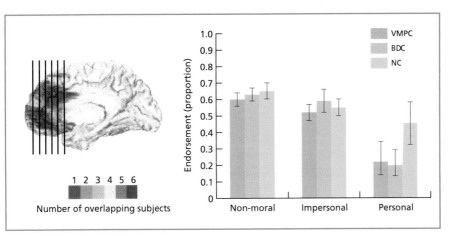

Patients with damage to the ventromedial/orbitofrontal prefrontal cortex (VMPC) are more likely to endorse actions that lead to injury of another person provided they lead to a greater good (e.g. killing one person to save five). Their performance is contrasted with other brain-damaged controls (BDC) and normal controls (NC). From Koenigs et al. (2007). Copyright © 2007 Nature Publishing Group. Reproduced with permission.

show low levels of skin conductance response to emotional stimuli (Koenigs et al., 2007).

Evaluation

Moral judgments concern the rights of individuals, such as not to be harmed and to be treated fairly. These judgments activate brain regions linked to emotional processing and also regions involved in cognitive appraisal and cognitive control. This is consistent with moral knowledge being represented at several levels: in terms of personal beliefs (or gut instincts), social norms, and reasoned pros and cons. Psychopaths know about moral norms but may neglect to follow them (they treat them as malleable). Patients with damage to the orbitofrontal/ventromedial prefrontal cortex tend to base moral judgments on a rational basis (analysis of costs and benefits), which is consistent with a role of this region in using emotions to guide decisions.

PUNISHMENT AS SOCIALLY SANCTIONED AGGRESSION

Social norms are not only concerned with what is right/wrong and appropriate/inappropriate but also with how one should respond to a perceived wrongdoing (e.g. eye-for-eye, turn the other cheek). The severity and nature of punishment are susceptible to significant cross-cultural variation, but variability need not imply randomness. For instance, the principle that the punishment should be in proportion to the misdemeanor may be culturally universal.

The ultimate form of punishment is death, and there is significant cross-cultural variability in the extent to which it is acceptable. For instance, in some cultures families may kill one of their own family members (e.g. if they have an affair) in order to preserve the family 'honor', even though such self-initiated acts are prohibited by the law. In some cultures, the law itself permits the killing of another and this may reflect the local attitudes of the population. In the USA, for instance, there is variability in the use of (and legality of) the death penalty, with greater use in the Southern and Western states. Nisbett and Cohen (1996) argue that there is a 'culture of honor' in these regions, in which violence is justified as retribution for a perceived wrong. They suggest that this reflects the cultural heritage of the immigrant populations who settled in these regions, which is perpetuated to the present day. (They note that the Western states were predominantly settled by migrants from the South.) In a series of surveys they show that the presence of a 'culture of honor' attitude is independent of economic and ethnic variations in these regions. These attitudes are reflected not only by the law (e.g. in the case of capital punishment) but also in the perceived legitimacy of violence in day-to-day interactions. For instance, people

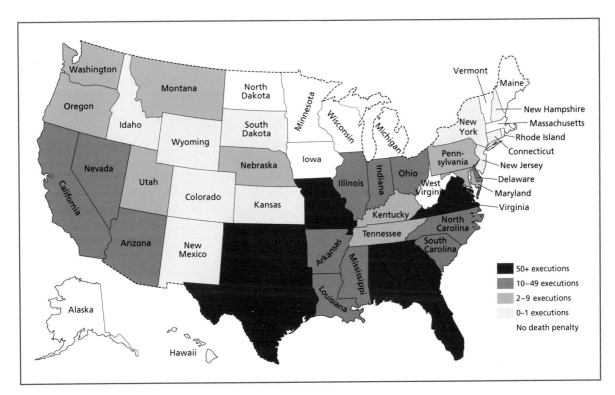

The use of capital punishment in the USA shows a divide between the Northern states and the Southern and Western states (the latter were predominantly migrated by settlers from the South). The map shows the number of executions since 1976. Nisbett and Cohen (1996) argue that Southern/Western states have a 'culture of honor' in which violence is justified in order to right a wrong but not that they are associated with more violence per se.

> from the Southern states are more likely to respond with violence to someone calling them 'pig' than someone from the Northern states, even when people are matched socio-economically and the questionnaire is administered in the same geographical location (Nisbett & Cohen, 1996). They are careful to note, however, that there is little evidence to suggest that there is greater acceptance of unprovoked violence (despite greater acceptance of violence as a form of retribution).

ANGER AND AGGRESSION

Anderson and Bushman (2002) define **aggression** as any behavior directed toward another individual that is carried out with the proximate (immediate) intent to cause harm. Violence would be one form of aggression associated with physical harm. Bullying is another form of aggression in which particular people are the repeated targets of aggression. When framed in this way, aggression appears to be dysfunctional by acting against group harmony, norms of fairness, and the welfare of others. However, aggression has a long evolutionary history and is regarded as a key aspect of animal behavior. All human societies tolerate, if not openly endorse, some acts of

KEY TERM

Aggression
Any behavior directed toward another individual that is carried out with the proximate (immediate) intent to cause harm.

aggression: for instance, as a means of punishment or retribution, as a means-to-an-end to overthrow an 'immoral' dictatorship, or even in certain competitive sports. Although societies tolerate some aggression, there will always be individuals (or groups of individuals) who cross the line of acceptability and it is in this sense that aggression can be considered as pathological. This section will describe the adaptive functions of aggression as well as pathological aspects of aggression.

Does aggression have a social function?

In non-human animals, instrumental aggression is generally considered to have a clear function: namely, social dominance (Hawley, 1999). Acts of aggression set up and maintain social hierarchies in which those at the top tend to have a leadership role (deciding what the group will do) and have privileged access to resources (such as food or mates). Those lower down the dominance hierarchy have more limited access to resources but may nevertheless benefit from group living (e.g. receiving the protection of higher status group members). Thus, the definition of aggression outlined above can be reconsidered and expanded: the proximate (immediate) intention of an aggressive act may be to cause harm but the ultimate (primary) intention may be to assert dominance over others or to defend ourselves, by defending status or well-being.

There is a fine line between social dominance and aggression. In many animal species, social dominance is achieved almost exclusively through aggression (and aggressive displays). Humans and other primates employ a variety of strategies (both pro-social and aggressive) to achieve social dominance. Competitive sports, for example, are used to establish dominance hierarchies (league tables) using socially permissible levels of aggression.

This particular spin on aggression accounts for one important observation: namely, that many aggressive acts in animals do not result in any physical harm at all (Lorenz, 1966). Other cues are used to determine social dominance in order to avoid physical injury. These include physical size and aggressive displays such as posturing (e.g. to make one appear bigger), vocalizations (e.g. roaring), and facial expressions (e.g. bared teeth, direct gaze). In addition, animals have evolved other display signals to say 'back off, you win', so-called **appeasement behaviors**. These may include distress displays (e.g. fear), averting gaze, or (in humans) saying 'sorry'.

In humans and other primates, displays relating to the emotion of anger (facial, vocal, bodily) are often regarded as crucial for demonstrating aggressive intentions (e.g. Berkowitz & Harmon-Jones, 2004). Anger is linked to situations in which one's goals are unfulfilled by someone else's improper actions. It may signal the fact that somebody else has violated a social norm or has challenged their social standing (e.g. their reputation, authority). Displays of anger may differ from other emotional displays in that they often do not elicit a matched response in the perceiver. Whereas seeing happiness may elicit happiness and seeing fear may elicit fear, seeing anger does not necessarily elicit anger. An angry face may sometimes trigger a fear response – a form of appeasement – rather than a reciprocated anger response (van Honk & Schutter, 2007). To give

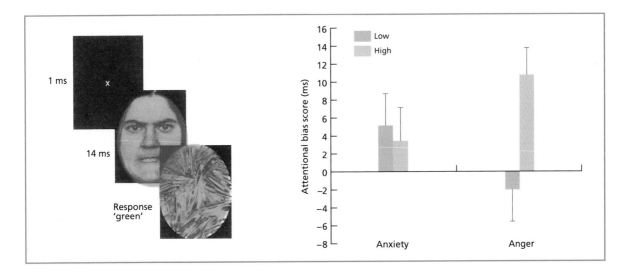

In this task 42 participants were divided into groups according to whether they were high/low in trait anxiety or high/low in trait angriness. The task involves presenting a colored face (subliminally) and then measuring naming times to a (supraliminal) colored display. Participants with high trait angriness are faster at naming the color after presentation of an angry face (relative to a neutral face). From van Honk et al. (2001b). Copyright © 2001 Elsevier. Reproduced with permission.

another example, in male rats a shock may produce a fear response when alone but an anger/aggression response in the presence of another male (Ulrich & Azrin, 1962).

Anger as an *emotion* is a response to a situation, but angriness as a *trait* may be a stable disposition that varies across individuals (Spielberger, Jacobs, Russell, & Crane, 1983). People with high trait angriness are biased towards interpreting the intentions of others in a hostile manner (e.g. it was a wrongdoing, not an inadvertent mistake) and may opt for confrontation over appeasement. Those with high trait angriness pay more attention towards angry faces (van Honk et al., 2001b). Those with high levels of *social* anxiety (rather than general anxiety) pay less attention to angry faces and may treat them as fear inducing (Putman, Hermans, & van Honk, 2004).

One of the most influential cognitive theories of aggression is the frustration–aggression model of Dollard, Doob, Miller, Mowrer, and Sears (1939) and later revised by Berkowitz (1989, 1990). In the original version of the model, aggression was regarded as a response to having one's goals thwarted, that is as a response to frustration (Dollard et al., 1939). In the later version, Berkowitz (1990) inserted anger as a mediator – that is, frustration from having one's goals thwarted can lead to feelings of anger and these feelings may generate aggressive acts. For Berkowitz, anger is more than an emotion; it is also an appraisal – namely, an appraisal that another person is to blame for one's unfulfilled goals through an improper or unfair action. Aggression may therefore serve a revenge motive in which one attempts to correct a perceived wrongdoing or restore fairness (Stillwell, Baumeister, & Del Priore, 2008). According to Berkowitz (1990), whether an aggressive thought is translated into an aggressive act depends on environmental cues and social norms about such behavior.

As in other species, there is evidence that human aggression is related to social dominance. According to Hawley, Little, and Rodkin (2007) moderate hostility and

There is a relationship between rates of homicide and income inequality of developed countries (but not between homicide rates and average income level). Some countries deviate more from the expected trend-line. For example, Finland and the USA have some of the highest rates of gun ownership compared to other countries, whereas Singapore has the lowest rate of gun ownership. From Wilkinson and Pickett (2009). Copyright © Dr Kate Pickett and Professor Richard Wilkinson, 2009. Reproduced by permission of Penguin Books Ltd.

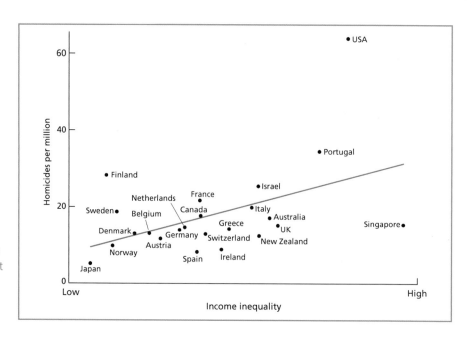

aggressive self-expression characterize highly ambitious, successful, and powerful people. However, it need not involve violence. Such people may tend to use a mix of coercive and pro-social methods to dominate the control of resources and the use of these two strategies tends to be positively correlated (Hawley, 2002, 2003). Aggressive acts are not always directed towards the person that has elicited anger, but may be displaced towards someone perceived to be of lower social status. This is known as **displaced aggression**. Examples of this include: men who are treated with hostility by their boss, who then shout at their wife or children when they commit a minor transgression; higher degrees of aggression in workplaces where managers behave unfairly; prisoners who are bullied, who then bully others lower down the hierarchy (Marcus-Newhall, Pedersen, Carlson, & Miller, 2000). Research suggests that violent acts are more common in societies in which there is greater inequality between the richest and poorest but the overall level of wealth is not important (Wilkinson & Pickett, 2009). Poverty does not cause anger and violence, but the perception of injustice and unfairness can. Mazur and Booth (1998) argue that anti-social actions often reflect attempts to dominate figures in authority or to prevail over a constraining environment.

SOCIAL LEARNING OF VIOLENCE

According to Bandura (1973): 'The specific forms that aggressive behavior takes, the frequency with which it is displayed, and the specific targets selected for attack are largely determined by social learning factors.' This conclusion was reached from a large number of studies, the most influential of which involved preschool children observing aggressive attacks on a Bobo doll – a child-sized inflatable figure with a weighted bottom so that it pops back up after being hit.

For example, Bandura, Ross, and Ross (1963) showed children films of an adult kicking, attacking with a hammer, and shouting at the doll (a control video showed non-aggressive behavior). Children exposed to the aggressive video tended to behave aggressively when introduced to the doll themselves. This effect was enhanced if they had seen the adult rewarded after the aggressive act but was diminished if they saw the adult punished (Bandura, 1965). Although not considered here, another important strand of evidence is the observation of violence on television (e.g. Johnson, Cohen, Smailes, Kasen, & Brook, 2002; Paik & Comstock, 1994).

Social learning theory places an emphasis on *behavior* rather than the underlying *intentions*. Children are able to make the distinction between aggression and playfighting. Similarly, they understand from a young age that dolls and inanimate things do not experience pain and other feelings. As such, it is unclear whether aggressive acts in these studies really reflect aggressive intentions to harm. This does not mean that social learning has no role, but as a general theory of aggression it is unsatisfactory (Anderson & Bushman, 2002).

The children attack the Bobo doll after seeing an adult behave aggressively towards it. From Bandura et al., (1961). Copyright © 1961 American Psychological Association. Reproduced with permission.

Biological basis of anger and aggression

Although the amygdala has been strongly implicated in the perception and generation of fear, it is also important in aggression. It is perhaps not surprising that aggression and fear should have partly overlapping neural substrates, given that both are linked to the **fight-or-flight response** (fight = aggression, flight = fear). Lesions of the amygdala disrupt social dominance hierarchies in primates (Rosvold et al., 1954) and can result in unusual tameness in situations in which a fight-or-flight response would be the norm (Kluver–Bucy syndrome; Kluver & Bucy, 1939). The specific role of the amygdala may be in the *regulation* of aggression (making an aggressive act more or less likely). Studies of reactive aggression in cats show that a defensive rage reaction can be elicited by direct stimulation of the dorsal **periaqueductal gray** region of the mid-brain, either electrically or chemically (Siegel et al., 1999). In addition, different sub-regions of the amygdala and hypothalamus were found to influence this behavior through both inhibitory and excitatory mechanisms (Siegel et al., 1999). In humans, amygdala lesions affect aggressive behavior but may either increase its likelihood (van Elst, Woermann, Lemieux, Thompson, & Trimble, 2000) or decrease it (Ramamurthi, 1988). This is consistent with a regulatory role in aggression and also with the view that there are different sub-regions of the amygdala that affect aggression in different ways.

The perception of anger has been linked to the ventral striatum and regions of the orbitofrontal/ventromedial prefrontal cortex that it projects to. Dougherty et al. (1999) asked participants to read stories from their own life that previously made them angry during fMRI and found that these stories trigger high ratings of angry feelings again, correlated with activity in these frontal regions. However, these regions have been linked to subjective feelings of emotion in general (Hornak et al., 2003; Koenigs et al., 2007) and it is unclear whether it is anger-specific. A selective deficit in recognizing anger has been reported following damage to the ventral striatal region of the basal ganglia (Calder et al., 2004), and the dopamine system in this region has been linked to the production of aggressive displays in rats (van Erp & Miczek, 2000). Lawrence, Calder, McGowan, and Grasby (2002) found that disruption of a certain class of dopamine receptors (D2) disrupts anger recognition in humans. Given the wider role that this region plays in motivation and reward-based processing, it has been argued that the ventral striatum dopamine system may be particularly important for *instrumental* forms of aggression, which are goal based and pre-meditated rather than a response to an immediate threat (e.g. Couppis & Kennedy, 2008). The latter may be more strongly linked to the amygdala–hypothalamus–periaqueductal gray interactions described earlier.

Finally, the androgen hormone **testosterone** has been linked to anger, aggression, and social

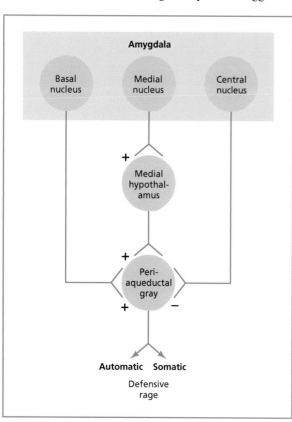

The amygdala modulates activity in the periaqueductal gray (also called central gray) region via direct excitatory (+) and inhibitory (–) connections and via the medial hypothalamus. The periaqueductal gray generates defensive autonomic and somatic ('fight') responses via brainstem nuclei. Adapted from Siegel et al. (1999).

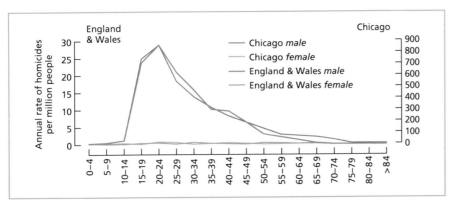

The age and sex distribution of convicted murderers shows a remarkably similar distribution between two different populations (Chicago USA vs. England & Wales), notwithstanding significant differences in the overall homicide rates (there is a 20-fold difference between them). What do you think is the relative contribution of biology and culture to these trends? From Cronin (1991). Copyright © 1991 Cambridge University Press. Reproduced with permission.

dominance. Testosterone has two types of effect on the brain. Firstly, it has an *organizing* effect during development. It stimulates the development of certain neural circuits and changes the sensitivity of others, making them more sensitive to testosterone. Testosterone is implicated in the production of predominantly male-related behaviors, and may account for sex-related differences in aggression (Mazur & Booth, 1998). Women do produce testosterone (via the adrenal glands and ovaries) but men produce far more of it (via the testes), with notable peaks prenatally and at puberty. Secondly, testosterone has an *activating* effect throughout the lifespan, that is, it binds to specific neural receptors and directly influences the functioning of certain neural circuits.

Participants given a testosterone injection show a greater response to angry faces, measured in terms of heart rate change, but not to neutral or happy faces, which was interpreted as an increased willingness to defend status (van Honk et al., 2001a). Resting levels of testosterone in men correlate with increased activity in the ventromedial prefrontal cortex when viewing angry faces (relative to neutral ones), but this negatively correlates with amygdala activity (Stanton, Wirth, Waugh, & Schultheiss, 2009). Ehrenkranz, Bliss, and Sheard (1974) studied prisoners divided into three categories: persistently physically aggressive, socially dominant but not physically aggressive, and neither physically aggressive nor socially dominant. Levels of testosterone were significantly higher in the first two groups relative to the third (but the first two groups were not different from each other). Other research links testosterone to the levels of violence of the crime committed by members of a prison population and the extent to which they violate rules in the prison (Dabbs, Carr, Frady, & Riad, 1995). In a non-prison sample of male army veterans, levels of testosterone were positively related to various self-report measures on antisocial behaviors (e.g. truancy at school and work) and violence (Dabbs & Morris, 1990). Socio-economic status (SES) was also found to be a moderating variable: testosterone had a greater influence on antisocial behavior in those from lower SES backgrounds.

Studies such as these establish a relationship between testosterone level and the level of aggression, dominance, and antisocial behavior. However, they do not establish the direction of cause and effect. Evidence suggests that the effects are bidirectional: high levels of testosterone are associated both with acts of competition and aggression *and* with the outcome of these acts. Testosterone levels increase prior to a competitive sporting engagement as a preparatory effect but after the engagement the testosterone levels of the losers drop and those of the winners increase or

A testosterone injection increases the physiological response to the sight of angry faces. This may bias people towards a 'fight' rather than a 'flight' response. From van Honk et al. (2001a). Copyright © 2001 American Psychological Association. Reproduced with permission.

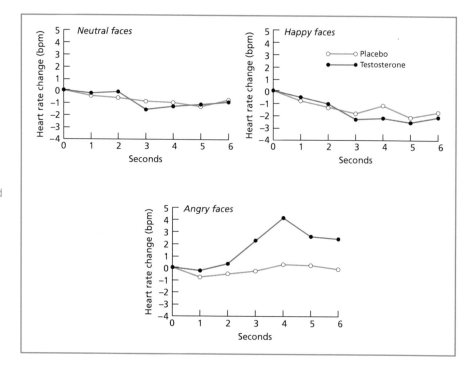

remain stable (Elias, 1981). Comparable results are found for non-physical competitions such as chess (Mazur, Booth, & Dabbs, 1992) and are even found when watching our own national team win or lose in the soccer World Cup (Bernhardt, Dabbs, Fielden, & Lutter, 1998) or when our political party wins or loses a general election (Stanton, Beehner, Saini, Kuhn, & LaBar, 2009). This modulation of testosterone by winning and losing may increase and decrease (respectively) the future motivation to take on more fights or challenges. A recent fMRI study shows that administration of testosterone (relative to a placebo) is associated with increased activity in the ventral striatum in a (non-social) game in which female participants played for money (Hermans et al., 2010). In social decision-making games, administration of testosterone relative to a placebo is associated with reduced levels of both trust (Bos, Terburg, & van Honk, 2010) and generosity (e.g. increases in the number of unfair offers made to others during the Ultimatum Game; Zak et al., 2009).

Pathological aspects of aggression

The terms psychopathy, **sociopathy,** and **antisocial personality disorder** (ASPD) are easily confused, not least because they mean different things to different researchers. The *Diagnostic and Statistical Manual of Mental Disorders* (DSM-IV) (American Psychiatric Association, 1994) proposed the term antisocial personality disorder to replace the older terms of psychopath and sociopath, which have pejorative connotations. As such, in their view, these are different labels for essentially the same thing. The DSM-IV definition of antisocial personality disorder is 'a pervasive pattern of disregard for, and violation of, the rights of others that begins in childhood or early adolescence and continues into adulthood' and it is diagnosed by behavior such as: a failure to conform to social norms; irritability and aggressiveness; impulsivity or failure to plan ahead; and having shallow or seemingly non-existent feelings. Such individuals comprise around 3–4% of the male population and less than 1% of the female population (Robins, Tipp, & Przybeck, 1991). However, others argue that a distinction should be made between the categories of psychopathy and ASPD/sociopathy. Hare and colleagues suggest that the DSM-IV diagnosis focuses too heavily on antisocial behavior and not enough on underlying personality traits (e.g. Hare, Hart, & Harpur, 1991). Their 'Psychopathy Checklist' (e.g. Hare, 1980) contains items that are similar to those used to diagnose

ASPD but also contains items that they believe capture non-behavioral psychopathic tendencies (e.g. glibness and superficial charm, grandiose sense of self-worth). Around 80–90% of inmates at a maximum security prison were found to meet the criteria for ASPD, whereas only 15–25% met the more conservative criteria from the Psychopathy Checklist (Hart & Hare, 1996). Blair (2003), in contrast, argues that a distinction between psychopathy and sociopathy/ASPD is motivated by the fact that only psychopaths show high levels of instrumental aggression whereas all groups show high levels of reactive aggression. In the studies reported below, I will take the approach of considering them separately, with psychopathy typically assessed via the Psychopathy Checklist and sociopathy/ASPD typically assessed via the DSM-IV criteria.

Psychopathy

Several studies have examined emotional processing in people diagnosed as psychopaths. Lykken (1957) was the first to note that these people do not show a normal fear-conditioned response to aversive stimuli (see also Flor, Birbaumer, Hermann, Ziegler, & Patrick, 2002). In addition they show reduced autonomic activity to fear and sadness in others (Blair et al., 1997). When presented with vignettes describing happiness, sadness, embarrassment, and guilt, a psychopathic group demonstrated particular problems with attributing guilt (Blair et al., 1995). Guilt tended to be described in terms of happiness or indifference. Blair (1995) put forward a cognitive model of psychopathy to account for this evidence in terms of an inability to respond appropriately to the distress cues of others, which normally acts as a 'violence inhibition mechanism'. In support of this, psychopaths show reduced amygdala activity in aversive conditioning (Veit et al., 2002), which is consistent with its role in detecting fear (or appeasement) in others. As noted previously, psychopaths perform normally on moral dilemmas (Cima et al., 2010), suggesting an awareness of right/wrong but an indifference to the consequences of an aggressive act.

Others have argued that prefrontal dysfunction is a core deficit in psychopathy (Gorenstein, 1982; Newman & Lorenz, 2002). The orbitofrontal cortex (OFC)

Psychopaths appear to be aware of social norms of right and wrong, but are willing to violate them. This may reflect an indifference to the consequences of aggression (e.g. they do not show a normal response to fear in others). Copyright © Sunset Boulevard/Corbis.

and ventromedial prefrontal cortex have strong bidirectional connections with the amygdala and are implicated in making emotionally guided decisions (Bechara et al., 1994; Bechara, Damasio, Damasio, & Lee, 1999). In a study of emotional memory, psychopathic individuals showed reduced medial OFC activity to emotional words, in addition to reduced amygdala responses (Kiehl et al., 2001). They also show less medial OFC activity during aversive conditioning (Birbaumer et al., 2005) and less medial OFC activity when making cooperation decisions relative to defect decisions in the Prisoner's Dilemma experiment (Rilling et al., 2007). The relationship between these differences and the actual symptoms of psychopathy is unclear: for example, patients with acquired lesions to these regions show a different behavioral pattern to psychopaths (see below). It is possible that a wider network of regions is disrupted in psychopathy (e.g. see Tiihonen et al., 2008). For example, functional imaging studies often show an *increase* in activity in lateral prefrontal regions, in addition to reduced orbitofrontal activity (e.g. Kiehl et al., 2001). This could be consistent with a deliberate attempt to suppress the influence of emotions, or a compensatory mechanism for a lack of context-appropriate emotional responses in the first place (Kiehl, 2008).

THE EXTRAORDINARY CASE OF PHINEAS GAGE

One of the most famous cases in the neuropsychological literature is that of Phineas Gage (Harlow, 1848/1993; Macmillan, 1986). On 13 September 1848, Gage was working on the Rutland and Burlington railroad. He was using a large metal rod (a tamping iron) to pack explosive charges into the ground when the charge accidentally exploded, pushing the tamping iron up through the top of his skull; it landed about 30 m behind him. The contemporary account noted that Gage was momentarily knocked over but that he then walked over to an ox-cart, made an entry in his time book, and went back to his hotel to wait for a doctor. He sat and waited half an hour for the doctor and greeted him with, 'Doctor, here is business enough for you!' (Macmillan, 1986).

Not only was Gage conscious after the accident, but he was able to walk and talk. Although this is striking in its own right, it is the cognitive consequences of the injury that have led to Gage's notoriety. Before the injury, Gage held a position of responsibility as a foreman and was described as shrewd and smart. After the injury, he was considered unemployable by his previous company; he was 'no longer Gage' (Harlow, 1848/1993). Gage was described as 'irreverent, indulging at times in grossest profanity ... manifesting but little deference for his fellows, impatient of restraint or advice when it conflicts with his desires ... devising many plans of future operation, which are no sooner arranged than they are abandoned in turn for others'. After various temporary jobs, including a stint in Barnum's Museum, he died of epilepsy (a secondary consequence of his injury) in San Francisco, some 12 years after his accident.

Where was Phineas Gage's brain lesion? This question was answered by an MRI reconstruction of Gage's skull, which found damage restricted to the frontal lobes, particularly the left orbitofrontal/ventromedial region and the left anterior region (Damasio, Grabowski, Frank, Galaburda, & Damasio, 1994). Research suggests that this region is crucial for certain aspects of decision making, planning, and social regulation of behavior, all of which appeared to have been disrupted in Gage. Other areas of the lateral prefrontal cortex are likely to have been spared.

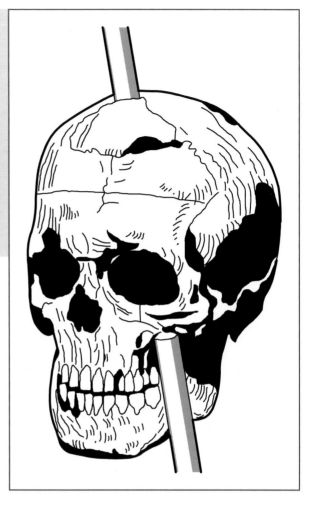

Acquired sociopathy/ASPD following damage to the orbitofrontal lobe

When testing a number of their patients with frontal lobe lesions, Damasio and colleagues (1990) made an interesting observation, namely, that many of their patients met a published American Psychiatric Association criterion for sociopathy (now called ASPD). The term *acquired sociopathy* is used to refer to those individuals who did not exhibit such symptoms prior to their brain injury. One of the earliest and most famous neurological cases in the literature is that of Phineas Gage (Macmillan, 1986). After a lesion to the left frontal lobe, Gage

A large metal rod was blown through the skull of Phineas Gage as a result of an explosion. Remarkably, he survived. However, his social behavior was profoundly changed.

was noted to be 'irreverent, indulging at times in grossest profanity'. Patient MGS is described as a modern case of Phineas Gage (Dimitrov, Phipps, Zahn, & Grafman, 1999). He had been decorated for service in Vietnam, with more than ten medals and a Purple Heart. Following a head injury, he was demoted for incompetent behavior. After an honorary discharge, he was noted to be sarcastic, lacking in tact (e.g. inappropriate disclosure of sexual history), moody, and unable to manage his own finances.

More detailed analysis has revealed that this kind of behavior arises specifically after lesions of the orbitofrontal cortex (particularly bilaterally) and also following lesions of parts of the medial surface of the frontal lobes, including the ventral region of the anterior cingulate (Hornak et al., 2003). Patients with lesions limited to the lateral prefrontal cortex do not show this socially disrupted behavior. Hornak et al. (2003) conducted a variety of assessments on patients with different lesions to the frontal cortex, including asking the patients about their own experiences of emotion (e.g. whether they feel angry more/less often) and asking a close relative about their social behavior using a questionnaire. Patients with bilateral orbitofrontal damage reported changes to subjective emotional experience relative to patients with lateral prefrontal damage, although the direction of change was variable (some reported less emotion, others more). Similar results were obtained with this group for the ratings of social behavior given by a relative. These patients tended to be rated as:

- less likely to notice when other people were sad/angry/disgusted;
- less likely to respond to emotions in others (e.g. through comfort or reassurance);
- less cooperative and more impatient;
- less close to their family and have problems with close relationships.

Tranel, Bechara, and Denberg (2002) draw a somewhat different conclusion about the anatomical location within the prefrontal cortex needed to disrupt social functioning. They suggest that damage to the *right* ventromedial/orbitofrontal cortex, but not the left, gives rise to symptoms of acquired sociopathy. This was found on a range of objective tests (e.g. anticipatory skin conductance responses to risky decisions) as well as being apparent in assessments of daily social living.

Patients with lesions to the orbitofrontal cortex (OFC) and/or the ventromedial cortex (including regions of the anterior cingulate cortex, ACC) show subjective changes in emotional intensity (top) and disturbances in social behavior as rated by a relative (bottom). Figures from Hornak et al. (2003). Copyright © 2003 Oxford University Press. Reproduced with permission.

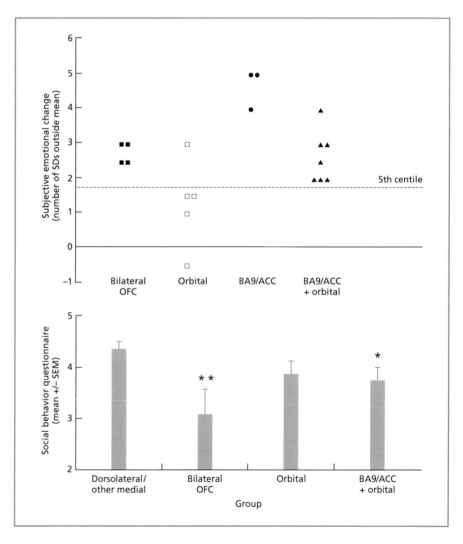

Double dissociations between acquired sociopathy and executive functions have been reported (Bechara, Damasio, Tranel, & Anderson, 1998), and also different neural substrates are implicated (orbitofrontal in sociopathy and lateral frontal in executive functions). The patients' difficulties appear to lie in the control of behavior specifically from emotional cues, rather than from a general problem in executive functions. Subjective emotional experiences of both a moral (e.g. pride, guilt) and a basic nature could be described as serving a motivational function on behavior: we are motivated to do 'right' because we want to achieve pride, joy, happiness and we want to avoid guilt, shame, or distress. In the absence of this emotional compass the quality of social interactions will almost certainly deteriorate.

Conduct disorder and bullying in childhood

Psychopathy and ASPD/sociopathy both require presentation in adulthood for diagnosis. However, comparable traits in childhood include a diagnosis of conduct disorder and also bullying behavior. Decety, Michalska, Akitsuki, and Lahey (2009) conducted an fMRI study of youths with aggressive conduct disorder. When watching another person in pain, both the conduct disorder and control groups activated regions of the brain linked to pain processing (including the insula and anterior cingulate cortex). However, the conduct disorder group activated it more, suggesting, paradoxically, a heightened neural empathic response. In addition, the conduct disorder group activated other regions not found in controls, including the ventral striatum. One possibility is that pain-producing acts (i.e. acts of aggression) are rewarding and this over-rides the tendency to be driven by empathy. Bullying is distinguished from other forms of aggressive behavior in that hostile acts are targeted towards specific individuals. Bullies appear to have good social cognition when tested formally. On tests of understanding deception, emotional reasoning, and theory of mind, bullies (aged 7–10 years) score as high as their non-bully peers and significantly higher than the victims of bullying (Sutton, Smith, & Swettenham, 1999). This suggests that many bullies have good social intelligence coupled with an ability (or willingness) to ignore the effects that their actions have on victims.

Evaluation

Whilst it would be relatively easy to draw up lists of social and cultural factors (variations in gun law, inequality levels and SES, 'culture of honor' mentality) and biological factors (testosterone, young male predominance, genetic predispositions) relating to aggression, this would give a misleading picture of the field. The reason why it is misleading is that these factors are not separate but intertwined: for instance, individual differences in aggression in adults are affected by gene–environment interactions (Caspi et al., 2002), testosterone–SES interactions (Dabbs & Morris, 1990), and so on. One reason why they are intertwined is because aggression and displays of anger serve adaptive social functions (establishing dominance, maintaining social order). Some individuals, however, do exhibit extreme forms of aggression. In cases of adult psychopathy this is related to reward-based, goal-directed motivations for aggression rather than a lack of understanding of social norms. This may involve inhibiting distress cues of others or an additional difficulty in learning fear-related associations (by the amygdala). In cases of acquired orbitofrontal damage, it may reflect an inability to use emotional cues to guide social behavior, resulting in antisocial trends but without high levels of goal-directed aggression.

KEY TERMS

Responsibility
The extent to which
someone can be held
to account for his/her
actions.

Neuroethics
The application
of neuroscientific
findings to ethical
issues.

CONTROL AND RESPONSIBILITY: 'IT WASN'T ME. IT WAS MY BRAIN'

Another key concept when considering antisocial behavior is that of **responsibility** – the extent to which someone can be held to account for his/her actions (e.g. Eastman & Campbell, 2006). This is related to the degree of control that people have over their behavior. Accidentally hitting somebody is not considered an antisocial act, but intentionally hitting someone is. Responsibility is typically deemed to vary across individuals (e.g. children vs. adults) and across situations (e.g. defensive, provoked, unprovoked). Imagine you find your lover in bed with someone else. Is it permissible to respond violently? In many countries, the law would offer a minimal punishment for such an act based on the assumption that the cheated lover had reduced control over his/her violent actions – a so-called crime of passion. These legal systems make assumptions about underlying psychological mechanisms – namely that there are some situations in which emotions drive antisocial behavior with minimal ability to over-ride it. It is, of course, an empirical question as to whether this assumption is true. These kinds of issues are explored in the emerging field of **neuroethics** (e.g. Farah, 2005; Moreno, 2003).

The ability to control behavior is central to the commonsense notion of 'free will' (or agency): namely, that we have a sense of being able to select between different courses of action. Controlled actions are intentional rather than accidental or reflexive (e.g. such as moving one's hand away from a flame). Control and responsibility, however, are not the same thing: a person with paranoid schizophrenia who attempts to kill their neighbor for 'tuning into their thoughts' would have control over this action but have diminished responsibility; a drunk driver who kills a pedestrian would have diminished control but would still be considered responsible for this action as the potential consequences could be foreseen.

The concept of controlled behavior is most easily explained by reference to its anti-thesis: automatic behavior (Schneider & Shiffrin, 1977). Whereas controlled behavior is considered slow, conscious, and based on reason, automatic behavior is considered fast, often unconscious, and based on intuition. Intuition, in this context, may refer either to stereotyped behaviors that have been performed numerous times in the past (so-called schemas) or to our gut reactions (e.g. based on emotions). Traditionally, these two sources of decision making (controlled vs. automatic) have been regarded as being in opposition to each other, but with controlled behavior having the upper hand such that it is able to over-ride automatic behavior (Miller & Cohen, 2001; Norman & Shallice, 1986). This controlling behavior is also referred to as the 'executive functions' of the brain and has been most closely associated with the prefrontal cortex (Fuster, 1989; Goldberg, 2001; Stuss & Benson, 1986). Controlled behavior may be achieved by two basic means. Firstly, the prefrontal cortex may hold in mind the current and future goals of the individual as well as maintaining in mind the elements of the current 'problem' (e.g. who did what to whom in a moral dilemma-type task). This is related to the notion of a working memory (Baddeley & Della Sala, 1996; Goldman-Rakic, 1996). Secondly, the prefrontal cortex may provide a biasing signal that either activates or inhibits the functioning of other regions of the brain (Miller & Cohen, 2001; Shallice & Burgess, 1996). This mechanism ensures that our behavior is not solely governed by automatic processes.

However, the distinction between automatic versus controlled is by no means clear cut. Behavior is rarely solely one or the other. It has also been argued that much of our control over behavior occurs unconsciously (e.g. Suhler & Churchland, 2009). For example, in the social domain many of our actions are influenced by biases that we are completely unaware of. This raises important philosophical questions about whether we really are in control of, and responsible for, our actions (Doris, 1998). Consider the following findings from the social psychology literature:

- Participants are more likely to interrupt a (staged) conversation between the experimenter and another person when they have been primed, in a previous task, with words relating to rudeness than those primed with words relating to politeness or not primed at all (Bargh, Chen, & Burrows, 1996).
- Participants who have been primed by an irrelevant task (containing words relating to old age) subsequently walk more slowly to get an elevator than those primed by words relating to young age (Bargh et al., 1996).
- People are more likely to litter in a particular place when it contains graffiti than when it does not (Keizer, Lindenberg, & Steg, 2008).
- If you find a dime on the street, you are more likely to help a passer-by who accidentally drops some papers (Isen & Levin, 1972).

Studies such as these raise important questions about the notion of control. Is our sense of control just a post hoc justification for the unconscious decisions we make? Can control occur unconsciously and, if so, are we responsible for such actions? Suhler and Churchland (2009) argue that we can be considered to be in control of our actions (and responsible for them) even if we are not consciously aware of all the information that enters into the decision. Their approach is to define control and responsibility relative to a normative model of brain function, rather than attempting to link it to prescribed ideas about what controlled behavior is (i.e. that one needs to be consciously aware of the basis of one's decisions). According to them, if normative control turns out to be unconscious, then so be it. They propose that lack of control and responsibility would then be inferred by disruptions of this normative neurobiological model. This represents a philosophical rather than a pragmatic approach to the issue, given the absence of an agreed-upon normative model of control or even a definition of abnormal (statistically rare, qualitatively different, etc.).

Eastman and Campbell (2006) take a different approach to Suhler and Churchland (2009) by addressing the question of how the neuroscience of aggression can be interpreted by the law. They note that all the functional and structural imaging studies of psychopathy are essentially correlational in nature. It is not clear what, if anything, is causally related to the committing of a criminal act. Psychopaths presumably have similar brain structure both when committing an antisocial act and when behaving responsibly. What may differ between these situations is the interaction with an external influence at the time of the act.

Control over actions may wax and wane over time. DeWall, Baumeister, Stillman, and Gailliot (2007) asked participants to complete a task involving a high degree of control (watching a video but ignoring words that appeared unexpectedly at the bottom of the screen) and compared them to another group shown the same stimuli but given no control instructions. Their prediction was that having to exert sustained control over this task would reduce their ability to exert control over a subsequent aggressive impulse. After this task, they either received an insult or praise from the experimenter about a short essay they had written earlier ('This is one of the worst essays I've ever read' vs. 'Excellent! No comments'). Finally,

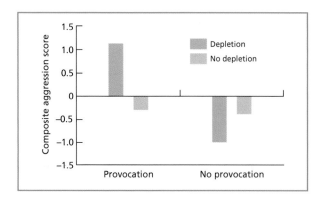

Participants who received an insult (provocation condition) *and* had previously exerted high levels of control on an unrelated task (depletion condition) chose louder and longer blasts of noise for the experimenter to receive. This suggests that our ability to exert self-control over aggression is diminished if we have previously had to exert control on a different task. From DeWall et al. (2007). Copyright © 2007 Elsevier. Reproduced with permission.

they played a competitive task against the experimenter in which the loser would receive a blast of loud noise. The participant could choose the level and duration of noise. Participants who had previously exerted high control and were then given the insult gave louder and longer blasts of noise to the experimenter. This suggests that our ability to control aggression is not fixed but is related to prior cognitive activity. Whether our responsibility for the aggressive act can be said to wax and wane is harder to answer.

Evaluation

The notions of control and responsibility are closely linked because both imply an element of intentionality or 'free will' over actions. They differ insofar as responsibility operates at the social level (i.e. the extent to which others hold you to account for your actions) whereas control operates at the level of the individual (Goodenough, 2004). The extent to which structural or functional brain difference could be used to argue for diminished responsibility is unclear due to several factors: establishing causation; establishing 'normality'; the extent to which normal control is itself influenced by unconscious (and uncontrollable) biases.

SUMMARY AND KEY POINTS OF THE CHAPTER

- Moral emotions (e.g. guilt, shame, moral disgust, anger) consist of emotions arising as a result of an appraisal of behavior (either one's own or somebody else's) relative to some normative standard of behaving. They tend to involve not only regions specialized for emotions but also regions implicated in higher cognition (lateral prefrontal cortex) and mentalizing.
- Moral judgments of right/wrong are underpinned by different sources of knowledge: an emotional reaction (or gut instinct; e.g. based on empathic concern for others), consensual norms (e.g. the law), and reasoned decisions (e.g. based on comparisons between costs and benefits). These different sources of knowledge may sometimes conflict with each other (requiring effortful control), and different individuals may give different weightings to the importance of these different factors.
- Patients with lesions to the orbitofrontal cortex and/or ventromedial prefrontal cortex may exhibit antisocial tendencies. They fail to respond appropriately to emotional cues, and may eschew emotional information in favor of more rational judgments in moral dilemmas.
- Feelings of anger are associated with lack of goal attainment due to the perceived improper/unfair actions of other people. Expressions of anger signal disapproval and possible aggressive intentions.

- Aggression (normal and pathological) has been linked to brain structures such as the amygdala (involved in regulating aggression), periaqueductal gray (involved in generating a reactive 'fight' response to a threat), the ventral striatum (linked to anger), and the orbitofrontal/ventromedial prefrontal cortex (contextual modulation of emotions).
- Responsibility is related to the notion of intentional control of actions and 'free will'. Although important legally, it is unclear how it can be translated into neuroscientific theory.

EXAMPLE ESSAY QUESTIONS

- Are morals different from other kinds of social norm?
- What is the role of testosterone on aggression?
- Why does damage to the orbitofrontal cortex (and ventromedial frontal cortex) give rise to antisocial behavior?
- How can studies in neuroscience shed light on the legal definition of responsibility?

RECOMMENDED FURTHER READING

- Hauser, M. D. (2006). *Moral minds*. London: Abacus. A popular science account, with a stronger focus on philosophy than neuroscience.
- Sinnott-Armstrong, W. (2008). *Moral psychology. Volume 3: The neuroscience of morality, emotion, brain disorders and development*. Boston, MA: MIT Press. Includes an up-to-date overview of all topics covered in this chapter.

CHAPTER 11

CONTENTS

Social learning during infancy   259

The social brain in childhood: Understanding self,
understanding others   269

The adolescent brain   277

Summary and key points of the chapter   281

Example essay questions   282

Recommended further reading   282

Developmental social neuroscience

11

The structure and function of the brain, and the neurons within it, are not resolutely fixed at birth according to a predetermined blueprint. Being part of a socially enriched environment affects both the structural and functional development of the brain (e.g. Branchi et al., 2006; Chugani et al., 2001). Cells and pathways in the brain may whither or grow depending on the quality and quantity of social interactions during development. As such, brain-level explanations – such as those being developed in the field of developmental social neuroscience – offer an exciting opportunity for resolving the **nature–nurture debate**. The brain is the organ in which gene-based influences ('nature') and environment-based influences ('nurture') come together. Traditionally within developmental science this has been construed as a series of interactions between a child's ability to understand and engage with the environment coupled with the suitability of the environment for providing the appropriate inputs. For example, the eminent developmental psychologist Jean Piaget (1896–1980) considered development as progressing through various stages: the structure of the stages was construed as largely predetermined, but successful passage to the next stage required appropriate interactions between the child and the environment.

Within this nature–nurture debate it is important not to confuse **phylogenetic development** (i.e. of species) with **ontogenetic development** (i.e. of individuals). It is widely accepted that many of the evolutionary developments that distinguish humans from other primates lie in the domain of social intelligence, and in this sense we can say that many facets of human social behavior are innate. However, this does not mean that human infants enter the world with a mature understanding of the social world, or that innate is the same as 'present from birth' (the term for this is congenital), and nor does it mean that developing good social skills is inevitable. The question of what the initial 'start-up kit' consists of in human infants remains open for debate, as do the many complex gene–environment interactions that occur during development that lead to adult social competence.

This chapter considers social development in three broad stages: infancy (birth to 18 months), childhood (18 months to puberty), and adolescence (puberty to adulthood). The focus of the chapter will be on the development of social cognition (empathy, extracting social cues from faces, theory of mind, etc.) and what we can learn from combining behavioral studies with the methods of cognitive neuroscience (fMRI, EEG, etc.). As such, there is not a strong emphasis on findings from genetics. However, the chapter will consider the relationship between early social competencies that can be considered to be innate (e.g. mimicry of tongue protrusion, preference for face-like stimuli, emotion contagion) and the development of related skills found in older children and adults (e.g. empathy, goal-based imitation, face-specific processing).

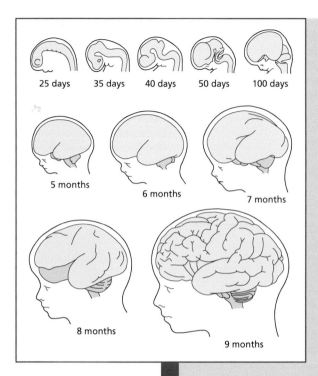

25 days · 35 days · 40 days · 50 days · 100 days

5 months · 6 months · 7 months

8 months

9 months

The embryonic and fetal development of the human brain.

Synapse formation has a rise-and-fall pattern. It peaks soon after birth, but different cortical regions differ greatly in the time taken to fall again to adult synaptic levels. From Huttenlocher and Dabholkar (1997). Reprinted with permission of John Wiley & Sons, Inc.

BUILDING A BRAIN

The nervous system derives from a set of cells arranged in a hollow cylinder, the **neural tube**. By around 5 weeks after conception, the neural tube has organized into a set of bulges and convolutions that will go on to form various parts of the brain. Closer to the hollow of the neural tube are several proliferative zones in which neurons and glial cells are produced by division of proliferating cells (neuroblasts and glioblasts). Purves (1994) estimates that the fetal human brain must add 250,000 neurons per minute at certain periods in early development. The newly formed neurons must then migrate outwards towards the region where they will be employed in the mature brain. At birth, the head makes up around a quarter of the length of the infant. Although the brain itself is small (450 g) relative to adult human size (1400 g), it is large in comparison to remote human ancestors and living primates (a newborn human brain is about 75% of that of an adult chimpanzee). The vast majority of neurons are formed prior to birth, so the expansion in brain volume during postnatal development is due to factors such as: the growth of synapses, dendrites, and axon bundles; the proliferation of glial cells; and the myelination of nerve fibers.

Huttenlocher and Dabholkar (1997) measured the synaptic density in various regions of human cortex at different ages. In all cortical areas studied to date, there is a characteristic rise and then fall in synapse formation (synaptogenesis). In primary visual and primary auditory cortex the peak density is between 4 and 12 months, at which point it is 150% above adult levels, but this falls to adult levels between 2 and 4 years.

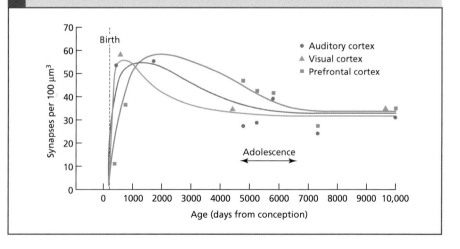

In prefrontal cortex, the peak is reached after 12 months but does not return to adult levels until 10–20 years old. Why does the number of synapses rise and then fall during the course of development? It is not necessarily the case that more synapses reflects more efficient functioning. During development, a process of fine-tuning the brain to the demands of the environment renders some connections redundant.

SOCIAL LEARNING DURING INFANCY

Infancy is normally defined as the period from birth to 18 months, beyond which the child is normally capable of language production. During this period the infant's social world is dominated by its parents or other caregivers, for which the infant will establish a powerful emotional bond. (Attachment and relationships were discussed in Chapter 8.) The infant learns to recognize these key figures via sight, sound, and smell. It learns to read the social cues that they provide (e.g. emotional expressions) and to understand their significance via shared activities (e.g. joint attention, turn-taking).

Recognizing others

The infant's first vision of the social world occurs at birth. However, other senses, particularly hearing, are able to acquire socially relevant information from within the womb. The newborn comes into the world already knowing the voice of the mother (DeCasper & Fifer, 1980), and her native language (Mehler, Bertoncini, Barriere, & Jassikgerschenfeld, 1978). This is demonstrated using an experimental set-up in which the infant's sucking switches on a tape recording: it sucks more strongly to hear a familiar recording. Infants even show a preference for listening to a story that the mother had frequently read aloud during her late pregnancy relative to an unfamiliar story (DeCasper & Spence, 1986). However, this behavioral evidence does not prove that they possess adult-like voice recognition from birth. In adults, there is a region in the upper part of the superior temporal sulcus that responds more to voices than to other kinds of auditory stimuli (Belin et al., 2000). Grossmann, Oberecker, Koch, and Friederici (2010) used near infrared spectroscopy to examine the emergence of this specialization and found that the region responds selectively to voices in 7-month-olds but not 4-month-olds. Other evidence suggests that at 7 months of age infants can discriminate between happy, angry, and neutral prosody and ERP evidence suggests that they are able to link these to the appropriate facial expressions (Grossmann, Striano, & Friederici, 2005).

Far more is known about face recognition by infants than voice recognition. Infants have an early preference for face-like stimuli relative to control stimuli. Infants within an hour of birth can show greater tracking of a moving face-like stimulus over a non-face-like stimulus (Johnson et al., 1991). This evidence has been used to propose an innate device for orienting attention towards stimuli that are likely to be faces (Morton & Johnson, 1991). However, this falls short of claiming that the newborn brain contains a representation of a standard face. Macchi

Cassia et al. (2004) manipulated photographs of real faces by inverting the internal features (such that the eyes are below the mouth) but such that the head itself retains the normal orientation. Newborns show a preference for a 'top-heavy' face but not necessarily the correct arrangement of features. At 3 months of age, infants show preferential looking at faces of their own race relative to different races (Bar-Haim, Ziv, Lamy, & Hodes, 2006) and prefer to look at faces that are of the same sex as their primary caregiver (Quinn, Yahr, Kuhn, Slater, & Pascalis, 2002).

Newborns in the first hour of life were sat on the experimenter's lap and shown a hand-held stimulus. Upon fixating the stimulus, the experimenter slowly moved it to the side (90°). The extent to which the baby tracked the stimulus (head and eye turns) was measured. From Johnson et al. (1991). Copyright © 1991 Elsevier. Reproduced with permission.

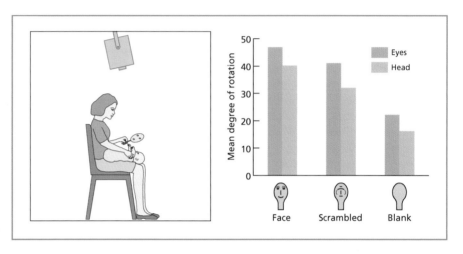

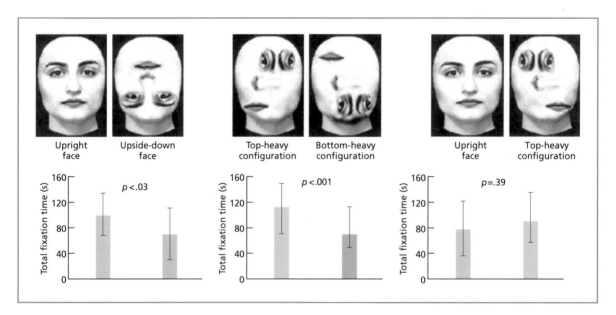

In this study, pairs of face-like stimuli are presented side by side (with side of presentation counter-balanced). The preferential looking time for one stimulus relative to another is measured. Infants 1–3 days old tend to prefer top-heavy face-like configurations, although they do not necessarily distinguish possible from impossible faces. From Macchi Cassia et al. (2004). Copyright © 2004 Association for Psychological Science. Reprinted by permission of SAGE Publications.

At what age can an infant recognize specific faces? Infants show a preference for the mother, relative to a stranger, as early as 2–3 days old and this occurs when odor cues are eliminated and the stranger has a similar hair style (Bushnell, Sai, & Mullin, 1989). However, early face recognition of the mother appears to depend on being able to link the mother's face with the mother's voice, which was already familiar from the prenatal stage (Sai, 2005). As such the mother's face may be learned differently to other faces. Long-term recognition of familiar people from a variety of views is possible by 3 months (Pascalis, de Haan, Nelson, & de Schonen, 1998). Before 3 months, studies have tended to rely only on short-term familiarity using habituation but nevertheless demonstrate the ability to distinguish between a recently exposed picture of a face and a completely novel face (Turati, Cassia, Simion, & Leo, 2006). Language may play a role in individuating faces even in infancy. Infants at 6 months old can discriminate different monkey faces (as assessed by habituation) but fail to do so at 9 months. However, if infants are trained to individuate them by pairing each face with a different name, then the tendency to individuate is retained at 9 months (Scott & Monesson, 2009). Pairing each face with the same category label 'monkey' leads to the normal decline in individuation.

Tzuorio-Mazoyer et al. (2002) conducted a functional imaging study of face processing in infants and noted that the regions implicated in adult face processing are activated but that there are also other regions activated. For instance, regions that are functionally specialized for language in adults respond to faces in infants. Brain damage in the fusiform region at only 1 day post-birth can lead to severe difficulties in face recognition (Farah, Rabinowitz, Quinn, & Liu, 2000). This suggests that this region may be committed to the processing of faces prior to birth even if it takes many years to reach an adult level of specialization. Studies outside infancy also suggest a slow maturation. Golarai et al. (2007) compared face perception (relative to objects) in children (7–11 years), teenagers, and adults using fMRI. Whilst all three groups responded similarly to objects, the fusiform face area was

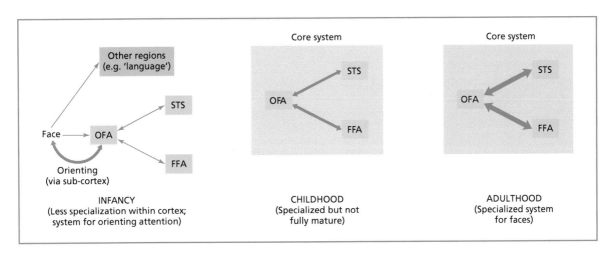

Functional imaging studies of infant face processing suggests that faces engage a wider network of regions than found in children and adults, in addition to showing activity in the 'core' system. During childhood this system becomes more specialized but the spatial extent of the activation in regions such as the fusiform face area (FFA) is far less than in adults. OFA = occipital face area; STS = superior temporal sulcus.

three times larger in adults relative to children. It would be interesting to know if this reflects cumulative experience or a property of the region itself – adults know more faces than children, but our knowledge of objects does not accumulate in the same way.

At around 5 months of age a child is able to hold and manipulate an object. As such, their interactions with other people shift away from face-to-face interactions (**dyadic interactions**) to joint attention-based interactions in which infant and adult may engage each other in looking at other objects or people (**triadic interactions**). Kaye and Fogel (1980) recorded mother–baby sessions at several points in time. At 6 weeks 70% of the session time was based on face-to-face interactions, reducing to 33% at about 6 months. The ability to recognize and respond to eye gaze cues is crucial for establishing joint attention. Newborns prefer to look at eyes making direct contact relative to averted eyes (Farroni et al., 2002), and faces presented with a direct gaze are subsequently recognized better by 4-month-old infants (Farroni, Massaccesi, Menon, & Johnson, 2007). However, the youngest age at which an infant is capable of following gaze to a peripheral location is around 3 months (Farroni, Mansfield, Lai, & Johnson, 2003). These mechanisms may enable joint attention to develop. However, merely following the gaze of another person is not sufficient for making the link between behavior and psychological states (i.e. that seeing is a source of knowing). Up until 10 months an infant will follow the head-turns of someone who has their eyes closed, but beyond this age they only follow head-turns when the eyes are open (Brooks & Meltzoff, 2005). This suggests that before this point they have little understanding of the significance of attention-orienting behavior, even though they can engage in it. A source localization study of ERP components to eye gaze in infants revealed sources of activity in occipital regions and prefrontal regions but *not* reliably in the region that is most closely linked to eye gaze processing in adults – namely the superior temporal sulcus (Johnson et al., 2005). This suggests an early use of general processes (e.g. in visual perception, motor coordination) in gaze detection that becomes more specialized during later development. Another EEG study (Mundy, Card, & Fox, 2000) suggests a difference between brain mechanisms for *responding* to joint attention cues (more posterior) and *initiating* such cues by looking back and forth between adult and object (more anterior).

By about 7 months of age, infants are generally able to distinguish between dynamic emotional expressions (Soken & Pick, 1999). However, this need not imply that they use an adult-like system to achieve this. For example, infants can be far better at discriminating certain emotions when posed by their mother rather than by a stranger (Kahana-Kalman & Walker-Andrews, 2001), suggesting that emotion recognition is not separable from face recognition in early development. Also, infant ERP studies show an influence of emotion on far later components of the ERP signal than expected from the adult literature (Leppanen, Moulson, Vogel-Farley, & Nelson, 2007; Nelson & DeHaan, 1996).

Social referencing becomes possible by linking gaze cues and facial expressions to learn about the potential emotional significance of objects (Klinnert et al., 1983). Recall from Chapter 4 that social referencing is based on a learned (classically conditioned) association between an adult's emotional response and attention directed towards an object, enabling an infant to learn whether something is 'good', 'safe', 'forbidden', etc. This kind of behavior comes online at around 12 months, although it continues to mature subsequently. For example, infants of this age respond appropriately to where the adult focuses their emotion even if the infant is attending elsewhere (Moses, Baldwin, Rosicky, & Tidball, 2001).

ADAPTING THE METHODS OF COGNITIVE NEUROSCIENCE FOR INFANTS AND CHILDREN

Methods such as fMRI and EEG are generally considered suitable for infants and children. One advantage of using these methods in younger people is that they do not necessarily require a verbal or motor response to be made.

Functional MRI and structural MRI

Gaillard, Grandon, and Xu (2001) provide an overview of some of the considerations needed. If one wants to compare across different ages, then the most significant problem is that the structural properties of the brain change during development. Although the volume of the brain is stable by about 5 years of age, there are differences in white and gray matter volumes until adulthood (Reiss, Abrams, Singer, Ross, & Denckla, 1996). The hemodynamic response function is relatively stable after 7 years of age but differs below this age (Marcar, Strassle, Loenneker, Schwarz, & Martin, 2004). Both the differences in brain structure and blood flow make it harder to compare activity in the same region across different ages. Younger children also find it harder to keep still in the scanner and this motion can disrupt the reliability of the magnetic resonance signal.

Near-infrared spectroscopy (NIRS)

One relatively new method that is now being used in developmental social neuroscience is **near-infrared spectroscopy (NIRS)** (for a summary see Lloyd-Fox, Blasi, & Elwell, 2010). This measures the amount of oxygenated blood and is – like fMRI – a hemodynamic method. Unlike fMRI it accommodates a good degree of movement and is more portable. The infant can sit upright on their parent's lap. However, it has poorer spatial resolution and does not normally permit whole-head coverage.

ERP/EEG

When working with young subjects using ERP/EEG, a limiting factor is the child's willingness to tolerate the electrodes, the task, and the time commitment required (Thomas & Casey, 2003). Children and adults can show quite different patterns of ERP (e.g. in terms of latency, amplitude, or scalp distribution), even for tasks that both groups find easy (Thomas & Nelson, 1996). These could reflect either age-related cognitive differences (i.e. the same task can be performed in different ways at different ages) or non-cognitive differences (e.g. the effects of skull thickness, cell packing density, or myelination).

KEY TERM

Near-infrared spectroscopy (NIRS)
A hemodynamic method that measures blood oxygenation, normally in one brain region.

Adults and children show very different visual ERP waveforms, despite having equivalent behavioral performance. Adapted from Thomas and Nelson (1996).

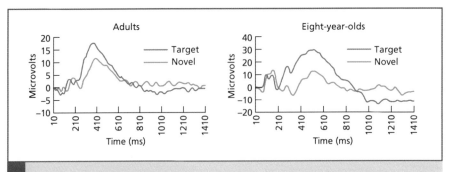

TMS

Current ethical and safety guidelines (Wassermann, 1996) do not recommend repetitive TMS to children except for compelling therapeutic purposes (e.g. treatment of depression).

Imitation, turn-taking and early communication

As noted in previous chapters, many researchers limit the use of the term imitation to those situations in which an individual reproduces the goals and intentions of another person rather than copying their motor actions. The latter, motor mimicry, appears to be present from birth and goal-based imitation appears to emerge after the first year. Newborn infants will mimic simple actions such as tongue protrusion (Meltzoff & Moore, 1977, 1983) – that is, they demonstrate an understanding that a seen tongue being protruded corresponds to their own, unseen, motor ability to do the same. Meltzoff and Moore (1977) concluded that, 'the ability to use intermodal equivalences is an innate ability of humans' (p. 78). At 3 months of age (but not 1 month) an infant shows signs of being able to discriminate when its own actions (facial, vocal, hands) are being imitated by an adult (Striano, Henning, & Stahl, 2005). They gaze more intently at this condition relative to control conditions in which the adult acts expressively but non-imitatively.

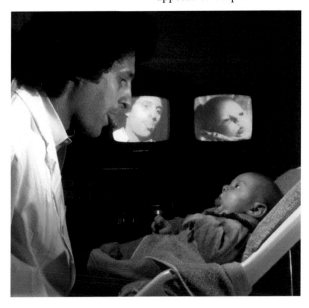

This 23-day-old infant imitates the tongue protrusion of the experimenter, suggesting an understanding of the link between seen actions of another and their own unseen actions. Photo by Andrew N. Meltzoff and E. Ferorelli. Used with permission from Andrew N. Meltzoff.

A parallel finding in the auditory–vocal domain is that (newborn) infants are susceptible to emotional contagion – that is, if other infants around them are crying then they too will be more likely to cry or show other distress cues (Simner, 1971). An infant is more likely to cry in response to the recorded cries of another infant than to a recording of their own previous crying (Dondi, Simion, & Caltran, 1999) and are more likely to do so than in response to an equally loud non-human sound (Simner, 1971).

Both motor mimicry and emotional contagion are assumed to reflect automatic reactions over which an infant has no control. Nevertheless, infants may use this behavior to learn the principle that other people are like themselves. Meltzoff and Decety (2003) refer to it as the 'like me' hypothesis. This matching of *physical* behavior (actions) is assumed to be transformed, during development, into a matching of *mental* behavior (thoughts, feelings, etc.) leading to the emergence of empathy and mind-reading. By 14 months of age, infants show evidence of being able to imitate on the basis of understanding the intentions of others rather than mimicry of the actions of others. In the study of Gergely et al. (2002), described in Chapter 3, infants respond on the basis of how an adult would have responded (should their arms be free) rather than how they actually did respond. Carpenter, Akhtar, and Tomasello (1998) found that infants aged 14–18 months would imitate acts differentially according to whether they were intentional versus accidental. Each action sequence had two parts and was performed on the same object. However, one part of the action would be accompanied by the experimenter saying 'whoops!' and the other part by saying 'there!' (signaling accidental and intentional actions, respectively). The infants were more likely to imitate the intentional actions than the accidental ones, suggesting an understanding of goals as distinct from actions. At 9 months of age, EEG recordings of infants' motor systems (mu rhythms) suggests that they discriminate actions in which a goal can be inferred relative to motorically similar but ambiguous actions (Southgate, Johnson, El Karoui, & Csibra, 2010).

Imitation/mimicry represents one form of early two-way interaction between infant and adult. Another example of a two-way interaction is **turn-taking** activities, in which adult and infant take it in turns to engage in an activity (although not necessarily the same activity). For example, the vocalizations of infants may occur in bursts followed by pauses, and the vocalizations of the adult may follow the same pattern but such that they occur during the infants' silent phases (Schaffer, 1984). Similarly, during feeding the jiggling behavior of the mother and the sucking behavior of the infant might obey the same turn-taking pattern (Kaye & Wells, 1980). In these cases, the infant has the opportunity to learn about principles underlying communication and reciprocity. By about 8 months, infants engage in simple interactive games such as peek-a-boo or passing an object back and forth between the infant and adult (Trevarthen & Hubley, 1978) – that is, there is a shift from one-sided interactions (e.g. infant takes object) to two-sided interactions (e.g. infant takes object, infant gives object back, infant takes object, etc.).

Gesture-based communication may occur before the production of the first spoken words. Carpenter, Nagell, and Tomasello (1998) examined protodeclarative pointing during infancy. Protodeclarative pointing carries the meaning 'look at that!'. It is important, theoretically, because it implies that the infant understands that seeing is a route to knowing, and that a different viewpoint leads to different knowledge. The infant and an experimenter interacted with a toy but then a second experimenter entered with a more interesting toy. Crucially, the second experimenter could be seen by the infant but not by the first experimenter. Gestures

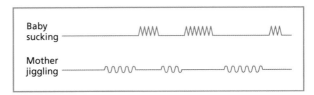

Simple motor interactions between adult and infant may have a turn-taking characteristic. For example, in the pattern of jiggling and sucking during breast-feeding the infant waits for the offset of the mother's jiggling. Other examples include turn-taking vocalizations between infant and adult, or the infant producing hand gestures in response to adult speech. These simple interactions are considered to be early forms of cooperation and communication. From Schaffer (1996). Reproduced with permission from Wiley-Blackwell.

by the infant to engage the attention of the first experimenter were seen to emerge at around 12–13 months of age. In terms of possible neural substrates, the level of EEG activity in frontal regions at 14 months of age predicts protodeclarative pointing at 18 months (Henderson, Yoder, Yale, & McDuffie, 2002). This suggests that this region may be important for initiating shared attention.

Although language development is typically studied from a 'cognitive' rather than 'social' developmental perspective, the social element is crucial given that communicative acts *are* social interactions. Note, for instance, that a difficulty in communication is one of the basic markers for autism (American Psychiatric Association, 1994). Kuhl (2007) has argued that normal language learning is gated by the social brain: that is, the degree to which an infant has developed an adequate understanding of social interactions will determine the pace and richness of speech acquisition. The quality of the interactions between infant and parent is also likely to be crucial, as well as the infant's own readiness. Tomasello and Todd (1983) videotaped a number of sessions of mother–child interactions between 12 and 18 months. They calculated the amount of time that the mother and child spent in activities requiring joint attention, and found that this correlates with overall vocabulary size at 18 months. Different types of attentional engagement predicted learning of particular types of words: the infants of mothers who *followed* their child's attention tended to learn more object names, whereas infants of mothers who *directed* their child's attention tended to learn more social words.

Kuhl, Tsao, and Liu (2003) studied phoneme discrimination by infants raised in English-speaking communities. At 10–12 months of age, infants tune-in to the phonemes of their language and this is achieved by reducing (tuning-out) their ability to detect the phonemes of other languages. Infants were exposed

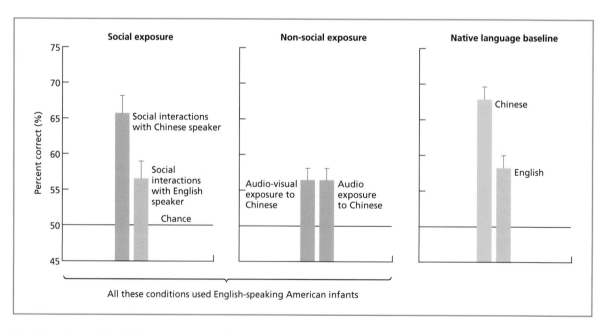

Social contact with a Chinese speaker enhances the discrimination of Chinese phonemes by English-speaking American infants (10–12 months old). The same effect is not found for comparable language exposure via CDs (i.e. audio only) or DVDs (i.e. audio-visual). Adapted from Kuhl et al. (2003).

to Chinese phonemes either via social interaction (over 12 sessions of 25 minutes) with a Chinese speaker or using the equivalent material presented non-socially via DVD (i.e. audio-visual) or CD (i.e. audio only). Only those infants exposed to Chinese socially showed an ability to make Chinese phonemic distinctions (and did not, in fact, differ from infants raised in monolingual Chinese cultures). One suggestion is that this learning depends on the level of triadic interactions between infant, speaker, and object, which is found in face-to-face interactions but not from DVDs and CDs (Mills & Conboy, 2009).

Adults, across most cultures, adjust the qualities of their speech when talking to infants using so-called **motherese** or **infant-directed speech**. This has several features, including higher pitch overall, more pitch variability, and a slower tempo with elongated vowels (Garnica, 1977). Infants show a preference for this kind of speech and one possible origin of the preference is that it is related to happy-sounding prosody (Singh, Morgan, & Best, 2002). Whilst adults tend to increase the pitch of their voice both when talking to pets and when talking to infants, they only exaggerate the pronunciation of vowels when talking to their infants (Burnham, Kitamura, & Vollmer-Conna, 2002). ERP studies comparing motherese with normal speech showed that infants at 6 and 13 months had larger amplitudes of certain ERP components to motherese relative to normal speech, notably over a component at 600–800 ms, which has been linked to attentional processes rather than word recognition (Zangl & Mills, 2007). Preschool children with autism do not fully distinguish motherese from normal speech, either in terms of preference or electrophysiological measures (Kuhl, Coffey-Corina, Padden, & Dawson, 2005).

KEY TERMS

Motherese
The prosodic aspects of human speech (e.g. increased pitch, intonation) that occur when speaking to infants.

Infant-directed speech
A more formal term for motherese.

Evaluation

Although the infant's expressive and communicative abilities are limited, he/she is both an avid consumer and user of social cues. Within the first year of life the infant is able to recognize others and respond to affective cues in faces and voices, and respond to attention-orienting cues (e.g. eye gaze). However, the infant is not a passive learner. He/she will engage in interactions (pointing, turn-taking, etc.) and is capable of manipulating the attention of others. Many of these behaviors suggest that

Motherese (or infant-directed speech) may increase attention to words and is associated with changes in neural processing, measured here using EEG (showing the mean amplitude of the ERP component at 600–800 ms). The infants listened to both familiar and unfamiliar words presented either in adult-directed or infant-directed speech. From Zangl and Mills (2007).

the infant is behaving like a 'rational' being (Gergely & Csibra, 2003) in that there appears to be an element of *interpretation* of social cues (e.g. responding to head turns only when the eyes are opened). The neuroscience evidence for this period, such as it is, suggests that the neural substrates supporting social behavior are far less specialized than found in adults (or older children).

THE DEVELOPMENT OF SELF-RECOGNITION AND SELF-AWARENESS

Lewis and Brooks-Gunn (1979) conducted an important study examining how children recognize themselves. Mirrors provide multiple cues to self-recognition: the person in the mirror looks like me, and the person in the mirror follows my movements. These different cues can be pulled apart using video cameras and monitors, such that there is a live stream (equivalent to the mirror condition) or a recorded stream so that the person looks like them but does not follow the movements. Between 3 and 8 months, infants show some signs of self-recognition based on shared movement cues. Between 8 and 12 months they show clear signs of recognizing themselves using movement cues, but they do not recognize their own face in a non-live video stream and they do not pass the mark test (i.e. rubbing a lipstick mark off their forehead). This suggests that they do not fully connect the image in the mirror to themselves. Between 12 and 24 months they can recognize themselves based on appearance even if the movements are not in synchrony, but it is only by 21–24 months that children pass the mark test. This suggests that bodily self-recognition develops incrementally up to the age of 2 years.

Infants and children seem to use different cues to recognize themselves at different developmental stages. Only at 2 years do they use mirrors to rub off a paint smudge on their forehead, implying a full understanding that the image out there corresponds to their own body.

Lewis and Carmody (2008) investigated the neural basis of self-representation by measuring structural changes (using structural MRI) in different brain regions in infants/children aged between 5 and 30 months (note: not longitudinally). They had three tasks: mirror-self-recognition (based on the mark test), use of personal pronouns (e.g. me, mine), and use of pretend play (e.g. feeding a doll with a spoon). Of the regions considered, only the temporo-parietal junction showed an association with self-representation (and the largest correlation was with mirror-self-recognition). As well as being part of the 'mentalizing' network (e.g. Saxe, 2006) this region is involved in putting oneself in another bodily perspective (Arzy, Thut, Mohr, Michel, & Blanke, 2006) and lesions can induce out-of-body experiences (Blanke et al., 2004). Recognizing oneself in a mirror may require a loosening of the sense of embodiment by projecting oneself onto an external location.

Other research suggests that children may not have a full understanding of the continuity of the self over time until 4 years. Povinelli and Simon (1998) conducted a version of the mark test in which children were video-recorded in two different locations at two time points – 1 week previously and a few minutes ago. Both 4- and 5-year-olds searched for the mark when shown the video of themselves a few minutes ago but not when shown the video taken 1 week ago. In contrast, 3-year-olds did not distinguish between these time points: they tended not to search for the mark in either condition. The authors interpret this as a difficulty in understanding that 'the self' can be duplicated in time and a struggle to represent the 'past self' and 'present self' as separate. This is consistent with other lines of evidence. As adults, our earliest recollections date from the period between 3 and 4 years old, and we remember far fewer events before the age of 6 years than would be expected from simple forgetting (Pillemer & White, 1989). The inability to recall episodes from our first few years is termed **childhood amnesia** or infantile amnesia. Perner and Ruffman (1995) claim that until 3–4 years children have *knowledge* about their past but they are unable to recall events as having been experienced by themselves (i.e. as episodic memories as opposed to facts). This shift is assumed to reflect changes in self-awareness (perhaps linked to theory-of-mind development) rather than changes in basic memory processes.

KEY TERM

Childhood amnesia
The inability of children and adults to recall episodes from the first few years of life.

THE SOCIAL BRAIN IN CHILDHOOD: UNDERSTANDING SELF, UNDERSTANDING OTHERS

The childhood period from the end of infancy (at 18 months) to adolescence (at puberty) is characterized by an increasingly complex and diverse set of social interactions. The child is increasingly influenced by his/her peers and behavior becomes more guided by the norms and skills of the surrounding culture. In short, this period is concerned with a process of socialization and enculturation. There is a correlation, in primate species, between brain size, social group size, and the length of the childhood period (Joffe, 1997). This long period of immaturity gives human children extra social learning opportunities before achieving adulthood.

In general, there have been fewer neuroscience-based investigations during the childhood period relative to infancy, and many of the studies described below are from the developmental psychology literature. These can be considered relevant for the following reason: In order to make sense of the large neuroscientific body of literature in *adults* on topics such as theory of mind and empathy, it is important to understand the developmental precursors of these abilities. For instance, apparently domain-specific mechanisms found in adulthood may have their developmental origins in more general processes. The questions of 'how they got there' and 'how they work' are unlikely to be independent.

Developing a theory of mind

Meltzoff and Moore (1977) argue that one of the mechanisms that is innate is a simple 'body scheme' that enables different parts of the body to be activated by vision (of others) and by touch and action (by oneself). This self–other link provides a foundation for the infants' understanding that others are 'like me' and, according to Meltzoff and Decety (2003), 'infant imitation is the seed and the adult theory of mind is the fruit' (see also Meltzoff, 2007). By 18 months, the imitative behavior of toddlers reproduces the intentions of adults by following what people *meant* to do and ignoring, say, accidents (Meltzoff, 1995). At 18 months, most infants engage in pretend play and understand that when their mother uses the banana as a telephone she is not mistaken but is deliberately substituting one object for another (Leslie, 1987). Leslie (1987) argues that there has been a 'decoupling' between ideas (what people think) and behavior (what people do). An absence of pretend play at 18 months is an early indicator of autism, along with a lack of protodeclarative pointing discussed earlier (Baird et al., 2000).

If this kind of behavior is present at the end of infancy and start of toddlerhood, why does it take until 4 years to pass 'classic' theory-of-mind tasks such as the Sally–Anne test discussed in Chapter 6 (Baron-Cohen et al., 1985)? One possibility is that aspects of non-social cognition need to catch up. These tasks require reasonable language abilities: following a narrative and remembering a sequence of events. However, from the age of 2 years onwards children use mental state words such as 'want', 'wish', and 'pretend' in an appropriate context, such as 'I thought it was an alligator. Now I know it's a crocodile' (Shatz, Wellman, & Silber, 1983). It may also be related to the particular nature of the mental state being tapped by theory-of-mind tasks (i.e. false beliefs) compared to imitation tasks (i.e. intentions). False belief tasks require second-order intentionality (e.g. beliefs about beliefs; Dennett, 1983), which could emerge later than first-order intentionality. False belief tasks also require inhibiting a strongly competing response (e.g. the child's own knowledge of the correct location) to perform correctly. Gergely and Csibra (2003) argue that from around 12 months infants understand that an action is a means to a goal (rather than the action being the goal), but they suggest that this falls short of being able to represent mental states as separate from current reality (as required in many theory-of-mind tests). Others suggest that the understanding of some mental states, such as desires, may develop earlier than others, such as beliefs (e.g. Wellman, 2002). This contrasts with the view that an understanding of mental states emerges as a result of maturation of a single over-arching theory-of-mind module.

There is some evidence of a functioning theory of mind before the age of 4 years, even on false belief tasks, when the other demands of the task are minimized. Clements and Perner (1994) showed 3-year-olds a version of the Sally–Anne task on a computer screen and recorded where they looked. The children tended to look at the location associated with the correct answer (i.e. what Sally believes) even though they gave the incorrect answer when prompted verbally (i.e. what they believe). This raises the possibility of an implicit (or unconscious) theory of mind present before the age of 4 years that becomes explicit after this age. Onishi and Baillargeon (2005) also used a looking-based measure to claim that infants can represent false beliefs as early as 15 months. In their task the adult saw a toy slice of watermelon placed in one of two boxes. A curtain was then lowered to obscure the adult's view. In the true belief condition, the object remains where it is. In the false belief condition, it is transferred to the other box. The curtain is then raised and the adult reaches into one

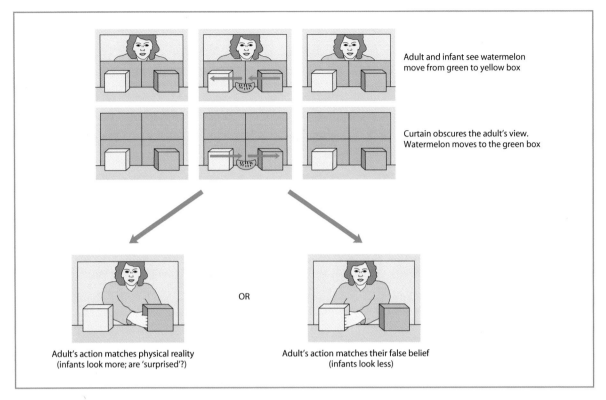

Adult and infant see watermelon move from green to yellow box

Curtain obscures the adult's view. Watermelon moves to the green box

OR

Adult's action matches physical reality (infants look more; are 'surprised'?)

Adult's action matches their false belief (infants look less)

Infants as young as 15 months show surprise (i.e. they have longer looking times) when an adult reaches for an object in a box that matches its true location but does not match the adult's presumed beliefs about the location. Is this evidence for an understanding of false beliefs several years before infants can explicitly reason about such mental states? From Onishi and Baillargeon (2005). Copyright © 2005 American Association for the Advancement of Science. Reproduced with permission.

of the two boxes (either consistent or inconsistent with their belief) and the infant's looking time is measured. In the false belief condition, longer looking times were found when the experimenter reached into the correct location (inconsistent with the false belief) relative to when the experimenter reached into the incorrect location (consistent with the false belief). The infants appeared to expect the experimenter to reach to where they believed the toy to be (but see Perner & Ruffman, 2005, for alternative accounts).

Evidence from neuroscience could perhaps be used to adjudicate between these positions. Are the 'core' regions implicated in adult theory of mind used as early as 18 months, with additional resources (language, executive functions) becoming more important later in development? At present there is very little evidence either way. At 4 years of age, EEG studies suggest that differences in regions such as the temporo-parietal junction and medial prefrontal cortex (both implicated in adult theory of mind) discriminate between children who fail and pass *explicit* measures of theory of mind, that is, those requiring an overt prediction of behavior (Sabbagh, Bowman, Evraire, & Ito, 2009). Do the same regions support performance on *implicit* measures of theory of mind (based on looking behavior) in younger children? Three-year-olds fail explicit measures of theory of mind but may look towards the correct answer on the very same task (Clements and Perner, 1994); neuroscience may be better equipped than behavioral measures to understand the origins of such a discrepancy.

Beyond the age of 4 years, theory of mind may continue to develop. Firstly, some typically developing children do not pass false belief tests at 4 years but around 90% do at 5 years and all do at 6 years of age (see Frith & Frith, 2003). Between 5 and 10 years children also become able to pass higher order belief reasoning of the sort 'John thinks that Mary thinks that...' (Perner & Wimmer, 1985). The non-social cognitive skills to perform explicit reasoning about mental states (including working memory and executive functions) may also continue to develop until adulthood, affecting performance on more challenging theory-of-mind tasks as discussed later (e.g. Dumontheil, Apperly, & Blakemore, 2010).

To return to the quote of Meltzoff and Decety (2003): Is imitation the seed and theory of mind the fruit? It would be fair to say that if imitation is the seed then it certainly is not the only ingredient in the recipe. Certain species of monkey show, at birth, some of the basic imitative abilities demonstrated by Meltzoff and Moore (1977), but their seed does not grow into a theory of mind (Ferrari et al., 2006). Conversely, congenitally blind individuals develop a similar brain network for performing theory-of-mind tasks even though they cannot engage in these forms of imitation (Bedny et al., 2009). Although we have a good understanding of what the basic ingredients are for developing a theory of mind, we lack a decent recipe for linking them together.

Development of empathy and pro-social behavior

As noted previously, 18-month-old children are more likely to imitate intentional versus accidental actions and this has been taken to imply that they have an understanding of the causes of actions (perhaps in terms of mental states) rather than copying motor programs. Another line of research suggests that children of this age also show pro-social, helping behavior that discriminates between accidental versus intentional behavior (Warneken & Tomasello, 2006). Young children help when an experimenter accidentally drops something or if they cannot open a cabinet door properly because their arms are full. However, they are less likely to help if the experimenter deliberately drops something or deliberately bumps into the cabinet door (Warneken & Tomasello, 2006). This occurs without an external reward being offered and without the need for parental encouragement (Warneken, Hare, Melis, Hanus, & Tomasello, 2007) and, in fact, giving rewards for helping may *reduce* the likelihood of future helping behavior (Warneken & Tomasello, 2008). Young children, like adults, may find the act of helping a reward in its own right. This intrinsic reward may be devalued when it is paired with an external award (i.e. the positive associations become linked with the external stimulus rather than their own internally generated behavior).

Young children may show concern and empathy for others and this may provide a motivation for helping in the absence of external rewards. Young children (18 months to 2 years) watched an adult doing a drawing and saw another adult grab the drawing and tear it up (Vaish, Carpenter, & Tomasello, 2009). In a control condition, one adult grabbed a blank piece of paper lying in front of the other adult. The children produced facial expressions of concern in the first scenario (as assessed by blind raters), and the more concerned they looked the greater was their tendency to step in and offer help to the adult.

It is unclear from this study whether helping behavior is driven by the need to alleviate the child's own personal distress (self-oriented) or a genuine concern for the adult (other-oriented). Children, like adults, may react differently when confronted with distress. Fabes, Eisenberg, Karbon, Troyer, and Switzer (1994) measured children's (6–8 years) physiological response (heart rate increase) and behavioral response to a crying infant in another room, listened to over an intercom. Children who were better able to regulate their own emotional response (as measured by a small increase in heart rate) spontaneously engaged in comforting behavior over the intercom. Children who were less able to regulate their own emotion (large increase in heart rate) showed signs of distress themselves, and their behavior showed irritation (rather than comforting) and avoidance. For instance, these children were more likely to switch the intercom off.

At 3 years of age, children show signs of discriminating who to cooperate with and who not to cooperate with. Olson and Spelke (2008) gave children a doll ('Reese') and then introduced, by way of a narrative, four more dolls. The dolls could either be family, friends, or strangers to Reese. They may have shared with Reese previously, or they may have shared with others recently. The child was then given some shells that 'Reese' had to distribute to the other dolls. When given four shells, children tended to distribute equally (i.e. one each) irrespective of status. However, when given fewer shells, then children tended to favor family and friends over strangers, those who had previously shared with them (direct reciprocity), and those who had shared with others (indirect reciprocity).

Other research based on the Ultimatum Game suggests that older children (aged 9–10 years) often apply rigid fairness norms (everything being shared equally), although adolescents and adults show more flexibility (Murnighan & Saxon, 1998). Crone and Westenberg (2009) argue that children might be basing their judgment on earlier-maturing emotion-processing regions and that over-riding these responses, in favor of self-interest, may require mature functioning of the dorsolateral prefrontal cortex.

Some recent studies have investigated children's empathy using EEG and fMRI. Light et al. (2009) measured EEG activity in 6–10-year-old children when performing a pleasurable task involving a pop-out toy. Changes in EEG activity were then correlated with individual differences in empathy, measured by the child's response (facial, vocal, bodily) to seeing the experimenter experience pain (trapping finger) followed by relief. Children showing high empathic concern on this measure tended to show greater EEG activity over prefrontal regions during the pop-out task, although the laterality difference shifted during the course of the task (shifting from right to left). Greimel et al. (2010) used fMRI to study changes in empathy from 8 to 27 years in which participants had to infer an emotional state from a face or judge their own emotional response to a face (relative to a control task of judging the width of a face). Interestingly, accuracy was unaffected by age but the brain regions supporting this task did shift developmentally. For instance, activity in the inferior frontal gyrus (linked to the human mirror neuron system) increased over age and was associated with differences between self and other perspectives (the left side

When 3-year-old children are asked to distribute shells, on a doll's behalf, to other dolls they show evidence of kin favoritism, direct reciprocity (giving to those who have given to you), and indirect reciprocity (giving to those who have given to others) when an equal distribution of resources is not possible. From Olson and Spelke (2008). Copyright © 2008 Elsevier. Reproduced with permission.

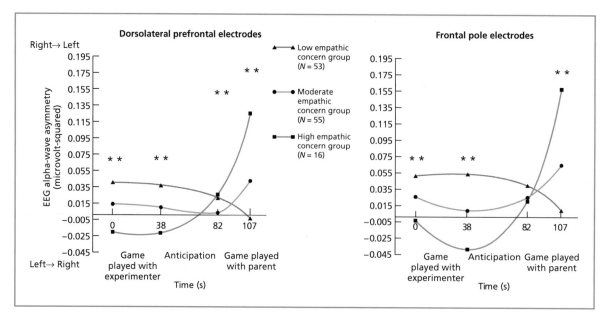

The level of empathic concern exhibited by children (measured as a dispositional trait) is related to frontal lobe activity, measured using EEG, when engaged in an emotionally arousing game in which a pop-up toy is played with by the experimenter and then, following a pause, is played with again by the parent. Note that the highly empathic group shows an asymmetry between the left and right hemispheres in different phases of the study. Graphs from Light et al. (2009). Reproduced with permission from Wiley-Blackwell.

showed an age-related increase in the self-versus-other contrast, whereas the right side showed the opposite). It is interesting to note that both of these studies suggest developmental differences in prefrontal functioning related to empathy either across individuals of the same age (Light et al., 2009) or developmentally at different ages (Greimel et al., 2010).

Role of peers, family, and culture

Culture is not normally considered to influence the development of mind-reading, as assessed by theory-of-mind tasks. For example, Callaghan et al. (2005) noted that passing false belief tests emerged between 3 and 5 years in a variety of cultures (rural Peru, Samoa, rural Canada, urban India, and urban Thailand). However, this does not necessarily imply equivalence in terms of underlying neural mechanisms. For instance, there is evidence, from fMRI, of different neural substrates for mentalizing in childhood (9 years) between English-speaking American children and bilingual Japanese children (Kobayashi, Glover, & Temple, 2006). It remains unclear whether these differences relate to differences in family dynamics or cultural/linguistic differences.

One aspect of the environment that *does* affect the development of mentalizing occurs at the level of the family, rather than at the level of culture (i.e. society). Children who live in families that engage in extensive talk about the feelings and thoughts of others tend to pass theory-of-mind tests at an earlier age (Dunn, Brown, Slomkowski, Tesla, & Youngblade, 1991) and are better at detecting the feeling states of others from vignettes when tested several years later at age 6, even after taking into account verbal ability (Dunn, Brown, & Beardsall, 1991). One may wonder about the direction of cause and effect in these studies; perhaps these parents have better genes for developing a social brain rather than providing their children with a richer mentalizing environment. However, the children of parents *trained* to discuss mental states show a larger improvement in 'mind-reading' tasks, including false belief, suggesting that the effect is environmental (Lohmann & Tomasello, 2003). Another important environmental factor within a family is the presence of a sibling. Jenkins and Astington (1996) found that children with a brother or sister performed better on theory-of-mind tasks, even after controlling for language ability, compared to children who had no siblings. The presence of another child in the house (either younger or older) provides a rich opportunity for learning about others' feelings, beliefs, and desires.

Harris (1995) provided a summary of studies of behavioral genetics that aim to partition variability in a trait (e.g. personality) according to environmental and genetic influences. Whilst the studies clearly demonstrated an important role for both genetic and environmental effects, the striking conclusion of her review was that most of the environmental effects are related to peer influences rather than family influences. One explanation for this finding is that there are crucial *within-family* differences. Behavioral genetic studies assume that children growing up in the same family have the same family environment, but of course this need not be true. Parents may treat different children differently. However, Harris (1995) rejected this explanation in favor of an alternative account: namely, that childhood socialization is influenced more by peers than by family. In her group socialization account, cultural information (values, fashions, norms) is transmitted from the parents' generation to the child's generation but, crucially, peer-level interactions act as a filter in determining which ideas are accepted and which are rejected. For example, migrant children tend to take on the language characteristics (e.g. accent) of their peers rather than parents, and are also more likely to endorse the social norms of their peers.

A child's awareness of his/her social identity is formed in the childhood years. Although children can reflect on internal states (e.g. hunger, wanting) from a young age, they do not attribute stable traits to themselves (shyness, kindness, etc.) until around the age of 7 years (Eder, 1990). The tendency to describe others in terms of psychological traits rather than in terms of their behavior increases throughout childhood. However, it is not until adolescence that they make a significant number of *comparisons* of psychological traits (e.g. 'X is kinder than me') when asked to describe others (Barenboim, 1981). One of the earliest developing aspects of a child's social identity comes from an awareness of his/her gender and the cultural norms that are bound up with this distinction. From an early age, adults treat boys and girls differently. The same baby will be treated very differently by strangers if it is dressed in blue versus pink (Will, Self, & Datan, 1976). Lloyd and Duveen (1990) examined the choice of toys played with by toddlers (18 months to 3½ years) when in the presence of one of their friends. Boys showed a preference for 'boys' toys' (as rated by adults) but this was strongly affected by the gender of the peer who was present. They showed a small preference for 'boys' toys' when a girl friend was present but a huge preference for 'boys' toys' in the presence of a male friend. Girls were also affected by the gender of their playmate but their behavior was more flexible. They were more likely to play with 'girls' toys' in the presence of a girl friend but were more likely to play with 'boys' toys' in the presence of a boy. Recall that boys never showed a preference for 'girls' toys' even in the presence of a girl,

Two models of socialization. In (a) the child acquires its social and cultural knowledge via inter-individual interactions. In (b), proposed by Harris (1995), the child acquires its basic social and cultural knowledge from its parents' generation but then selects and transforms this via peer–peer interactions. In this model, social norms from the parents will tend to be endorsed by the child if, and only if, the child's peers endorse it too. From Harris (1995). Copyright © 1955 American Psychological Association. Reproduced with permission.

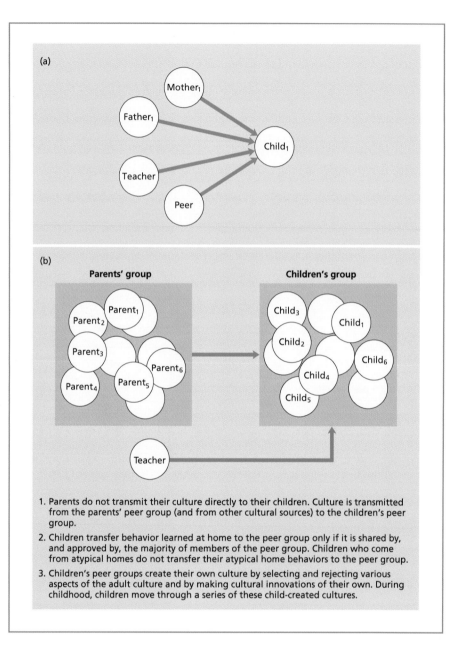

1. Parents do not transmit their culture directly to their children. Culture is transmitted from the parents' peer group (and from other cultural sources) to the children's peer group.

2. Children transfer behavior learned at home to the peer group only if it is shared by, and approved by, the majority of members of the peer group. Children who come from atypical homes do not transfer their atypical home behaviors to the peer group.

3. Children's peer groups create their own culture by selecting and rejecting various aspects of the adult culture and by making cultural innovations of their own. During childhood, children move through a series of these child-created cultures.

suggesting that boys' play behavior is more determined by their own gender at this age, whereas girls' play behavior is more determined by their social setting.

Evaluation

During the preschool and early school years, the child comes to understand that people (including themselves) are embedded in a social and cultural context comprising norms (e.g. fairness, right/wrong), beliefs, and roles (e.g. go to work, raise a family).

They also become aware that different people have different traits (e.g. kindness, shyness) and that they too have a particular social identity (based on their traits, sex, beliefs).

THE ADOLESCENT BRAIN

Adolescence is formally defined as the period between onset of puberty and adulthood and is characterized by significant hormonal changes and changes to physical appearance. For adolescents an over-arching social concern involves the cultivation of their social identity (how they see themselves) and their reputation (how they are perceived by others). This involves a need for social comparison between themselves and others in terms of their social standing (e.g. popularity, dominance) but also in terms of their adherence to other social norms (e.g. Sebastian, Viding, Williams, & Blakemore, 2010). Adherence to age-appropriate popular norms (e.g. fashion or music trends) is one way of appearing popular or, at least, guarding against unpopularity. Changes in social cognition go hand-in-hand with structural and functional changes that occur in the brain during adolescence.

The success of many contemporary fashion trends (from mods and rockers – to punks, new wave, and Goths) is driven by teenagers. What changes in the adolescent brain and their social world might explain this?

Structural changes in the adolescent brain

The region of the brain described as undergoing the most structural change during adolescence is the prefrontal cortex. These changes occur somewhat differently in the white matter and gray matter. White matter density increases steadily during the first two decades of life, stabilizing during late adolescence (Sowell et al., 1999). This is generally attributed to **myelination** – the increase in the fatty sheath that surrounds axons, which increases the speed of information transmission. The prefrontal cortex is one of the last areas to achieve adult levels of myelination. Changes in gray matter take a rather different course, showing an inverted-U-shaped function characterized by a peak and then a decline. The volume of gray matter in the prefrontal cortex peaks before the onset of adolescence (12 years for boys, 10 years for girls), before decreasing during adolescence to adult levels (Giedd et al., 1999). The maturation tends to occur earliest for posterior regions of the frontal lobes, and latest for anterior regions (Gogtay et al., 2004). These changes in gray matter are likely to be associated with the density of synaptic connections (Huttenlocher & Dabholkar, 1997).

KEY TERMS

Adolescence
The period between onset of puberty and adulthood, characterized by significant hormonal changes and changes to physical appearance.

Myelination
The increase in the fatty sheath that surrounds axons, which increases the speed of information transmission.

Studies of animals – both rodents and non-human primates – have highlighted important changes to another brain system during adolescence, namely in terms of dopaminergic inputs to the orbitofrontal/ventromedial prefrontal cortex through the ventral striatal 'reward' system (Casey, Getz, & Galvan, 2008; Ernst & Spear, 2009). Specifically, there is a developmental increase in dopaminergic input to the frontal regions in early adolescence due to changing patterns of receptor binding (Tseng & O'Donnell, 2005), shifting towards greater predominance of dopaminergic activity within the ventral striatum in later adolescence (Ernst & Spear, 2009). Ernst, Pine, and Hardin (2006) have suggested that these changes lead to a dominance of reward-related over punishment-related motivations. Whereas the reward-related system is linked to changes in the dopaminergic striatal–frontal pathway, the punishment-related system is linked to amygdala functioning. During adolescence, the amygdala shows a reduced response to stress, measured by stress-induced gene regulation in rats (Kellogg, Awatramani, & Piekut, 1998), and functional imaging in humans shows altered amygdala response to fear expressions relative to adults (Killgore & Yurgelun-Todd, 2010). Ernst et al. (2006) argue that this imbalance between reward/punishment-related cues in adolescence, coupled with immature control (due to late maturation of prefrontal cortex), leads to greater risk-related behavior. It may also tend to lead to greater independence as the adolescent 'follows his/her own heart' rather than norms prescribed by their parents. Casey et al. (2008) propose a similar model that highlights the different rates of maturation of reward-related regions (e.g. the ventral striatum) and prefrontal regions (although this model does not consider the amygdala in detail). The faster maturation of the ventral striatal pathway than the prefrontal cortex during adolescence is assumed to give rise to increased risk taking and a greater propensity for rewarding social affiliations (friends, romantic partners).

Gaining control over the social world

Consistent with the structural changes occurring in prefrontal regions, several functional imaging studies have shown that adolescents show different patterns of brain activity in frontal areas in socially relevant tasks. Wang, Lee, Sigman, and Dapretto (2006) conducted an irony comprehension task on children/early

Many neuroanatomically based models explain adolescent behavior in terms of an imbalance between late-maturing regions of the prefrontal cortex and earlier-maturing regions such as the ventral striatum. This leads to increased reward-based behavior (greater sociality, thrill seeking) coupled with less cognitive control (greater risk taking). From Ernst and Spear (2009).

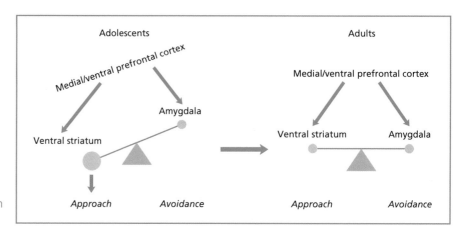

adolescents (9–14 years) and adults. The participants had to decide whether a speaker was sincere or ironic using both prosodic cues (e.g. sarcastic voice) and contextual cues (integrating prior information). The younger group were more likely to activate regions of the prefrontal cortex (both medially and laterally) than the adult group on this task, which may reflect greater effort and a less intuitive response. A similar finding was obtained by Blakemore, den Ouden, Choudhury, and Frith (2007). They compared judgments about mind-reading scenarios (e.g. 'You are at the cinema and have trouble seeing the screen – do you move to another seat?') with physical scenarios (e.g. 'A huge tree suddenly comes crashing down in a forest – does it make a loud noise?'). Adults showed greater activity than adolescents in the right superior temporal sulcus (STS) region in response to mind-reading scenarios, whereas adolescents showed greater activity than adults in the medial prefrontal cortex region in response to mind-reading scenarios. This result is intriguing, given the claim that the medial prefrontal cortex may serve a general function in representing self in relation to other (Amodio & Frith, 2006), whereas the right STS region has been linked more specifically to representing mental states (Saxe, 2006). However, differences in activation need to be interpreted carefully. An increase in activity in frontal regions by adolescents could reflect the fact that these regions work less efficiently (less efficiency generating more effort and more activity), rather than reflecting a greater importance for conducting the task.

Although children from the age of 4 years upwards tend to pass 'theory-of-mind' tasks, competency on some of these tasks may not reach adult levels until late in adolescence. Dumontheil et al. (2010) developed a more challenging theory-of-mind test (although based on the same principle as the Sally–Anne task) for older participants. Participants were given an array of compartments with objects on them that were all visible to the participant, but a 'Director' was positioned on the other side of the array and could only see a sub-set of the objects due to an occlusion. The Director would call out instructions to the participant (e.g. 'move the small ball to the left') and the participant would have to respond appropriately by taking into account the Director's perspective. There was found to be a significant difference between older adolescents (14–17 years old) and adults (19 years and above) on this task, but not on a control task that did not require perspective taking. The authors argue that although a basic understanding of theory of mind comes online at around 4 years of age (if not before) it may require a mature system of executive functioning in order for it to be optimized, and this may not come online until the end of adolescence. It would also be interesting to see how individual differences in social behavior may be related to developing expertise in theory of mind. For example, would an adolescent who is very pre-occupied with his/her reputation (e.g. 'what does person X think of me?') perform better than someone who is less concerned but of the same age?

A similar conclusion to Dumontheil et al. (2010) was reached in a study of the Ultimatum Game (described in Chapter 7) with adolescents between 9 and 18 years old (Guroglu, van den Bos, & Crone, 2009). When the responder had the opportunity to reject an offer, there was an age-related increase during adolescence in the amount of offer that the proposer (the participant) proposed. This was interpreted as a greater developmental inclination to take on board the perspective of the other person.

Much of the adolescent concern for reputation and motivation for risk taking may be related to the need to develop relationships with others. This is spurred on,

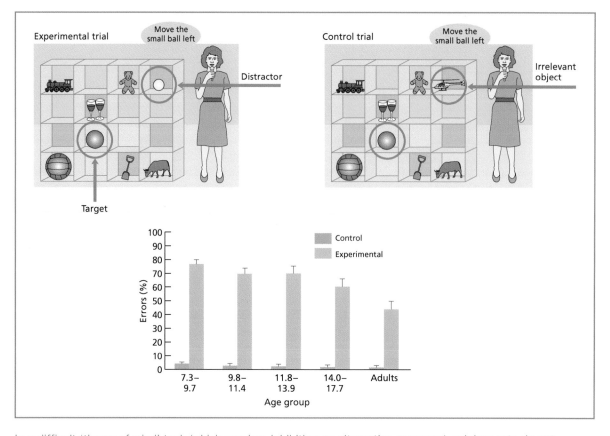

In a difficult 'theory-of-mind' task (which requires inhibiting an alternative response), adolescents do not perform as well as adults. Thus, whilst competence in these tasks emerges at an early age (from 4 years), sophisticated performance develops more slowly and depends on domain-general resources such as executive functions. From Dumontheil et al. (2010). Copyright © 2009 The authors. Journal compilation copyright © 2010 Blackwell Publishing Ltd. Reproduced with permission from Wiley-Blackwell.

at least in part, by reaching sexual maturity. The presence of intimate and confiding relationships has a positive influence on health and stress, and the absence of them during adolescence is related to vulnerability to mental health problems (Pine, Cohen, Gurley, Brook, & Ma, 1998). Social rejection, in the cyberball game (discussed in Chapter 8), leads to greater negative feelings in adolescence relative to adulthood (Sebastian et al., 2010). There is evidence that the increased risk taking found in adolescents is far greater when measured in the presence of peers than when similar measures are taken alone (Gardner & Steinberg, 2005). Experimentation with alcohol and drugs by this age group is typically a 'social' activity insofar as it normally occurs in social gatherings rather than via individual discovery. Alcohol reduces anxiety, which may facilitate certain social interactions (e.g. Varlinskaya & Spear, 2002). Of course, many recreational drugs stimulate the reward systems of the brain (e.g. Koob, 1992), as do social interactions themselves. By pairing the two together, the social interaction itself may be perceived as being more rewarding than in the absence of such a stimulant. It would be easy to dismiss the interaction between adolescence and drug/alcohol use as being related to social norms rather than developmental changes in the brain. However, adolescent rats show comparable effects in the absence of any social norms for alcohol use. In adolescent rats, but not adult rats, low doses of alcohol facilitate social interactions (Varlinskaya & Spear, 2002). If a sober adolescent rat is exposed to an intoxicated sibling then alcohol drinking by the sober rat is enhanced, and in a dose-dependent manner – the more the sibling has consumed, the more he/she will consume (Hunt, Holloway, & Scordalakes, 2001).

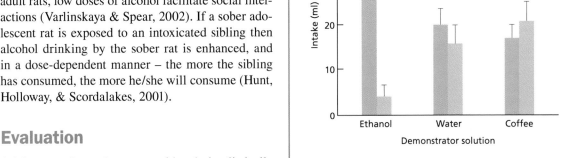

Evaluation

Adolescence is no longer considered simplistically as half-adult/half-child but rather as a distinct phase in development that is qualitatively different to both childhood and adulthood. The prefrontal cortex reaches maturity during adolescence, but other structures relating to reward (e.g. ventral striatum) and punishment (e.g. amygdala) may also mature during this period and – crucially – not necessarily at the same rate.

If a sober adolescent rat is exposed to an intoxicated sibling then he/she will voluntarily consume more alcohol. In this experiment, a sibling (the 'demonstrator') was administered a solution of coffee (COF), ethanol (EtOH), or water (H_2O). The demonstrator was then allowed to interact with the other rat, the 'observer' (who would be able to smell the coffee/ethanol on the demonstrator's breath). This observer was then removed and, in isolation, was given two cups containing ethanol and coffee. The intake of each (out of a maximum of 50 ml) was recorded. From Hunt et al. (2001). Copyright © 2001 John Wiley & Sons, Inc. All rights reserved. Reproduced with permission.

SUMMARY AND KEY POINTS OF THE CHAPTER

- Infants show an innate preference for face-like stimuli (e.g. for a top-heavy configuration), but this falls short of innate knowledge of the exact layout of a face. These stimuli do not selectively engage regions that, in adults, have been proposed to be face specific. This suggests that specialization emerges slowly over time.
- Between 12 and 18 months, infants show evidence of being able to understand the relationship between actions and intentions (e.g. imitating intentional actions more than accidental ones) and to understand the relationship between seeing and knowing (e.g. by directing an adult's attention to an object that the infant can see but the adult cannot). It is unclear whether this is evidence of a 'theory of mind' at this early age or whether these skills are precursors to the later development of this ability (as measured on *explicit* judgments of mental states).
- Evidence from EEG and fMRI suggests that the neural substrates of social cognition (e.g. on tasks of empathy, face processing, or mindreading) differ between children, adolescents, and adults. This can even occur when behavioral performance across these age groups does not differ: that is, seemingly mature levels of performance are not necessarily supported by a mature neural architecture.

- The preschool and early school years could be considered as a process of enculturation in which the child learns about social norms (e.g. fairness, reciprocity), engages in peer-based interactions, and develops an understanding of his/her own social identity.
- Structural changes in the brain during adolescence consist of late maturation of prefrontal cortex (important for the control of behavior) and changes in the reward-based systems of the brain. This combination may give rise to increased risk taking by this age group. Adolescents also show increased concern with reputation (how others perceive them) and reputation management (adherence to peer-endorsed fashions), which may relate to development changes in the functioning of certain prefrontal regions (e.g. medial prefrontal cortex).

EXAMPLE ESSAY QUESTIONS

- How is face recognition in infancy and childhood related to that found in adults?
- Does the development of early language and gestural communication depend on knowledge of the social world?
- What kind of basic abilities and social interactions act as developmental precursors to the emergence of mature forms of empathy and theory of mind?
- What physical changes occur in the brain during adolescence and how might this relate to changes in social functioning at this age?

RECOMMENDED FURTHER READING

- De Haan, M., & Gunnar, M. R. (2009). *Handbook of developmental social neuroscience*. New York: Guilford Press. Up-to-date but not all areas are covered in detail (e.g. theory of mind, social norms).

- Zelazo, P. D., Chandler, M., & Crone, E. (2010). *Developmental social cognitive neuroscience*. New York: Psychology Press. An excellent collection of papers.

References

Adam, E. K., Hawkley, L. C., Kudielka, B. M., & Cacioppo, J. T. (2006). Day-to-day dynamics of experience–cortisol associations in a population-based sample of older adults. *Proceedings of the National Academy of Sciences of the USA, 103*(45), 17058–17063.

Adams, J. M., & Jones, W. H. (1997). The conceptualisation of marriage commitment: An integrative analysis. *Journal of Social and Personal Relationships, 11,* 1177–1196.

Adams, R. B., Gordon, H. L., Baird, A. A., Ambady, N., & Kleck, R. E. (2003). Effects of gaze on amygdala sensitivity to anger and fear faces. *Science, 300*(5625), 1536–1536.

Adams, R. B., & Kleck, R. E. (2003). Perceived gaze direction and the processing of facial displays of emotion. *Psychological Science, 14*(6), 644–647.

Adolphs, R. (1999). Social cognition and the human brain. *Trends in Cognitive Sciences, 3,* 469–479.

Adolphs, R. (2002). Neural systems for recognizing emotion. *Current Opinion in Neurobiology, 12,* 169–177.

Adolphs, R., Damasio, H., Tranel, D., Cooper, G., & Damasio, A. R. (2000). A role for somatosensory cortices in the visual recognition of emotion as revealed by three-dimensional lesion mapping. *Journal of Neuroscience, 20*(7), 2683–2690.

Adolphs, R., Gosselin, F., Buchanan, T. W., Tranel, D., Schyns, P., & Damasio, A. R. (2005). A mechanism for impaired fear recognition after amygdala damage. *Nature, 433,* 68–72.

Adolphs, R., Tranel, D., & Buchanan, T. W. (2005). Amygdala damage impairs emotional memory for gist but not details of complex stimuli. *Nature Neuroscience, 8*(4), 512–518.

Adolphs, R., Tranel, D., & Damasio, A. R. (1998). The human amygdala in social judgment. *Nature, 393*(6684), 470–474.

Adolphs, R., Tranel, D., Damasio, H., & Damasio, A. (1994). Impaired recognition of emotion in facial expressions following bilateral damage to the human amygdala. *Nature, 372,* 669–672.

Aguirre, G. K., Zarahn, E., & D'Esposito, M. (1998). The variability of human BOLD hemodynamic response. *NeuroImage, 8,* 360–369.

Ainsworth, M. D. S., Blehar, M. C., Waters, E., & Wahl, S. (1978). *Patterns of attachment.* Hillsdale, NJ: Lawrence Erlbaum Associates, Inc.

Alexander, G. E., & Crutcher, M. D. (1990). Functional architecture of basal ganglia circuits: Neural substrates of parallel processing. *Trends in Neurosciences, 13,* 266–271.

Alley, T. R. (1988). Physiognomy and social perception. In T. R. Alley (Ed.), *Social and applied aspects of perceiving faces.* Hillsdale, NJ: Lawrence Erlbaum Associates, Inc.

Allison, T., Puce, A., & McCarthy, G. (2000). Social perception from visual cues: Role of the STS region. *Trends in Cognitive Sciences, 4,* 267–278.

Allport, G. W. (1954). *The nature of prejudice.* Reading, MA: Addison-Wesley.

Allport, G. W. (1968). The historical background of modern social psychology. In G. Lindzey & E. Aronson (Eds.), *Handbook of social psychology.* New York: McGraw Hill.

Amaral, D. G., Price, J. L., Pitkanen, A., & Carmichael, S. T. (1992). Anatomical organization of the primate amygdaloid complex. In J. P. Aggleton (Ed.), *The amygdala.* New York: Wiley-Liss.

American Psychiatric Association (1994). *Diagnostic and Statistical Manual of Mental Disorders* (4th edition, DSM–IV). Washington, DC: American Psychiatric Association.

Amodio, D. M. (2008). The social neuroscience of intergroup relations. *European Review of Social Psychology, 19,* 1–54.

Amodio, D. M., & Devine, P. G. (2006). Stereotyping and evaluation in implicit race bias: Evidence for independent constructs and unique effects on behavior. *Journal of Personality and Social Psychology, 91*(4), 652–661.

Amodio, D. M., & Frith, C. D. (2006). Meeting of minds: The medial frontal cortex and social cognition. *Nature Reviews Neuroscience, 7*(4), 268–277.

Amodio, D. M., Harmon-Jones, E., & Devine, P. G. (2003). Individual differences in the activation and control of affective race bias as assessed by startle eyeblink response and self-report. *Journal of Personality and Social Psychology, 84*(4), 738–753.

Amodio, D. M., Harmon-Jones, E., Devine, P. G., Curtin, J. J., Hartley, S. L., & Covert, A. E. (2004). Neural signals for the detection of unintentional race bias. *Psychological Science, 15*(2), 88–93.

Amodio, D. M., Kubota, J. T., Harmon-Jones, E., & Devine, P. G. (2006). Alternative mechanisms for regulating racial responses according to internal vs. external cues. *Social Cognitive and Affective Neuroscience, 1,* 26–36.

Anders, S., Eippert, F., Weiskopf, N., & Veit, R. (2008). The human amygdala is sensitive to the valence of pictures and sounds irrespective of arousal: An fMRI study. *Social Cognitive and Affective Neuroscience, 3*(3), 233–243.

Anderson, C. A., & Bushman, B. J. (2002). Human aggression. *Annual Review of Psychology, 53*, 27–51.

Anderson, J. R., Myowa-Yamakoshi, M., & Matsuzawa, T. (2004). Contagious yawning in chimpanzees. *Proceedings of the Royal Society of London, Series B. 271*, S468–S470.

Anderson, S. W., Bechara, A., Damasio, H., Tranel, D., & Damasio, A. R. (1999). Impairment of social and moral behavior related to early damage in human prefrontal cortex. *Nature Neuroscience, 2*, 1032–1037.

Apperly, I. A. (2008). Beyond simulation-theory and theory-theory: Why social cognitive neuroscience should use its own concepts to study 'theory of mind'. *Cognition, 107*(1), 266–283.

Apperly, I. A., Samson, D., Carroll, N., Hussain, S., & Humphreys, G. (2006). Intact first- and second-order false belief reasoning in a patient with severely impaired grammar. *Social Neuroscience, 1*(3/4), 334–348.

Aron, A., Aron, E. N., & Smollan, D. (1992). Inclusion of other in the self scale and the structure of interpersonal closeness. *Journal of Personality and Social Psychology, 63*(4), 596–612.

Arzy, S., Thut, G., Mohr, C., Michel, C. M., & Blanke, O. (2006). Neural basis of embodiment: Distinct contributions of temporoparietal junction and extrastriate body area. *Journal of Neuroscience, 26*, 8074–8081.

Asch, S. E. (1951). Effects of group pressure upon the modification and distortion of judgements. In H. Guetzkow (Ed.), *Groups, leadership and men*. Pittsburgh: Carnegie Press.

Asch, S. E. (1956). Studies of independence and conformity: 1. A minority of one against a unanimous majority. *Psychological Monographs, 70*, 1–70.

Asperger, H. (1944). 'Autistic psychopathy' in childhood. In U. Frith (Ed.), *Autism and Asperger syndrome*. Cambridge, UK: Cambridge University Press.

Attwell, D., & Iadecola, C. (2002). The neural basis of functional brain imaging signals. *Trends in Neurosciences, 25*, 621–625.

Auyeung, B., Baron-Cohen, S., Ashwin, E., Knickmeyer, R., Taylor, K., & Hackett, G. (2009). Fetal testosterone and autistic traits. *British Journal of Psychology, 100*, 1–22.

Axelrod, R., & Hamilton, W. D. (1981). The evolution of cooperation. *Science, 211*, 1390–1396.

Baddeley, A., & Della Sala, S. (1996). Working memory and executive control. *Philosophical Transactions of the Royal Society of London, Series B, 298*, 1397–1404.

Baird, G., Charman, T., Baron-Cohen, S., Cox, A., Swettenham, J., Wheelwright, S., & Drew, A. (2000). A screening instrument for autism at 18 months of age: A 6 year follow-up study. *Journal of the American Academy of Child and Adolescent Psychiatry, 39*, 694–702.

Baird, G., Simonoff, E., Pickles, A., Chandler, S., Loucas, T., Meldrum, D., & Charman, T. (2006). Prevalence of disorders of the autism spectrum in a population cohort of children in South Thames: The Special Needs and Autism Project (SNAP). *Lancet, 368*(9531), 210–215.

Bale, T. L., Davis, A. M., Auger, A. P., Dorsa, D. M., & McCarthy, M. M. (2001). CNS region-specific oxytocin receptor expression: Importance in regulation of anxiety and sex behavior. *Journal of Neuroscience, 21*(7), 2546–2552.

Ballantyne, A. O., Spilkin, A. M., Hesselink, J., & Trauner, D. A. (2008). Plasticity in the developing brain: Intellectual, language and academic functions in children with ischaemic perinatal stroke. *Brain, 131*, 2975–2985.

Bandura, A. (1965). Influence of models reinforcement contingencies on the acquisition of imitative responses. *Journal of Personality and Social Psychology, 1*(6), 589–595.

Bandura, A. (1973). *Aggression: A social learning analysis*. Englewood Cliffs, NJ: Prentice-Hall.

Bandura, A. (2002). Reflexive empathy: On predicting more than has ever been observed. *Behavioral and Brain Sciences, 25*(1), 24.

Bandura, A., Barbaranelli, C., Caprara, G. V., & Pastorelli, C. (1996). Mechanisms of moral disengagement in the exercise of moral agency. *Journal of Personality and Social Psychology, 71*, 364–374.

Bandura, A., Ross, D., & Ross, S. A. (1961). Transmission of aggression through imitation of aggressive models. *Journal of Abnormal and Social Psychology, 63*, 575–582.

Bandura, A., Ross, S. A., & Ross, D. (1963). Imitation of film-mediated aggressive models. *Journal of Abnormal Psychology, 66*(1), 3.

Banerjee, K., Huebner, B., & Hauser, M. D. (2011). Intuitive moral judgments are robust across demographic variation in gender, education, politics, and religion: A large-scale web-based study. *Journal of Cognition and Culture, 10*, 253–281.

Bar-Haim, Y., Ziv, T., Lamy, D., & Hodes, R. M. (2006). Nature and nurture in own-race face processing. *Psychological Science, 17*, 159–163.

Barenboim, C. (1981). The development of person perception in childhood and adolescence: From behavioral comparisons to psychological constructs to psychological comparisons. *Child Development, 52*(1), 129–144.

Bargh, J. A., Chen, M., & Burrows, L. (1996). Automaticity of social behavior: Direct effects of trait construct and stereotype activation on action. *Journal of Personality and Social Psychology, 71*(2), 230–244.

Barker, A. T., Jalinous, R., & Freeston, I. L. (1985). Non-invasive magnetic stimulation of human motor cortex. *Lancet, 1*, 1106–1107.

Baron, R. A., & Richardson, D. R. (1994). *Human agression*. New York: Plenum Press.

Baron-Cohen, S. (1995a). The Eye-Direction Detector (EDD) and the Shared Attention Mechanism (SAM): Two cases for evolutionary psychology. In C. Moore & P. Dunham (Eds.), *The role of joint attention in development.* Hillsdale, NJ: Lawrence Erlbaum Associates, Inc.

Baron-Cohen, S. (1995b). *Mind-blindness: An Essay on autism and theory of mind.* Cambridge, MA: MIT Press.

Baron-Cohen, S. (2002). The extreme male brain theory of autism. *Trends in Cognitive Sciences, 6*, 248–254.

Baron-Cohen, S. (2009). Autism: The Empathizing-Systemizing (E-S) Theory. In *Year in Cognitive Neuroscience 2009* (Vol. 1156, pp. 68–80). Oxford: Blackwell Publishing.

Baron-Cohen, S., Ashwin, E., Ashwin, C., Tavassoli, T. and Chakrabarti, B. (2009). Talent in autism: Hyper-systemizing, hyper-attention to detail and sensory hyper-sensitivity. *Philosophical Transactions of the Royal Society, Series B, 364* (1522), 1377–1383.

Baron-Cohen, S., Campbell, R., Karmiloff-Smith, A., Grant, J., & Walker, J. (1995). Are children with autism blind to the mentalistic significance of eyes? *British Journal of Developmental Psychology, 13*, 379–398.

Baron-Cohen, S., & Cross, P. (1992). Reading the eyes: Evidence for the role of perception in the development of theory of mind. *Mind and Language, 6*, 166–180.

Baron-Cohen, S., Leslie, A., & Frith, U. (1985). Does the autistic child have a 'theory of mind'? *Cognition, 21*, 37–46.

Baron-Cohen, S., Leslie, A. M., & Frith, U. (1986). Mechanical, behavioral and intentional understanding of picture stories in autistic children. *British Journal of Developmental Psychology, 4*, 113–125.

Baron-Cohen, S., Richler, J., Bisarya, D., Gurunathan, N., & Wheelwright, S. (2003). The systemizing quotient: An investigation of adults with Asperger syndrome or high-functioning autism, and normal sex differences. *Philosophical Transactions of the Royal Society of London, Series B, 358*, 361–374.

Baron-Cohen, S., & Wheelwright, S. (2004). The empathy quotient: An investigation of adults with Asperger syndrome or high functioning autism and normal sex differences. *Journal of Autism and Developmental Disorders, 34*, 163–175.

Baron-Cohen, S., Wheelwright, S., Hill, J., Raste, Y., & Plumb, I. (2001). The 'Reading the Mind in the Eyes' test revised version: A study with normal adults, and adults with Asperger syndrome or high-functioning autism. *Journal of Child Psychology and Psychiatry and Allied Disciplines, 42*(2), 241–251.

Baron-Cohen, S., Wheelwright, S., Stone, V., & Rutherford, M. (1999). A mathematician, a physicist and a computer scientist with Asperger syndrome: Performance on psychology and folk physics tests. *Neurocase, 5*, 475–483.

Barraclough, N. E., Xiao, D., Baker, C. I., Oram, M. W., & Perrett, D. I. (2005). Integration of visual and auditory information by superior temporal sulcus neurons responsive to the sight of actions. *Journal of Cognitive Neuroscience, 17*, 377–391.

Bartels, A., & Zeki, S. (2000). The neural basis of romantic love. *NeuroReport, 11*, 3829–3834.

Bateson, M., Nettle, D., & Roberts, G. (2006). Cues of being watched enhance cooperation in a real-world setting. *Biology Letters, 2*, 412–414.

Batson, C. D. (1991). *The altruism question: Toward a social-psychological answer.* Hillsdale, NJ: Lawrence Erlbaum Associates, Inc.

Batson, C. D. (2009). These things called empathy: Eight related but distinct phenomena. In J. Decety & W. J. Ickes (Eds.), *The social neuroscience of empathy.* Cambridge, MA: MIT Press.

Batson, C. D., Batson, J. G., Griffitt, C. A., Barrientos, S., Brandt, J. R., Sprengelmeyer, P., & Bayly, M. J. (1989). Negative-state relief and the empathy altruism hypothesis. *Journal of Personality and Social Psychology, 56*(6), 922–933.

Batson, C. D., Duncan, B. D., Ackerman, P., Buckley, T., & Birch, K. (1981). Is empathic emotion a source of altruistic motivation? *Journal of Personality and Social Psychology, 40*(2), 290–302.

Batson, C. D., Dyck, J. L., Brandt, J. R., Batson, J. G., Powell, A. L., McMaster, M. R., & Griffitt, C. (1988). 5 studies testing 2 new egoistic alternatives to the empathy altruism hypothesis. *Journal of Personality and Social Psychology, 55*(1), 52–77.

Batson, C. D., & Shaw, L. L. (1991). Encouraging words concerning the evidence for altruism. *Psychological Inquiry, 2*, 159–168.

Baumeister, R. F., Twenge, J. M., & Nuss, C. K. (2002). Effects of social exclusion on cognitive processes: Anticipated aloneness reduces intelligent thought. *Journal of Personality and Social Psychology, 83*(4), 817–827.

Baxter, M. G., & Murray, E. A. (2002). The amygdala and reward. *Nature Reviews Neuroscience, 3*(7), 563–573.

Beauregard, M., & Paquette, V. (2006). Neural correlates of a mystical experience in Carmelite nuns. *Neuroscience Letters, 405*, 186–190.

Bechara, A., Damasio, A. R., Damasio, H., & Anderson, S. W. (1994). Insensitivity to future consequences following damage to human prefrontal cortex. *Cognition, 50*, 7–15.

Bechara, A., Damasio, H., Damasio, A. R., & Lee, G. P. (1999). Different contributions to the human amygdala and ventromedial prefrontal cortex to decision making. *Journal of Neuroscience, 19*, 5437–5481.

Bechara, A., Damasio, H., Tranel, D., & Anderson, S. W. (1998). Dissociation of working memory from decision making within the human prefrontal cortex. *Journal of Neuroscience, 18*, 428–437.

Bechara, A., Tranel, D., Damasio, H., Adolphs, R., Rockland, C., & Damasio, A. R. (1995). Double dissociation of conditioning and declarative knowledge relative to the amygdala and hippocampus in humans. *Science, 269*(5227), 1115–1118.

Beckett, C., Castle, J., Rutter, M., & Sonuga-Barke, E. J. (2010). VI. Institutional deprivation, specific cognitive functions, and scholastic achievement: English and Romanian adoptee (ERA) study findings. *Monographs of the Society for Research in Child Development, 75*(1), 125–142.

Bedny, M., Pascual-Leone, A., & Saxe, R. R. (2009). Growing up blind does not change the neural bases of Theory of Mind. *Proceedings of the National Academy of Sciences of the USA, 106*(27), 11312–11317.

Beer, J. S., Stallen, M., Lombardo, M. V., Gonsalkorale, K., Cunningham, W. A., & Sherman, J. W. (2008). The Quadruple Process model approach to examining the neural underpinnings of prejudice. *NeuroImage, 43*(4), 775–783.

Belin, P., Zatorre, R. J., Lafaille, P., Ahad, P., & Pike, B. (2000). Voice-selective areas in human auditory cortex. *Nature, 403*, 309–312.

Belsky, J., & Ravine, M. (1987). Temperament and attachment security in the Strange Situation: An empirical rapprochement. *Child Development, 58*, 787–795.

Bentin, S., Allison, T., Puce, A., Perez, E., & McCarthy, G. (1996). Electrophysiological studies of face perception in humans. *Journal of Cognitive Neuroscience, 8*, 551–565.

Bentin, S., & Deouell, L. Y. (2000). Structural encoding and identification in face processing: ERP evidence for separate mechanisms. *Cognitive Neuropsychology, 17*, 35–54.

Berg, J., Dickhaut, J., & McCabe, K. (1995). Trust, reciprocity and social history. *Games and Economic Behavior, 10*, 122–142.

Berkowitz, L. (1989). Frustration aggression hypothesis: Examination and reformulation. *Psychological Bulletin, 106*(1), 59–73.

Berkowitz, L. (1990). On the formation and regulation of anger and aggression: A cognitive-neoassociationistic analysis. *American Psychologist, 45*(4), 494–503.

Berkowitz, L., & Harmon-Jones, E. (2004). Toward an understanding of the determinants of anger. *Emotion, 4*(2), 107–130.

Bernhardt, P. C., Dabbs, J. M., Fielden, J. A., & Lutter, C. D. (1998). Testosterone changes during vicarious experiences of winning and losing among fans at sporting events. *Physiology and Behavior, 65*, 59–62.

Berry, D. S., & Brownlow, S. (1989). Were the physiognomists right: Personality-correlates of facial babyishness. *Personality and Social Psychology Bulletin, 15*(2), 266–279.

Berry Mendes, W. (2009). Assessing autonomic nervous system activity. In E. Harmon-Jones & J. S. Beer (Eds.), *Methods in social neuroscience*. New York: Guilford Press.

Berthoz, S., Armony, J. L., Blair, R. J. R., & Dolan, R. J. (2002). An fMRI study of intentional and unintentional (embarrassing) violations of social norms. *Brain, 125*, 1696–1708.

Birbaumer, N., Viet, R., Lotze, M., Erb, M., Hermann, C., Grodd, W., & Flor, H. (2005). Deficient fear conditioning in psychopathy: A functional magnetic resonance imaging study. *Archives of General Psychiatry, 62*(7), 799–805.

Bird, C. M., Casteli, F., Malik, O., Frith, U., & Husain, M. (2004). The impact of extensive medial frontal lobe damage on 'theory of mind' and cognition. *Brain, 127*, 914–928.

Blackhart, G. C., Nelson, B. C., Knowles, M. L., & Baumeister, R. F. (2009). Rejection elicits emotional reactions but neither causes immediate distress nor lowers self-esteem: A meta-analytic review of 192 studies on social exclusion. *Personality and Social Psychology Review, 13*, 269–309.

Blackmore, S. (1999). *The meme machine*. Oxford: Oxford University Press.

Blair, K., Marsh, A. A., Morton, J., Vythilingam, M., Jones, M., Mondillo, K., Pine, D. C., Drevets, W. C., & Blair, J. R. (2006). Choosing the lesser of two evils, the better of two goods: Specifying the roles of ventromedial prefrontal cortex and dorsal anterior cingulate in object choice. *Journal of Neuroscience, 26*(44), 11379–11386.

Blair, R. J. R. (1995). A cognitive developmental approach to morality: Investigating the psychopath. *Cognition, 57*, 1–29.

Blair, R. J. R. (1996). Morality in the autistic child. *Journal of Autism and Developmental Disorders, 26*, 571–579.

Blair, R. J. R. (2003). Neurobiological basis of psychopathy. *British Journal of Psychiatry, 182*, 5–7.

Blair, R. J. R., & Cipolotti, L. (2000). Impaired social response reversal: A case of acquired 'sociopathy'. *Brain, 123*, 1122–1141.

Blair, R. J. R., Jones, L., Clark, F., & Smith, M. (1997). The psychopathic individual: A lack of responsiveness to distress cues? *Psychophysiology, 34*, 192–198.

Blair, R. J. R., Sellars, C., Strickland, I., Clark, F., Williams, A. O., Smith, M., & Jones, L. (1995). Emotion attributions in the psychopath. *Personality and Individual Differences, 19*(4), 431–437.

Blakemore, S. J., den Ouden, H., Choudhury, S., & Frith, C. (2007). Adolescent development of the neural circuitry for thinking about intentions. *Social Cognitive and Affective Neuroscience, 2*(2), 130–139.

Blanke, O., Landis, T., Spinelli, L., & Seeck, M. (2004). Out-of-body experience and autoscopy of neurological origin. *Brain, 127*, 243–258.

Blanke, O., Mohr, C., Michel, C. M., Pascual-Leone, A., Brugger, P., Seeck, M., Landis, T., & Thut, G. (2005).

Linking out-of-body experience and self processing to mental own-body imagery at the temporoparietal junction. *Journal of Neuroscience, 25,* 550–557.

Bloch, M. (2008). Why religion is nothing special but is central. *Philosophical Transactions of the Royal Society of London, Series B, 363,* 2055–2061.

Blood, A. J., & Zatorre, R. J. (2001). Intensely pleasurable responses to music correlate with activity in brain regions implicated in reward and emotion. *Proceedings of the National Academy of Science, USA, 98,* 11818–11823.

Bodamer, J. (1947). Die prosopagnosie. *Archiv fuer Psychiatrie und Zeitschrift fuer Neurologie, 179,* 6–54.

Bolger, D. J., Perfetti, C. A., & Schneider, W. (2005). Cross-cultural effect on the brain revisited: Universal structures plus writing system variation. *Human Brain Mapping, 25,* 92–104.

Bolhuis, J. J. (1990). Mechanisms of avian imprinting: A review. *Biological Reviews, 66,* 303–345.

Borg, J. S., Lieberman, D., & Kiehl, K. A. (2008). Infection, incest, and iniquity: Investigating the neural correlates of disgust and morality. *Journal of Cognitive Neuroscience, 20*(9), 1529–1546.

Boria, S., Fabbri-Destro, M., Cattaneo, L., Sparaci, L., Sinigaglia, C., Santelli, E., Cossu, G., & Rizzolatti, G. (2009). Intention understanding in autism. *PloS One, 4*(5).

Bos, P. A., Terburg, D., & van Honk, J. (2010). Testosterone decreases trust in socially naive humans. *Proceedings of the National Academy of Sciences of the USA, 107,* 9991–9995

Bottini, G., Corcoran, R., Sterzi, R., Paulesu, E., Schenone, P., Scarpa, P., Frackowiak, R. S. J., & Frith, C. D. (1994). The role of the right hemisphere in the interpretation of figurative aspects of language: A positron emission tomography activation study. *Brain, 117,* 1241–1253.

Botvinick, M. M., Braver, T. S., Barch, D. M., Carter, C. S., & Cohen, J. D. (2001). Conflict monitoring and cognitive control. *Psychological Review, 108*(3), 624–652.

Bourgeois, P., & Hess, U. (2008). The impact of social context on mimicry. *Biological Psychology, 77*(3), 343–352.

Bowlby, J. (1969). *Attachment and loss: Volume 1. Attachment.* London: Hogarth Press.

Boyer, P. (2008). Religion: Bound to believe? *Nature, 455,* 1038–1039.

Boyer, P., Robbins, P., & Jack, A. I. (2005). Varieties of self-systems worth having. *Consciousness and Cognition, 14*(4), 647–660.

Branchi, I., D'Andrea, I., Fiore, M., Di Fausto, V., Aloe, L., & Alleva, E. (2006). Early social enrichment shapes social behavior and nerve growth factor and brain-derived neurotrophic factor levels in the adult mouse brain. *Biological Psychiatry, 60*(7), 690–696.

Brooks, R., & Meltzoff, A. N. (2005). The development of gaze following and its relation to language. *Developmental Science, 8,* 535–543.

Brown, R. (1995). *Prejudice.* Oxford: Blackwell.

Bruce, V., & Young, A. W. (1986). Understanding face recognition. *British Journal of Psychology, 77,* 305–327.

Buccino, G., Lui, F., Canessa, N., Patteri, I., Lagravinese, G., Benuzzi, F., Porro, C. A., & Rizzolatti, G. (2004). Neural circuits involved in the recognition of actions performed by nonconspecifics: An fMRI study. *Journal of Cognitive Neuroscience, 16,* 114–126.

Bufalari, I., Aprile, T., Avenanti, A., Di Russo, F., & Aglioti, S. M. (2007). Empathy for pain and touch in the human somatosensory cortex. *Cerebral Cortex, 17,* 2553–2561.

Bulbulia, J. (2004). The cognitive and evolutionary psychology of religion. *Biology and Philosophy, 19,* 655–686.

Burnham, D., Kitamura, C., & Vollmer-Conna, U. (2002). What's new, pussycat? On talking to babies and animals. *Science, 296*(5572), 1435–1435.

Bush, G., Luu, P., & Posner, M. I. (2000). Cognitive and emotional influences in anterior cingulate cortex. *Trends in Cognitive Sciences, 4,* 215–222.

Bush, G., Vogt, B. A., Holmes, J., Dale, A. M., Greve, D., Jenike, M. A., & Rosen, B. R. (2002). Dorsal anterior cingulate cortex: A role in reward-based decision making. *Proceedings of the National Academy of Sciences of the USA, 99*(1), 523–528.

Bushnell, I. W. R., Sai, F., & Mullin, J. T. (1989). Neonatal recognition of the mother's face. *British Journal of Developmental Psychology, 7,* 3–15.

Buss, D. M. (1989). Sex-differences in human mate preferences: Evolutionary hypothesis tested in 37 cultures. *Behavioral and Brain Sciences, 12*(1), 1–14.

Buttelmann, D., Carpenter, M., Call, J., & Tomasello, M. (2007). Encultured chimpanzees imitate rationally. *Developmental Science, 10,* F31–F38.

Byrne, R. W., & Corp, N. (2004). Neocortex size predicts deception rates in primates. *Proceedings of the Royal Society of London, Series B, 271,* 1693–1699.

Cacioppo, J. T., & Berntson, G. G. (1992). Social psychological contributions to the decade of the brain: Doctrine of multi-level analysis. *American Psychologist, 47,* 1019–1028.

Cacioppo, J. T., & Hawkley, L. C. (2009). Perceived social isolation and cognition. *Trends in Cognitive Sciences, 13,* 447–454.

Cacioppo, J. T., Norris, C. J., Decety, J., Monteleone, G., & Nusbaum, H. (2009). In the eye of the beholder: Individual differences in perceived social isolation predict regional brain activation to social stimuli. *Journal of Cognitive Neuroscience, 21,* 83–92.

Cahill, L., Prins, B., Weber, M., & McGaugh, J. L. (1994). Beta-adrenergic activation and memory for emotional events. *Nature, 371,* 702–704.

Cahill, L., Weinberger, N. M., Roozendaal, B., & McGaugh, J. L. (1999). Is the amygdala a locus of 'conditioned fear'? Some questions and caveats. *Neuron, 23,* 227–228.

Calder, A. J., Keane, J., Lawrence, A. D., & Manes, F. (2004). Impaired recognition of anger following damage to the ventral striatum. *Brain, 127,* 1958–1969.

Calder, A. J., Keane, J., Manes, F., Antoun, N., & Young, A. W. (2000). Impaired recognition and experience of disgust following brain injury. *Nature Neuroscience, 3*(11), 1077–1078.

Calder, A. J., & Young, A. W. (2005). Understanding the recognition of facial identity and facial expression. *Nature Reviews Neuroscience, 6*(8), 641–651.

Calder, A. J., Young, A. W., Rowland, D., Perrett, D. I., Hodges, J. R., & Etcoff, N. L. (1996). Facial emotion recognition after bilateral amygdala damage: Differentially severe impairment of fear. *Cognitive Neuropsychology, 13,* 699–745.

Call, J., & Tomasello, M. (2008). Does the chimpanzee have a theory of mind? 30 years later. *Trends in Cognitive Sciences, 12,* 187–192.

Callaghan, T., Rochat, P., Lillard, A., Claux, M. L., Odden, H., Itakura, S., Tapanya, S., & Singh, S. (2005). Synchrony in the onset of mental state reasoning: Evidence from five cultures. *Psychological Science, 16,* 378–384.

Calvert, G. A., Hansen, P. C., Iversen, S. D., & Brammer, M. J. (2001). Detection of audio-visual integration sites in humans by application of electrophysiological criteria to the BOLD effect. *NeuroImage, 14,* 427–438.

Campbell, R., Heywood, C., Cowey, A., Regard, M., & Landis, T. (1990). Sensitivity to eye gaze in prosopagnosic patients and monkeys with superior temporal sulcus ablation. *Neuropsychologia, 28,* 1123–1142.

Campbell, R., Landis, T., & Regard, M. (1986). Face recognition and lip reading: A neurological dissociation. *Brain, 109,* 509–521.

Canli, T., Zhao, Z., Desmond, J. E., Kang, E. J., Gross, J., & Gabrieli, J. D. E. (2001). An fMRI study of personality influences on brain reactivity to emotional stimuli. *Behavioral Neuroscience, 115*(1), 33–42.

Cannon, W. B. (1927). The James–Lange theory of emotions: A critical examination and an alternative theory. *American Journal of Psychology, 39,* 106–124.

Capgras, J., & Reboul-Lachaux, J. (1923). L'illusion des sosies dans un delire sytematise chronique. *Bulletin de la Societe Clinique de Medecine Mentale, 2,* 6–16.

Cardinal, R. N., Parkinson, J. A., Hall, J., & Everitt, B. J. (2002). Emotion and motivation: The role of the amygdala, ventral striatum, and prefrontal cortex. *Neuroscience and Biobehavioral Reviews, 26*(3), 321–352.

Carpenter, M., Akhtar, N., & Tomasello, M. (1998). Fourteen-through 18-month-old infants differentially imitate intentional and accidental actions. *Infant Behavior and Development, 21,* 315–330.

Carpenter, M., Nagell, K., & Tomasello, M. (1998). Social cognition, joint attention, and communicative competence from 9 to 15 months of age. *Monographs of the Society for Research in Child Development, 63*(4).

Carr, L., Iacoboni, M., Dubeau, M. C., Mazziotta, J. C., & Lenzi, G. L. (2003). Neural mechanisms of empathy in humans: A relay from neural systems for imitation to limbic areas. *Proceedings of the National Academy of Sciences of the USA, 100*(9), 5497–5502.

Carre, J. M., & McCormick, C. M. (2008). In your face: Facial metrics predict aggressive behaviour in the laboratory and in varsity and professional hockey players. *Proceedings of the Royal Society of London, Series B, 275*(1651), 2651–2656.

Carter, C. S. (1998). Neuroendocrine perspectives on social attachment and love. *Psychoneuroendocrinology, 23,* 779–818.

Carter, C. S., DeVries, A. C., & Getz, L. L. (1995). Physiological substrates of mammalian monogamy: The Prairie vole model. *Neuroscience and Biobehavioral Review, 19,* 303–314.

Carter, C. S., MacDonald, A. M., Botvinick, M., Ross, L. L., Stenger, V. A., Noll, D., & Cohen, J. D. (2000). Parsing executive processes: Strategic vs. evaluative functions of the anterior cingulate cortex. *Proceedings of the National Academy of Sciences of the USA, 97,* 1944–1948.

Casey, B. J., Getz, S., & Galvan, A. (2008). The adolescent brain. *Developmental Review, 28*(1), 62–77.

Caspi, A., McClay, J., Moffitt, T. E., Mill, J., Martin, J., Craig, I. W., Taylor, A., & Poulton, R. (2002). Role of genotype in the cycle of violence in maltreated children. *Science, 297*(5582), 851–854.

Castelli, F., Frith, C., Happe, F., & Frith, U. (2002). Autism, Asperger syndrome and brain mechanisms for the attribution of mental states to animated shapes. *Brain, 125,* 1839–1849.

Castelli, F., Happe, F., Frith, U., & Frith, C. D. (2000). Movement and mind: A functional imaging study of perception and interpretation of complex intentional movements. *NeuroImage, 12,* 314–325.

Cavada, C., Company, T., Tejedor, J., Cruz-Rizzolo, R. J., & Reinoso-Suarez, F. (2000). The anatomical connections of the macaque monkey orbitofrontal cortex. A review. *Cerebral Cortex, 10*(3), 220–242.

Chance, M. (1967). The interpretation of some agonistic postures: The role of 'cut-off' acts and postures. *Symposium of the Zoological Society of London, 8,* 71–89.

Chandler, J., Griffin, T. M., & Sorensen, N. (2008). In the 'I' of the storm: Shared initials increase disaster donations. *Judgment and Decision Making Journal, 3*(5), 404–410.

Chapman, H. A., Kim, D. A., Susskind, J. M., & Anderson, A. K. (2009). In bad taste: Evidence for the oral origins of moral disgust. *Science, 323*(5918), 1222–1226.

Chartrand, T. L., & Bargh, J. A. (1999). The Chameleon effect: The perception–behavior link and social interaction. *Journal of Personality and Social Psychology,* *76*(6), 893–910.

Cheesman, J., & Merikle, P. M. (1984). Priming with and without awareness. *Perception and Psychophysics,* *36*(4), 387–395.

Cheng, Y., Chou, K. H., Decety, J., Chen, I. Y., Hung, D., Tzeng, O. J. L., & Lin, C. P. (2009). Sex differences in the neuroanatomy of human mirror-neuron system: A voxel-based morphometric investigation. *Neuroscience,* *158*(2), 713–720.

Cheng, Y. W., Lee, P. L., Yang, C. Y., Lin, C. P., Hung, D., & Decety, J. (2008). Gender differences in the mu rhythm of the human mirror-neuron system. *PloS One,* *3*(5).

Cheng, Y. W., Lin, C. P., Liu, H. L., Hsu, Y. Y., Lims, K. E., Hung, D., & Decety, J. (2007). Expertise modulates the perception of pain in others. *Current Biology,* *17*(19), 1708–1713.

Chisholm, K. (1998). A three year follow-up of attachment and indiscriminate friendliness in children adopted from Romanian orphanages. *Child Development,* *69*(4), 1092–1106.

Chittka, L., & Niven, J. (2009). Are bigger brains better? *Current Biology,* *19*(21), R995–R1008.

Cho, M. M., DeVries, A. C., Williams, J. R., & Carter, C. S. (1999). The effects of oxytocin and vasopressin on partner preferences in male and female prairie voles (Microtus ochrogaster). *Behavioral Neuroscience,* *113*(5), 1071–1079.

Chomsky, N. (1980). Rules and representations. *Behavioral and Brain Sciences,* *3*, 1–61.

Chugani, H. T., Behen, M. E., Muzik, O., Juhasz, C., Nagy, F., & Chugani, D. C. (2001). Local brain functional activity following early deprivation: A study of postinstitutionalized Romanian orphans. *NeuroImage,* *14*(6), 1290–1301.

Churchland, P. S., & Sejnowski, T. J. (1988). Perspectives on cognitive neuroscience. *Science,* *242*, 741–745.

Cialdini, R. B., Brown, S. L., Lewis, B. P., Luce, C., & Neuberg, S. L. (1997). Reinterpreting the empathy–altruism relationship: When one into one equals oneness. *Journal of Personality and Social Psychology,* *73*(3), 481–494.

Ciaramelli, E., Muccioli, M., Ladavas, E., & di Pellegrino, G. (2007). Selective deficit in personal moral judgment following damage to ventromedial prefrontal cortex. *Social Cognitive and Affective Neuroscience,* *2*(2), 84–92.

Cima, M., Tonnaer, F., & Hauser, M. D. (2010). Psychopaths know right from wrong but don't care. *Social Cognitive and Affective Neuroscience,* *5*(1), 59–67.

Civai, C., Corradi-Dell'Acqua, C., Gamer, M., & Rumiati, R. I. (2010). Are irrational reactions to unfairness truly emotionally-driven? Dissociated behavioural and emotional responses in the Ultimatum Game task. *Cognition,* *114*, 89–95.

Clark, A. (2003). *Natural-born cyborgs: Minds, technologies, and the future of human intelligence.* Oxford: Oxford University Press.

Clark, A. (2008). *Supersizing the mind: Embodiment, action and cognitive extension.* Oxford: Oxford University Press.

Clements, W. A., & Perner, J. (1994). Implicit understanding of belief. *Cognitive Development,* *9*, 377–395.

Clore, G. L., & Ortony, A. (2000). Cognition in emotion: Always, sometimes or never? In R. D. Lane & L. Nadel (Eds.), *Cognitive neuroscience of emotion.* Oxford: Oxford University Press.

Cohen, L., Lehericy, S., Chochon, F., Lemer, C., Rivaud, S., & Dehaene, S. (2002). Language-specific tuning of visual cortex functional properties of the Visual Word Form Area. *Brain,* *125*, 1054–1069.

Cole, S. W., Hawkley, L. C., Arevalo, J. M., Sung, C. Y., Rose, R. M., & Cacioppo, J. T. (2007). Social regulation of gene expression in human leukocytes [Electronic version]. *Genome Biology,* *8*(9).

Collins, D., Neelin, P., Peters, T., & Evans, A. (1994). Automatic 3D intersubject registration of MR volumetric data in standardized Talaraich space. *Journal of Computer Assisted Tomography,* *18*, 192–205.

Connor, R. C. (2007). Dolphin social intelligence: Complex alliance relationships in bottlenose dolphins and a consideration of selective environments for extreme brain size evolution in mammals. *Philosophical Transactions of the Royal Society of London, Series B, 362,* 587–602.

Connor, R. C., Wells, R., Mann, J., & Read, A. (2000). The bottlenose dolphin: Social relationships in a fission-fusion society. In J. Mann, R. Connor, P. Tyack, & H. Whitehead (Eds.), *Cetacean societies: Field studies of whales and dolphins.* Chicago: Chicago University Press.

Conway, M. A. (2005). Memory and the self. *Journal of Memory and Language,* *53*(4), 594–628.

Conway, M. A., & Pleydell-Pearce, C. W. (2000). The construction of autobiographical memories in the self-memory system. *Psychological Review,* *107*, 261–288.

Coricelli, G., Critchley, H. D., Joffily, M., O'Doherty, J. P., Sirigu, A., & Dolan, R. J. (2005). Regret and its avoidance: A neuroimaging study of choice behavior. *Nature Neuroscience,* *8*(9), 1255–1262.

Corkin, S. (2002). What's new with the amnesic patient HM? *Nature Reviews Neuroscience,* *3*, 153–160.

Cosmides, L. (1989). The logic of social exchange: Has natural selection shaped how humans reason? Studies with the Wason selection task. *Cognition,* *31*, 187–276.

Costa, P. T., & McCrae, R. R. (1985). *The NEO personality inventory manual.* Odessa, FL: Psychological Assessment Resources.

Costa, P. T., Terracciano, A., & McCrae, R. R. (2001, February). *Gender differences in personality traits across cultures: Robust and surprising findings.*

Paper presented at the 2nd Annual Meeting of the Society for Personality and Social Psychology, San Antonio, TX.

Couppis, M. H., & Kennedy, C. H. (2008). The rewarding effect of aggression is reduced by nucleus accumbens dopamine receptor antagonism in mice. *Psychopharmacology, 197*(3), 449–456.

Craig, A. D. (2009). How do you feel – now? The anterior insula and human awareness. *Nature Reviews Neuroscience, 10*(1), 59–70.

Creswell, C. S., & Skuse, D. H. (1999). Autism in association with Turner syndrome: Genetic implications for male vulnerability to pervasive developmental disorders. *Neurocase, 5*, 511–518.

Critchley, H. D., Elliott, R., Mathias, C. J., & Dolan, R. J. (2000). Neural activity relating to generation and representation of galvanic skin conductance responses: A functional magnetic resonance imaging study. *Journal of Neuroscience, 20*, 3033–3040.

Critchley, H. D., Mathias, C. J., Josephs, O., O'Doherty, J., Zanini, S., Dewar, B. K., Cipolotti, L., Shallice, T., & Dolan, R. J. (2003). Human cingulate cortex and autonomic control: Converging neuroimaging and clinical evidence. *Brain, 126*, 2139–2152.

Critchley, H. D., Wiens, S., Rotshtein, P., Ohman, A., & Dolan, R. J. (2004). Neural systems supporting interoceptive awareness. *Nature Neuroscience, 7*(2), 189–195.

Crone, E. A., & Westenberg, P. M. (2009). A brain-based account of developmental changes in social decision making. In M. D. Haan & M. R. Gunnar (Eds.), *Handbook of developmental social neuroscience.* New York: Guilford Press.

Cronin, H. (1991). *The ant and the peacock.* Cambridge, UK: Cambridge University Press.

Cunningham, W. A., Raye, C. L., & Johnson, M. K. (2004). Implicit and explicit evaluation: fMRI correlates of valence, emotional intensity, and control in the processing of attitudes. *Journal of Cognitive Neuroscience, 16*(10), 1717–1729.

Cushman, F., Young, L., & Hauser, M. (2006). The role of conscious reasoning and intuition in moral judgment: Testing three principles of harm. *Psychological Science, 17*(12), 1082–1089.

Custance, D. M., Whiten, A., & Bard, K. A. (1995). Can young chimpanzees (Pan troglodytes) imitate arbitrary actions? Hayes and Hayes (1952) revisited. *Behaviour, 132*, 837–859.

Dabbs, J. M., Carr, T. S., Frady, R. L., & Riad, J. K. (1995). Testosterone, crime, and misbehavior among 692 male prison-inmates. *Personality and Individual Differences, 18*(5), 627–633.

Dabbs, J. M., & Morris, R. (1990). Testosterone, social-class, and antisocial-behavior in a sample of 4,462 men. *Psychological Science, 1*(3), 209–211.

Damasio, A. R. (1994). *Descartes' error: Emotion, reason and the human brain.* New York: G. P. Putnam & Sons.

Damasio, A. R. (1999). *The feeling of what happens: Body and emotion in the making of consciousness.* New York: Harcourt.

Damasio, A. R. (2003). Feelings of emotion and the self. In J. Le Doux, J. Debiec, & H. Moss (Eds.), *Self: From Soul to Brain* (Vol. 1001, pp. 253-261). New York: New York Academy of Sciences.

Damasio, A. R., Grabowski, T. J., Bechara, A., Damasio, H., Ponto, L. L. B., Parvizi, J., & Hichwa, R. D. (2000). Subcortical and cortical brain activity during the feeling of self-generated emotions. *Nature Neuroscience, 3*(10), 1049–1056.

Damasio, A. R., Tranel, D., & Damasio, H. (1990). Individuals with sociopathic behavior caused by frontal damage fail to respond autonomically to social stimuli. *Behavioral Brain Research, 41*, 81–94.

Damasio, H., Grabowski, T., Frank, R., Galaburda, A. M., & Damasio, A. R. (1994). The return of Phineas Gage: Clues about the brain from the skull of a famous patient. *Science, 264*, 1102–1105.

Dapretto, M., Davies, M. S., Pfeifer, J. H., Scott, A. A., Sigman, M., Bookheimer, S. Y., & Iacoboni, M. (2006). Understanding emotions in others: Mirror neuron dysfunction in children with autism spectrum disorders. *Nature Neuroscience, 9*, 28–30.

Darwin, C. (1871). *The descent of man and selection in relation to sex.* London: John Murray.

Darwin, C. (1965). *The expression of the emotions in man and animals.* Chicago: University of Chicago Press. (Original work published 1872)

Dasgupta, N., & Greenwald, A. G. (2001). On the malleability of automatic attitudes: Combating automatic prejudice with images of admired and disliked individuals. *Journal of Personality and Social Psychology, 81*(5), 800–814.

Davidoff, J., Fonteneau, E., & Fagot, J. (2008). Local and global processing: Observations from a remote culture. *Cognition, 108*(3), 702–709.

Davis, M. (1980). A multi-dimensional approach to individual differences in empathy. *JCAS Catalog of Selected Documents in Psychology, 75*, 989–1015.

Davis, M. (1992). The role of the amygdala in fear-potentiated startle: Implications for animal-models of anxiety. *Trends in Pharmacological Sciences, 13*(1), 35–41.

Dawkins, R. (1976). *The selfish gene.* Oxford: Oxford University Press.

de Gelder, B. (2006). Towards the neurobiology of emotional body language. *Nature Reviews Neuroscience, 7*(3), 242–249.

de Quervain, D. J. F., Fischbacher, U., Treyer, V., Schelthammer, M., Schnyder, U., Buck, A., & Fehr, E. (2004). The neural basis of altruistic punishment. *Science, 305*(5688), 1254–1258.

De Renzi, E. (1986). Prosopagnosia in two patients with CT scan evidence of damage confined to the right hemisphere. *Neuropsychologia, 24*, 385–389.

de Veer, M. W., & Van den Bos, R. (1999). A critical review of methodology and interpretation of mirror self-recognition research in nonhuman primates. *Animal Behaviour, 58*, 459–468.

de Vignemont, F., & Frith, U. (2008). *Autism, morality and empathy.* Cambridge, MA: MIT Press.

de Vignemont, F., & Singer, T. (2006). The empathic brain: How, when and why? *Trends in Cognitive Sciences, 10*(10), 435–441.

de Waal, F. B. M. (2008). Putting the altruism back into altruism: The evolution of empathy. *Annual Review of Psychology, 59*, 279–300.

De Waal, F. B. M., & Luttrell, L. M. (1988). Mechanisms of social reciprocity in 3 primate species: Symmetrical relationship characteristics or cognition. *Ethology and Sociobiology, 9*(2–4), 101–118.

DeBruine, L. M. (2005). Trustworthy but not lust-worthy: Context-specific effects of facial resemblance. *Proceedings of the Royal Society of London, Series B, 272*(1566), 919–922.

DeCasper, A. J., & Fifer, W. P. (1980). Of human bonding: Newborns prefer their mothers' voices. *Science, 208*(4448), 1174–1176.

DeCasper, A. J., & Spence, M. J. (1986). Prenatal maternal speech influences newborns' perception of speech sounds. *Infant Behavior and Development, 9*, 133–150.

Decety, J., & Chaminade, T. (2003). When the self represents the other: A new cognitive neuroscience view on psychological identification. *Consciousness and Cognition, 12*, 577–596.

Decety, J., & Jackson, P. J. (2004). The functional architecture of human empathy. *Behavioral and Cognitive Neuroscience Reviews, 3*, 71–100.

Decety, J., & Jackson, P. L. (2006). A social-neuroscience perspective on empathy. *Current Directions in Psychological Science, 15*(2), 54–58.

Decety, J., & Lamm, C. (2007). The role of the right temporoparietal junction in social interaction: How low-level computational processes contribute to meta-cognition. *Neuroscientist, 13*(6), 580–593.

Decety, J., Michalska, K. J., Akitsuki, Y., & Lahey, B. B. (2009). Atypical empathic responses in adolescents with aggressive conduct disorder: A functional MRI investigation. *Biological Psychology, 80*(2), 203–211.

Dehaene, S., & Cohen, L. (2007). Cultural recycling of cortical maps. *Neuron, 56*, 384–398.

Dehaene, S., Dehaene-Lambertz, G., & Cohen, L. (1998). Abstract representations of numbers in the animal and human brain. *Trends in Neurosciences, 21*, 355–361.

Dehaene, S., Posner, M. I., & Tucker, D. M. (1994). Localisation of a neural system for error detection and compensation. *Psychological Science, 5*, 303–305.

Delgado, M. R., Frank, R. H., & Phelps, E. A. (2005). Perceptions of moral character modulate the neural systems of reward during the trust game. *Nature Neuroscience, 8*, 1611–1618.

Dennett, D. C. (1978). Beliefs about beliefs. *Behavioral and Brain Sciences, 1*, 568–570.

Dennett, D. C. (1983). Intentional systems in cognitive ethology: The panglossian paradigm defended. *Behavioral and Brain Sciences, 6*(3), 343–355.

Dennett, D. C. (2003). The self as a responding – and responsible – artifact. In J. Le Doux, J. Debiec, & H. Moss (Eds.), *Self: From soul to brain* (Vol. 1001, pp. 39–50). New York: New York Academy of Sciences.

Devine, P. G., & Elliot, A. J. (1995). Are racial stereotypes really fading: The Princeton trilogy revisited. *Personality and Social Psychology Bulletin, 21*(11), 1139–1150.

Devine, P. G., Plant, E. A., Amodio, D. M., Harmon-Jones, E., & Vance, S. L. (2002). The regulation of explicit and implicit race bias: The role of motivations to respond without prejudice. *Journal of Personality and Social Psychology, 82*(5), 835–848.

Devos, T., & Banaji, M. R. (2005). American = white? *Journal of Personality and Social Psychology, 88*(3), 447–466.

DeVries, A. C., DeVries, M. B., Taymans, S. E., & Carter, C. S. (1996). The effects of stress on social preferences are sexually dimorphic in prairie voles. *Proceedings of the National Academy of Sciences of the USA, 93*(21), 11980–11984.

DeWall, C. N., Baumeister, R. F., Stillman, T. F., & Gailliot, M. T. (2007). Violence restrained: Effects of self-regulation and its depletion on aggression. *Journal of Experimental Social Psychology, 43*(1), 62–76.

di Pellegrino, G., Fadiga, L., Fogassi, L., Gallese, V., & Rizzolatti, G. (1992). Understanding motor events: A neurophysiological study. *Experimental Brain Research, 91*, 176–180.

Diamond, J. (1997). *Guns, germs and steel.* London: Cape.

Diener, E. (1980). *Deindividuation: The absence of self-awareness and self-regulation in group members.* Hillsdale, NJ: Lawrence Erlbaum Associates, Inc.

Dijksterhuis, A. (2005). Why we are social animals: The high road to imitation as social glue. In S. Hurley & N. Chater (Eds.), *Perspectives on imitation: From neuroscience to social science* (Vol. 1). Cambridge, MA: MIT Press.

Dimberg, U., & Petterson, M. (2000). Facial reactions to happy and angry facial expressions: Evidence for right hemisphere dominance. *Psychophysiology, 37*, 693–696.

Dimberg, U., Thunberg, M., & Elmehed, K. (2000). Unconscious facial reactions to emotional facial expressions. *Psychological Science, 11*(1), 86–89.

Dimitrov, M., Phipps, M., Zahn, T. P., & Grafman, J. (1999). A thoroughly modern Gage. *Neurocase, 5*, 345–354.

Dinstein, I., Thomas, C., Behrmann, M., & Heeger, D. J. (2008). A mirror up to nature. *Current Biology, 18*(1), R13–R18.

Dion, K., Berscheid, E., & Walster, E. (1972). What is beautiful is good. *Journal of Personality and Social Psychology, 24*(3), 285–290.

Dittes, J. E., & Kelley, H. H. (1956). Effects of different conditions of acceptance upon conformity to group norms. *Journal of Abnormal and Social Psychology, 53*, 100–107.

Dolan, R. J. (2007). The human amygdala and orbitofrontal cortex in behavioural regulation. *Philosophical Transactions of the Royal Society of London, Series B, 362*, 787–799.

Dolcos, F., LaBar, K. S., & Cabeza, R. (2004). Interaction between the amygdala and the medial temporal lobe memory system predicts better memory for emotional events. *Neuron, 42*(5), 855–863.

Dollard, J., Doob, L., Miller, N., Mowrer, O., & Sears, R. (1939). *Frustration and aggression*, New Haven, CT: Yale University Press.

Dondi, M., Simion, F., & Caltran, G. (1999). Can newborns discriminate between their own cry and the cry of another newborn infant? *Developmental Psychology, 35*, 418–426.

Doris, J. M. (1998). Persons, situations, and virtue ethics: Moral psychology. *Nous, 32*(4), 504–530.

Dougherty, D. D., Shin, L. M., Alpert, N. M., Pitman, R. K., Orr, S. P., Lasko, M., Macklin, M. L., Fischman, A. J., & Rauch, S. L. (1999). Anger in healthy men: A PET study using script-driven imagery. *Biological Psychiatry, 46*(4), 466–472.

Downing, P. E., Chan, A. W., Peelen, M. V., Dodds, C. M., & Kanwisher, N. (2006). Domain specificity in visual cortex. *Cerebral Cortex, 16*, 1453–1461.

Downing, P. E., Jiang, Y. H., Shuman, M., & Kanwisher, N. (2001). A cortical area selective for visual processing of the human body. *Science, 293*, 2470–2473.

Duchaine, B. C., & Nakayama, K. (2006). Developmental prosopagnosia: A window to content-specific face processing. *Current Opinion in Neurobiology, 16*(2), 166–173.

Dumontheil, I., Apperly, I. A., & Blakemore, S. J. (2010). Online usage of theory of mind continues to develop in late adolescence. *Developmental Science, 13*, 331–338.

Dunbar, R. I. M. (1992). Neocortex size as a constraint on group size in primates. *Journal of Human Evolution, 20*, 469–493.

Dunbar, R. I. M. (1998). The social brain hypothesis. *Evolutionary Anthropology, 6*, 178–190.

Dunbar, R. I. M. (2004). Gossip in evolutionary perspective. *Review of General Psychology, 5*, 100–110.

Dunn, J., & Brophy, M. (2005). Communication, relationships, and individual differences in children's understanding of mind. In J. W. Astington & J. A. Baird (Eds.), *Why language matters for theory of mind*. Oxford: Oxford University Press.

Dunn, J., Brown, J., & Beardsall, L. (1991). Family talk about feeling states and children's later understanding of others' emotions. *Developmental Psychology, 27*, 448–455.

Dunn, J., Brown, J., Slomkowski, C., Tesla, C., & Youngblade, L. (1991). Young children's understanding of other people's feelings and beliefs: Individual differences and their antecedents. *Child Development, 62*, 1352–1366.

Eastman, N., & Campbell, C. (2006). Neuroscience and legal determination of criminal responsibility. *Nature Reviews Neuroscience, 7*(4), 311–318.

Eder, R. A. (1990). Uncovering young children's psychological selves: Individual and developmental differences. *Child Development, 61*(3), 849–863.

Ehrenkranz, J., Bliss, E., & Sheard, M. H. (1974). Plasma testosterone: Correlation with aggressive-behavior and social dominance in man. *Psychosomatic Medicine, 36*(6), 469–475.

Eisenberg, N., Fabes, R. A., Murphy, B., Karbon, M., Maszk, P., Smith, M., Oboyle, C., & Suh, K. (1994). The relations of emotionality and regulation to dispositional and situational empathy-related responding. *Journal of Personality and Social Psychology, 66*(4), 776–797.

Eisenberger, N. I., Jarcho, J. M., Lieberman, M. D., & Naliboff, B. D. (2006). An experimental study of shared sensitivity to physical pain and social rejection. *Pain, 126*, 132–138.

Eisenberger, N. I., Lieberman, M. D., & Williams, K. D. (2003). Does rejection hurt? An fMRI study of social exclusion. *Science, 302*(5643), 290–292.

Ekman, P. (1972). Universal and cultural differences in facial expression of emotion. In J. R. Cole (Ed.), *Nebraska symposium on motivation*. Lincoln: Nebraska University Press.

Ekman, P. (1992). An argument for basic emotions. *Cognition and Emotion, 6*, 169–200.

Ekman, P., & Friesen, W. V. (1976). *Pictures of facial affect*. Palo Alto, CA: Consulting Psychologists Press.

Ekman, P., Friesen, W. V., & Ellsworth, P. (1972). *Emotion in the human face: Guidelines for research and an integration of findings*. New York: Pergamon Press.

Elias, M. (1981). Serum cortisol, testosterone, and testosterone-binding globulin responses to competitive fighting in human males. *Aggressive Behavior, 7*, 215–224.

Ellis, H. D., & Lewis, M. B. (2001). Capgras delusion: A window on face recognition. *Trends in Cognitive Sciences, 5*, 149–156.

Ellis, H. D., & Young, A. W. (1990). Accounting for delusional misidentifications. *British Journal of Psychiatry, 157*, 239–248.

Ellis, H. D., Young, A. W., Quayle, A. H., & DePauw, K. W. (1997). Reduced autonomic responses to faces in Capgras delusion. *Proceedings of the Royal Society of London, Series B, 264*, 1085–1092.

Engel, A. K., Konig, P., & Singer, W. (1991). Direct physiological evidence for scene segmentation by temporal encoding. *Proceedings of the National Academy of Sciences of the USA, 88*, 9136–9140.

Engel, A. K., Moll, C. K. E., Fried, I., & Ojemann, G. A. (2005). Invasive recordings from the human brain: Clinical insights and beyond. *Nature Reviews Neuroscience, 6*, 35–47.

Epley, N., Akalis, S., Waytz, A., & Cacioppo, J. T. (2008). Creating social connection through inferential reproduction: Loneliness and perceived agency in gadgets, gods, and greyhounds. *Psychological Science, 19*(2), 114–120.

Erickson, K., Drevets, W., & Schulkin, J. (2003). Glucocorticoid regulation of diverse cognitive functions in normal and pathological emotional states. *Neuroscience and Biobehavioral Reviews, 27*(3), 233–246.

Ernst, M., Pine, D. S., & Hardin, M. (2006). Triadic model of the neurobiology of motivated behavior in adolescence. *Psychological Medicine, 36*(3), 299–312.

Ernst, M., & Spear, L. P. (2009). Reward systems. In M. D. Haan & M. R. Gunnar (Eds.), *Handbook of developmental social neuroscience*. New York: Guilford Press.

Eslinger, P. J., & Damasio, A. R. (1985). Severe disturbance of higher cognition after bilateral frontal ablation: Patient EVR. *Neurology, 35*, 1731–1741.

Everitt, B. J., & Robbins, T. W. (2005). Neural systems of reinforcement for drug addiction: From actions to habits to compulsion. *Nature Neuroscience, 8*(11), 1481–1489.

Fabes, R. A., Eisenberg, N., Karbon, M., Troyer, D., & Switzer, G. (1994). The relations of children's emotion regulation to their vicarious emotional responses and comforting behaviors. *Child Development, 65*(6), 1678–1693.

Falk, A., Fehr, E., & Fischbacher, U. (2005). Driving forces behind informal sanctions. *Econometrica, 73*(6), 2017–2030.

Farah, M. J. (2005). Neuroethics: The practical and the philosophical. *Trends in Cognitive Sciences, 9*(1), 34–40.

Farah, M. J., Rabinowitz, C., Quinn, G. E., & Liu, G. T. (2000). Early commitment of neural substrates for face recognition. *Cognitive Neuropsychology, 17*(1-3), 117–123.

Farrer, C., & Frith, C. D. (2002). Experiencing oneself vs another person as being the cause of an action: The neural correlates of the experience of agency. *NeuroImage, 15*, 596–603.

Farroni, T., Csibra, G., Simion, G., & Johnson, M. H. (2002). Eye contact detection in humans from birth. *Proceedings of the National Academy of Sciences of the USA, 99*, 9602–9605.

Farroni, T., Mansfield, E. M., Lai, C., & Johnson, M. H. (2003). Infants perceiving and acting on the eyes: Tests of an evolutionary hypothesis. *Journal of Experimental Child Psychology, 85*(3), 199–212.

Farroni, T., Massaccesi, S., Menon, E., & Johnson, M. H. (2007). Direct gaze modulates face recognition in young infants. *Cognition, 102*(3), 396–404.

Fazio, R. H., Jackson, J. R., Dunton, B. C., & Williams, C. J. (1995). Variability in automatic activation as an unobtrusive measure of racial-attitudes: A bona-fide pipeline. *Journal of Personality and Social Psychology, 69*(6), 1013–1027.

Fehr, E., & Fischbacher, U. (2004). Social norms and human cooperation. *Trends in Cognitive Sciences, 8*(4), 185–190.

Fehr, E., & Gachter, S. (2000). Cooperation and punishment in public goods experiments. *American Economic Review, 90*(4), 980–994.

Fehr, E., & Gachter, S. (2002). Altruistic punishment in humans. *Nature, 415*, 137–140.

Feldman Barrett, L. (2006). Are emotions natural kinds? *Perspectives on Psychological Science, 1*, 28–58.

Ferrari, P. F., Visalberghi, E., Paukner, A., Fogassi, L., Ruggiero, A., & Suomi, S. J. (2006). Neonatal imitation in rhesus macaques. *PLoS Biology, 4*, 1501–1508.

Ferstl, E. C., & von Cramon, D. Y. (2002). What does the frontomedian cortex contribute to language processing: Coherence or theory of mind? *NeuroImage, 17*, 1599–1612.

Festinger, L., Schacter, S., & Back, K. (1950). *Social pressures in informal groups: A study of human factors in housing*. New York: Harper.

Feys, J. (1991). Briefly induced belongingness to self and preference. *European Journal of Social Psychology, 21*(6), 547–552.

Fiedler, K., Messner, C., & Bluemke, M. (2006). Unresolved problems with the 'I', the 'A' and the 'T': A logical and psychometric critique of the Implicit Association Test (IAT). *European Review of Social Psychology, 17*, 74–147.

Fine, C., Lumsden, J., & Blair, R. J. R. (2001). Dissociation between 'theory of mind' and executive functions in a patient with early left amygdala damage. *Brain, 124*, 287–298.

Finger, E. C., Marsh, A. A., Kamel, N., Mitchell, D. G. V., & Blair, J. R. (2006). Caught in the act: The impact of audience on the neural response to morally and socially inappropriate behavior. *NeuroImage, 33*(1), 414–421.

Fischbacher, U., Gachter, S., & Fehr, E. (2001). Are people conditionally cooperative? Evidence from a public goods experiment. *Economics Letters, 71*(3), 397–404.

Fitch, W. T., Hauser, M. D., & Chomsky, N. (2005). The evolution of the language faculty: Clarifications and implications. *Cognition, 97*, 179–210.

Fitch, W. T., & Reby, D. (2001). The descended larynx is not uniquely human. *Proceedings of the Royal Society of London, Series B, 268*, 1669–1675.

Fletcher, P. C., Happe, F., Frith, U., Baker, S. C., Dolan, R. J., Frackowiak, R. S. J., & Frith, C. D. (1995). Other minds in the brain: A functional imaging study of 'theory of mind' in story comprehension. *Cognition, 57*, 109–128.

Flor, H., Birbaumer, N., Hermann, C., Ziegler, S., & Patrick, C. J. (2002). Aversive Pavlovian conditioning in psychopaths: Peripheral and central correlates. *Psychophysiology, 39*(4), 505–518.

Fodor, J. A. (1983). *The modularity of mind*. Cambridge, MA: MIT Press.

Fodor, J. A. (1992). A theory of the child's theory of mind. *Cognition, 44*, 283–296.

Fogassi, L., Ferrari, P. F., Gesierich, B., Rozzi, S., Chersi, F., & Rizzolatti, G. (2005). Parietal lobe: From action organization to intention understanding. *Science, 308*(5722), 662–667.

Forster, P. (2004). Ice ages and the mitochondrial DNA chronology of human dispersals: A review. *Philosophical Transactions of the Royal Society of London, Series B, 359*, 255–264.

Fox, C. J., Moon, S. Y., Iaria, G., & Barton, J. S. J. (2009). The correlates of subjective perception of identity and expression in the face network: An fMRI adaptation study. *NeuroImage, 44*, 569–580.

Fraedrich, E. M., Lakatos, K., & Spangler, G. (2010). Brain activity during emotion perception: The role of attachment representation. *Attachment and Human Development, 12*, 231–248.

Fraley, R. C., Waller, N. G., & Brennan, K. A. (2000). An item-response theory analysis of self-report measures of adult attachment. *Journal of Personality and Social Psychology, 78*, 350–365.

Fredrikson, M., Hursti, T., Salmi, P., Bojeson, S., Furst, C., Peterson, C., & Steineck, G. (1993). Conditioned nausea after cancer chemotherapy and autonomic system conditionability. *Scandinavian Journal of Psychology, 34*, 318–327.

Freud, S. (1922). *Group psychology and the analysis of the ego*. London: International Psychoanalytic Press.

Fridlund, A. J., & Cacioppo, J. T. (1986). Guidelines for human electromyographic research. *Psychophysiology, 23*, 567–589.

Frischen, A., Bayliss, A. P., & Tipper, S. P. (2007). Gaze cueing of attention: Visual attention, social cognition, and individual differences. *Psychological Bulletin, 133*(4), 694–724.

Frith, C. D. (1992). *The cognitive neuropsychology of schizophrenia*. Hove, UK: Psychology Press.

Frith, C. D. (2007). The social brain? In N. Emery, N. Clayton, & C. Frith (Eds.), *Social intelligence: From brain to culture*. Oxford: Oxford University Press.

Frith, C. D., Blakemore, S. J., & Wolpert, D. M. (2000). Explaining the symptoms of schizophrenia: Abnormalities in the awareness of action. *Brain Research Reviews, 31*(2-3), 357–363.

Frith, C. D., & Done, D. J. (1989). Experiences of alien control in schizophrenia reflect a disorder in the central monitoring of action. *Pschological Medicine, 19*, 521–530.

Frith, C. D., & Frith, U. (1999). Interacting minds: A biological basis. *Science, 286*, 1692–1695.

Frith, U. (1989). *Autism: Explaining the enigma*. Oxford: Blackwell.

Frith, U., & Frith, C. D. (2003). Development and neurophysiology of mentalising. *Philosophical Transactions of the Royal Society of London, Series B, 358*, 459–472.

Fritsch, G. T., & Hitzig, E. (1870). On the electrical excitability of the cerebrum. In G. V. Bonin (Ed.), *Some papers on the cerebral cortex*. Springfield, IL: Charles C. Thomas.

Fultz, J., Batson, C. D., Fortenbach, V. A., McCarthy, P. M., & Varney, L. L. (1986). Social evaluation and the empathy altruism hypothesis. *Journal of Personality and Social Psychology, 50*(4), 761–769.

Fusar-Poli, P., & Broome, M. R. (2007). Love and brain: From mereological fallacy to 'folk' neuroimaging. *Psychiatry Research, 154*, 285–286.

Fuster, J. M. (1989). *The prefrontal cortex: Anatomy, physiology, and neuropsychology of the frontal lobe* (2nd edition). New York: Raven Press.

Gaffan, D. (1992). Amygdala and the memory of reward. In J. P. Aggleton (Ed.), *The amygdala: Neurobiological aspects of emotion, memory and mental dysfunction*. New York: Wiley-Liss.

Gaillard, W. D., Grandon, C. B., & Xu, B. (2001). Developmental aspects of pediatric fMRI: Considerations for image acquisition, analysis and interpretation. *NeuroImage, 13*, 239–249.

Gallagher, H. L., Happe, F., Brunswick, N., Fletcher, P. C., Frith, U., & Frith, C. D. (2000). Reading the mind in cartoons and stories: An fMRI study of 'theory of mind' in verbal and nonverbal tasks. *Neuropsychologia, 38*, 11–21.

Gallagher, S. (2000). Philosophical conceptions of the self: Implications for cognitive science. *Trends in Cognitive Sciences, 4*(1), 14–21.

Gallagher, S. (2007). Simulation trouble. *Social Neuroscience, 2*(3–4), 353–365.

Gallegos, D. R., & Tranel, D. (2005). Positive facial affect facilitates the identification of famous faces. *Brain and Language, 93*, 338–348.

Gallese, V. (2001). The 'shared manifold' hypothesis: From mirror neurons to empathy. *Journal of Consciousness Studies, 8*, 33–50.

Gallese, V. (2003). The manifold nature of interpersonal relations: The quest for a common mechanism. *Philosophical*

Transactions of the Royal Society of London, Series B, 358, 517–528.

Gallese, V., & Goldman, A. (1998). Mirror neurons and the simulation theory of mind-reading. *Trends in Cognitive Sciences, 2*, 493–501.

Gallup, G. G. J. (1970). Chimpanzees: Self-recognition. *Science, 167*, 86–87.

Ganel, T., Valyear, K. F., Goshen-Gottstein, Y., & Goodale, M. A. (2005). The involvement of the 'fusiform face area' in processing facial expression. *Neuropsychologia, 43*, 1645–1654.

Gardner, M., & Steinberg, L. (2005). Peer influence on risk taking, risk preference, and risky decision making in adolescence and adulthood: An experimental study. *Developmental Psychology, 41*(4), 625–635.

Garnica, O. (1977). Some prosodic and paralinguistic features of speech to young children. In C. E. Snow & C. A. Ferguson (Eds.), *Talking to children: Language input and acquisition.* Cambridge, UK: Cambridge University Press.

Gauthier, I., Skudlarski, P., Gore, J. C., & Anderson, A. W. (2000). Expertise for cars and birds recruits brain areas involved in face recognition. *Nature Neuroscience, 3*, 191–197.

Gauthier, I., & Tarr, M. J. (1997). Becoming a 'Greeble' expert: Exploring mechanisms for face recognition. *Vision Research, 37*, 1673–1682.

Gauthier, I., Tarr, M. J., Anderson, A. W., Skudlarski, P., & Gore, J. C. (1999). Activation of middle fusiform 'face area' increases with expertise in recognizing novel objects. *Nature Neuroscience, 2*, 568–573.

Gazzola, V., Aziz-Zadeh, L., & Keysers, C. (2006). Empathy and the somatotopic auditory mirror system in humans. *Current Biology, 16*, 1824–1829.

Gehring, W. J., Goss, B., Coles, M. G. H., Meyer, D. E., & Donchin, E. (1993). A neural system for error detection and compensation. *Psychological Science, 4*, 385–390.

Gergely, G., Bekkering, H., & Kiraly, I. (2002). Rational imitation in preverbal infants. *Nature, 415*, 755–755.

Gergely, G., & Csibra, G. (2003). Teleological reasoning in infancy: The naive theory of rational action. *Trends in Cognitive Sciences, 7*, 287–292.

Gianaros, P. J., Horenstein, J. A., Cohen, S., Matthews, K. A., Brown, S. M., Flory, J. D., Critchley, H. D., Manuck, S. B., & Hariri, A. R. (2007). Perigenual anterior cingulate morphology covaries with perceived social standing. *Social Cognitive and Affective Neuroscience, 2*(3), 161–173.

Giedd, J. N., Blumenthal, J., Jeffries, N. O., Castellanos, F. X., Liu, H., Zijdenbos, A., Paus, T., Evans, A. C., & Rapoport, J. L. (1999). Brain development during childhood and adolescence: A longitudinal MRI study. *Nature Neuroscience, 2*, 861–863.

Gilbert, S. J., Bird, G., Brindley, R., Frith, C. D., & Burgess, P. W. (2008). Atypical recruitment of medial prefrontal cortex in autism spectrum disorders: An fMRI study of two executive function tasks. *Neuropsychologia, 46*(9), 2281–2291.

Gillath, O., Bunge, S. A., Shaver, P. R., Wendelken, C., & Mikulincer, M. (2005). Attachment-style differences in the ability to suppress negative thoughts: Exploring the neural correlates. *NeuroImage, 28*, 835–847.

Gillihan, S. J., & Farah, M. J. (2005). Is self special? A critical review of evidence from experimental psychology and cognitive neuroscience. *Psychological Bulletin, 131*(1), 76–97.

Gobrogge, K. L., Liu, Y., Jia, X. X., & Wang, Z. X. (2007). Anterior hypothalamic neural activation and neurochemical associations with aggression in pair-bonded male prairie voles. *Journal of Comparative Neurology, 502*(6), 1109–1122.

Gogtay, N., Giedd, J. N., Lusk, L., Hayashi, K. M., Greenstein, D., Vaituzis, A. C., Nugent, T. F., Herman, D. H., Clasen, L. S., Toga, A. W., Rapoport, J. L., & Thompson, P. M. (2004). Dynamic mapping of human cortical development during childhood through early adulthood. *Proceedings of the National Academy of Sciences of the USA, 101*(21), 8174–8179.

Golarai, G., Ghahremani, D. G., Whitfield-Gabrieli, S., Reiss, A., Eberhardt, J. L., Gabrieli, J. D. E., & Grill-Spector, K. (2007). Differential development of high-level visual cortex correlates with category-specific recognition memory. *Nature Neuroscience, 10*(4), 512–522.

Goldberg, E. (2001). *The executive brain: Frontal lobes and the civilised mind.* Oxford: Oxford University Press.

Goldman, A. (2006). *Simulating minds: The philosophy, psychology and neuroscience of mindreading.* Oxford: Oxford University Press.

Goldman-Rakic, P. S. (1996). The prefrontal landscape: Implications of functional architecture for understanding human mentation and the central executive. *Philosophical Transactions of the Royal Society of London, Series B, 351*, 1445–1453.

Gonzalez, A., Atkinson, L., & Fleming, A. S. (2009). Attachment and comparative psychobiology of mothering. In M. D. Haan & M. R. Gunnar (Eds.), *Handbook of developmental social neuroscience.* New York: Guilford Press.

Goodenough, O. R. (2004). Responsibility and punishment: Whose mind? A response. *Philosophical Transactions of the Royal Society of London, Series B, 359*(1451), 1805–1809.

Gopnik, M., & Wellman, H. (1992). Why the child's theory of mind really is a theory. *Mind and Language, 7*, 145–171.

Gorenstein, E. E. (1982). Frontal-lobe functions in psychopaths. *Journal of Abnormal Psychology, 91*(5), 368–379.

Gosselin, N., Peretz, I., Johnsen, E., & Adolphs, R. (2007). Amygdala damage impairs emotion recognition from music. *Neuropsychologia, 45*, 236–244.

Gould, S. J. (1991). Exaptation: A crucial tool for evolutionary psychology. *Journal of Social Issues, 47*, 43–65.

Grafman, J., Schwab, K., Warden, D., Pridgen, A., Brown, H. R., & Salazar, A. M. (1996). Frontal lobe injuries, violence, and aggression: A report of the Vietnam head injury study. *Neurology, 46*, 1231–1238.

Graham, J., Haidt, J., & Nosek, B. A. (2009). Liberals and Conservatives rely on different sets of moral foundations. *Journal of Personality and Social Psychology, 96*(5), 1029–1046.

Graham, M. D., Rees, S. L., Steiner, M., & Fleming, A. S. (2006). The effects of adrenalectomy and corticosterone replacement on maternal memory in postpartum rats. *Hormones and Behavior, 49*(3), 353–361.

Grandin, T. (1995). *Thinking in pictures: And other reports from my life with autism*. New York: Double Day.

Granqvist, P., Fredrikson, M., Unge, P., Hagenfeldt, A., Valind, S., Larhammar, D., & Larsson, M. (2005). Sensed presence and mystical experiences are predicted by suggestibility, not by the application of transcranial weak complex magnetic fields. *Neuroscience Letters, 379*(1), 1–6.

Graziano, M. S. (1999). Where is my arm? The relative role of vision and proprioception in the neuronal representation of limb position. *Proceedings of the National Academy of Sciences of the USA, 96*, 10418–10421.

Graziano, M. S., Cooke, D. F., & Taylor, C. S. R. (2000). Coding the location of the arm by sight. *Science, 290*, 1782–1786.

Greene, J. D. (2008). *The secret joke of Kant's soul*. Cambridge, MA: MIT Press.

Greene, J. D., Morelli, S. A., Lowenberg, K., Nystrom, L. E., & Cohen, J. D. (2008). Cognitive load selectively interferes with utilitarian moral judgment. *Cognition, 107*(3), 1144–1154.

Greene, J. D., Nystrom, L. E., Engell, A. D., Darley, J. M., & Cohen, J. D. (2004). The neural bases of cognitive conflict and control in moral judgment. *Neuron, 44*(2), 389–400.

Greene, J. D., Sommerville, R. B., Nystrom, L. E., Darley, J. M., & Cohen, J. D. (2001). An fMRI investigation of emotional engagement in moral judgment. *Science, 293*(5537), 2105–2108.

Greenwald, A. G., Banaji, M. R., Rudman, L. A., Farnham, S. D., Nosek, B. A., & Mellott, D. S. (2002). A unified theory of implicit attitudes, stereotypes, self-esteem, and self-concept. *Psychological Review, 109*(1), 3–25.

Greenwald, A. G., McGhee, D. E., & Schwartz, J. L. K. (1998). Measuring individual differences in implicit cognition: The Implicit Association Test. *Journal of Personality and Social Psychology, 74*(6), 1464–1480.

Greimel, E., Schulte-Ruther, M., Fink, G. R., Piefke, M., Herpertz-Dahlmann, B., & Konrad, K. (2010). Development of neural correlates of empathy from childhood to early adulthood: An fMRI study in boys and adult men. *Journal of Neural Transmission, 117*(6), 781–791.

Gross, C. G., Rocha-Miranda, C. E., & Bender, D. B. (1972). Visual properties of neurons in the inferotemporal cortex of the macaque. *Journal of Neurophysiology, 35*, 96–111.

Grossman, E., Donnelly, M., Price, R., Pickens, D., Morgan, V., Neighbor, G., & Blake, R. (2000). Brain areas involved in perception of biological motion. *Journal of Cognitive Neuroscience, 12*(5), 711–720.

Grossman, K. E. (1988). Longitudinal and systematic approaches to the study of biological high- and low-risk groups. In M. Rutter (Ed.), *Studies of psychosocial risk: The power of longitudinal data*. Cambridge, UK: Cambridge University Press.

Grossmann, T., Oberecker, R., Koch, S. P., & Friederici, A. D. (2010). The developmental origins of voice processing in the human brain. *Neuron, 65*(6), 852–858.

Grossmann, T., Striano, T., & Friederici, A. D. (2005). Infants' electric brain responses to emotional prosody. *NeuroReport, 16*(16), 1825–1828.

Gunnar, M. R., Morison, S. J., Chisholm, K., & Schuder, M. (2001). Salivary cortisol levels in children adopted from Romanian orphanages. *Development and Psychopathology, 13*(3), 611–628.

Gupta, U., & Singh, P. (1982). An exploratory study of love and liking and types of marriage. *Indian Journal of Applied Psychology, 19*, 92–97.

Guroglu, B., van den Bos, W., & Crone, E. A. (2009). Fairness considerations: Increasing understanding of intentionality during adolescence. *Journal of Experimental Child Psychology, 104*(4), 398–409.

Gutchess, A. H., Welsh, R. C., Boduroglu, A., & Park, D. C. (2006). Cultural differences in neural function associated with object processing. *Cognitive Affective and Behavioral Neuroscience, 6*(2), 102–109.

Guth, W., Schmittberger, R., & Schwarze, B. (1982). An experimental analysis of ultimatum bargaining. *Journal of Economics, Behavior, and Organizations, 3*, 367–388.

Guthrie, S. (1993). *Faces in the clouds: A new theory of religion*. New York: Oxford University Press.

Hadjikhani, N., & de Gelder, B. (2003). Seeing fearful body expressions activates the fusiform cortex and amygdala. *Current Biology, 13*(24), 2201–2205.

Hadjikhani, N., Joseph, R. M., Snyder, J., & Tager-Flusberg, H. (2006). Anatomical differences in the mirror neuron system and social cognition network in autism. *Cerebral Cortex, 16*(9), 1276–1282.

Haggard, P. (2008). Human volition: Towards a neuroscience of will. *Nature Reviews Neuroscience, 9*, 934–946.

Haggard, P., Clark, S., & Kalogeras, J. (2002). Voluntary action and conscious awareness. *Nature Neuroscience, 5*, 382–385.

Haidt, J. (2001). The emotional dog and its rational tail: A social intuitionist approach to moral judgment. *Psychological Review, 108*(4), 814–834.

Haidt, J. (2003). The moral emotions. In R. J. Davidson, K. R. Scherer, & H. H. Goldsmith (Eds.), *Handbook of affective sciences.* Oxford: Oxford University Press.

Haidt, J. (2007). The new synthesis in moral psychology. *Science, 316*(5827), 998–1002.

Hamann, S., & Canli, T. (2004). Individual differences in emotion processing. *Current Opinion in Neurobiology, 14*(2), 233–238.

Hamilton, A. F. D., Brindley, R. M., & Frith, U. (2007). Imitation and action understanding in autistic spectrum disorders: How valid is the hypothesis of a deficit in the mirror neuron system? *Neuropsychologia, 45*(8), 1859–1868.

Hamilton, W. D. (1964). The genetical evolution of social behaviour. Parts I and II. *Journal of Theoretical Biology, 7*, 1–52.

Han, S. H., & Northoff, G. (2008). Culture-sensitive neural substrates of human cognition: A transcultural neuroimaging approach. *Nature Reviews Neuroscience, 9*(8), 646–654.

Happe, F. (1995). Understanding minds and metaphors: Insights from the study of figurative language in autism. *Metaphor and Symbolic Activity, 10*, 275–295.

Happe, F. (1999). Autism: Cognitive deficit or cognitive style? *Trends in Cognitive Sciences, 3*, 216–222.

Happe, F., Ehlers, S., Fletcher, P., Frith, U., Johansson, M., Gillberg, C., Dolan, R., Frackowiak, R., & Frith, C. (1996). 'Theory of mind' in the brain. Evidence from a PET scan study of Asperger syndrome. *NeuroReport, 8*(1), 197–201.

Hare, B., Brown, M., Williamson, C., & Tomasello, M. (2002). The domestication of social cognition in dogs. *Science, 298*(5598), 1634–1636.

Hare, R. D. (1980). A research scale for the assessment of psychopathy in criminal populations. *Personality and Individual Differences, 1*, 111–119.

Hare, R. D., Hart, S. D., & Harpur, T. J. (1991). Psychopathy and the DSM-IV criteria for antisocial personality disorder. *Journal of Abnormal Psychology, 100*, 391–398.

Hare, T. A., O'Doherty, J., Camerer, C. F., Schultz, W., & Rangel, A. (2008). Dissociating the role of the orbitofrontal cortex and the striatum in the computation of goal values and prediction errors. *Journal of Neuroscience, 28*(22), 5623–5630.

Hariri, A. R., Bookheimer, S. Y., & Mazziotta, J. C. (2000). Modulating emotional responses: Effects of a neocortical network on the limbic system. *NeuroReport, 11*(1), 43–48.

Harlow, H. F. (1958). The nature of love. *American Psychologist, 13*, 673–685.

Harlow, J. M. (1993). Recovery from the passage of an iron bar through the head. *History of Psychiatry, 4*, 271–281. (Original work published 1848)

Harris, J. R. (1995). Where is the child's environment: A group socialization theory of development. *Psychological Review, 102*(3), 458–489.

Harris, L. T., & Fiske, S. T. (2006). Dehumanizing the lowest of the low: Neuroimaging responses to extreme outgroups. *Psychological Science, 17*(10), 847–853.

Hart, A. J., Whalen, P. J., Shin, L. M., McInerney, S. C., Fischer, H., & Rauch, S. L. (2000). Differential response in the human amygdala to racial outgroup vs ingroup face stimuli. *NeuroReport, 11*(11), 2351–2355.

Hart, S. D., & Hare, R. D. (1996). Psychopathy and antisocial personality disorder. *Current Opinion in Psychiatry, 9*(2), 129–132.

Hassin, R., & Trope, Y. (2000). Facing faces: Studies on the cognitive aspects of physiognomy. *Journal of Personality and Social Psychology, 78*(5), 837–852.

Hatfield, T., Han, J. S., Conley, M., Gallagher, M., & Holland, P. (1996). Neurotoxic lesions of basolateral, but not central, amygdala interfere with pavlovian second-order conditioning and reinforcer devaluation effects. *Journal of Neuroscience, 16*(16), 5256–5265.

Hauser, M. D. (2006). *Moral minds.* London: Abacus.

Hauser, M. D. (2009). The possibility of impossible cultures. *Nature, 460*, 190–196.

Hauser, M. D., Chomsky, N., & Fitch, W. T. (2002). The faculty of language: What is it, who has it, and how did it evolve? *Science, 298*, 1569–1579.

Hawkley, L. C., Masi, C. M., Berry, J. D., & Cacioppo, J. T. (2006). Loneliness is a unique predictor of age-related differences in systolic blood pressure. *Psychology and Aging, 21*(1), 152–164.

Hawley, P. H. (1999). The ontogenesis of social dominance: A strategy-based evolutionary perspective. *Developmental Review, 19*(1), 97–132.

Hawley, P. H. (2002). Social dominance and prosocial and coercive strategies of resource control in preschoolers. *International Journal of Behavioral Development, 26*(2), 167–176.

Hawley, P. H. (2003). Prosocial and coercive configurations of resource control in early adolescence: A case for the well-adapted Machiavellian. *Merrill-Palmer Quarterly Journal of Developmental Psychology, 49*(3), 279–309.

Hawley, P. H., Little, T. D., & Rodkin, P. C. (2007). *Aggression and adaptation: The bright side to bad behavior.* Hillsdale, NJ: Lawrence Erlbaum Associates, Inc.

Haxby, J. V., Hoffman, E. A., & Gobbini, M. I. (2000). The distributed human neural system for face perception. *Trends in Cognitive Sciences, 4*(6), 223–233.

Haynes, J. D., & Rees, G. (2006). Decoding mental states from brain activity in humans. *Nature Reviews Neuroscience, 7*, 523–534.

Hazan, C., & Shaver, P. (1987). Romantic love conceptualized as an attachment process. *Journal of Personality and Social Psychology, 52*, 511–524.

Healy, S. D., & Rowe, C. (2007). A critique of comparative studies of brain size. *Proceedings of the Royal Society of London, Series B, 274*(1609), 453–464.

Heberlein, A. S., & Adolphs, R. (2007). Neurobiology of emotion recognition: Current evidence for shared substrates. In E. Harmon-Jones & P. Winkielman (Eds.), *Social neuroscience*. New York: Guilford Press.

Heberlein, A. S., Padon, A. A., Gillihan, S. J., Farah, M. J., & Fellows, L. K. (2008). Ventromedial frontal lobe plays a critical role in facial emotion recognition. *Journal of Cognitive Neuroscience, 20*(4), 721–733.

Heider, F., & Simmel, M. (1944). An experimental study of apparent behavior. *American Journal of Psychology, 57*, 243–259.

Hein, G., Silani, G., Preuschoff, K., Batson, C. D., & Singer, T. (2010). Neural responses to ingroup and outgroup members' suffering predict individual differences in costly helping. *Neuron, 68*(1), 149–160.

Heinrichs, M., Baumgartner, T., Kirschbaum, C., & Ehlert, U. (2003). Social support and oxytocin interact to suppress cortisol and subjective responses to psychosocial stress. *Biological Psychiatry, 54*, 1389–1398.

Henderson, L. M., Yoder, P. J., Yale, M. E., & McDuffie, A. (2002). Getting to the point: Electrophysiological correlates of protodeclarative pointing. *International Journal of Developmental Neuroscience, 20*, 449–458.

Henrich, J., Boyd, R., Bowles, S., Camerer, C., Fehr, E., Gintis, H., McElreath, R., Alvard, M., Barr, A., Ensminger, J., Henrich, N. S., Hill, K., Gil-White, F., Gurven, M., Marlowe, F. W., et al. (2005). 'Economic man' in cross-cultural perspective: Behavioral experiments in 15 small-scale societies. *Behavioral and Brain Sciences, 28*(6), 795–855.

Henrich, J., Ensminger, J., McElreath, R., Barr, A., Barrett, C., Bolyanatz, A., Cardenas, J. C., Gurven, M., Gwako, E., Henrich, N., Lesorogol, C., Marlowe, F. W., Tracer, D., & Ziker, J. (2010). Markets, religion, community size, and the evolution of fairness and punishment. *Science, 327*(5972), 1480–1484.

Henrich, J., McElreath, R., Barr, A., Ensminger, J., Barrett, C., Bolyanatz, A., Cardenas, J. C., Gurven, M., Gwako, E., Henrich, N., Lesorogol, C., Marlowe, F. W., Tracer, D., & Ziker, J. (2006). Costly punishment across human societies. *Science, 312*(5781), 1767–1770.

Herman, L. H. (2002). Vocal, social and self-imitation by bottlenosed dolphins. In K. Dautenhahn & C. L. Nehaniv (Eds.), *Imitation in animals and artifacts*. Cambridge, MA: MIT Press.

Hermans, E. J., Bos, P. A., Ossewaarde, L., Ramsey, N. F., Fernandez, G., & van Honk, J. (2010). Effects of exogenous testosterone on the ventral striatal BOLD response during reward anticipation in healthy women. *NeuroImage, 52*, 277–283.

Herrmann, E., Call, J., Hernàndez-Lloreda, M. V., Hare, B., & Tomasello, M. (2007). Humans have evolved specialized skills of social cognition: The cultural intelligence hypothesis. *Science, 317*(5843), 1360–1366.

Hess, U. (2009). Facial EMG. In E. Harmon-Jones & J. S. Beer (Eds.), *Methods in social neuroscience*. New York: Guilford Press.

Hess, U., & Blairy, S. (2001). Facial mimicry and emotional contagion to dynamic emotional facial expressions and their influence on decoding accuracy. *International Journal of Psychophysiology, 40*, 129–141.

Heyes, C. (2010). Where do mirror neurons come from? *Neuroscience and Biobehavioral Reviews, 34*(4), 575–583.

Heyes, C. M., & Galef, B. G. (1996). *Social learning in animals: The roots of culture*. San Diego: Academic Press.

Hietanen, J. K. (1999). Does your gaze direction and head orientation shift my visual attention? *NeuroReport, 10*(16), 3443–3447.

Hihara, S., Notoya, T., Tanaka, M., Ichinose, S., Ojima, H., Obayashi, S., Fujii, N., & Iriki, A. (2006). Extension of corticocortical afferents into the anterior bank of the intraparietal sulcus by tool-use training in adult monkeys. *Neuropsychologia, 44*, 2636–2646.

Hihara, S., Obayashi, S., Tanaka, M., & Iriki, A. (2003). Rapid learning of sequential tool use by macaque monkeys. *Physiology and Behavior, 78*, 427–434.

Hill, E. L., & Frith, U. (2003). Understanding autism: Insights from mind and brain. *Philosophical Transactions of the Royal Society of London, Series B, 358*, 281–289.

Hill, R. A., & Dunbar, R. I. M. (2003). Social network size in humans. *Human Nature: An Interdisciplinary Biological Perspective, 14*, 53–72.

Hoffman, E., & Haxby, J. (2000). Distinct representations of eye gaze and identity in the distributed human neural system for face perception. *Nature Neuroscience, 3*, 80–84.

Hofmann, W., Gawronski, B., Gschwendner, T., Le, H., & Schmitt, M. (2005). A meta-analysis on the correlation between the Implicit Association Test and explicit self-report measures. *Personality and Social Psychology Bulletin, 31*(10), 1369–1385.

Hoge, R. D., & Pike, G. B. (2001). Quantitative measurement using fMRI. In P. Jezzard, P. M. Matthews, & S. M. Smith (Eds.), *Functional MRI*. Oxford: Oxford University Press.

Hogg, M. A., & Vaughan, G. M. (2011). *Social psychology* (6th edition). Harlow, UK: Pearson.

Holmes, N. P., Calvert, G. A., & Spence, C. (2007). Tool use changes multisensory interactions in seconds: Evidence from the crossmodal congruency task. *Experimental Brain Research, 183*, 465–476.

Holmes, W. G., & Sherman, P. W. (1982). The ontogeny of kin recognition in 2 species of ground-squirrels. *American Zoologist, 22*(3), 491–517.

Hooker, C. I., Paller, K. A., Gitelman, D. R., Parrish, T. B., Mesulam, M. M., & Reber, P. J. (2003). Brain networks for analyzing eye gaze. *Cognitive Brain Research, 17*(2), 406–418.

Hoorens, V., & Nuttin, J. M. (1993). Overvaluation of own attributes: Mere ownership or subjective frequency. *Social Cognition, 11*(2), 177–200.

Hornak, J., Bramham, J., Rolls, E. T., Morris, R. G., O'Doherty, J., Bullock, P. R., & Polkey, C. E. (2003). Changes in emotion after circumscribed surgical lesions of the orbitofrontal and cingulate cortices. *Brain, 126*, 1691–1712.

Hornak, J., Rolls, E. T., & Wade, D. (1996). Face and voice expression identification inpatients with emotional and behavioural changes following ventral frontal lobe damage. *Neuropsychologia, 34*(4), 247–261.

Horner, V., & Whiten, A. (2005). Causal knowledge and imitation/emulation switching in chimpanzees (Pan troglodytes) and children. *Animal Behavior, 64*, 851–859.

Hosobuchi, Y., Adams, J. E., & Linchitz, R. (1977). Pain relief by electrical-stimulation of central gray-matter in humans and its reversal by naloxone. *Science, 197*, 183–186.

Huber, D., Veinante, P., & Stoop, R. (2005). Vasopressin and oxytocin excite distinct neuronal populations in the central amygdala. *Science, 308*(5719), 245–248.

Hughes, C., Russell, J., & Robbins, T. W. (1994). Evidence for executive dysfunction in autism. *Psychological Medicine, 27*, 209–220.

Humphrey, N. K. (1976). The social function of intellect. In P. Bateson & R. A. Hinde (Eds.), *Growing points in ethology*. Cambridge, UK: Cambridge University Press.

Hunt, G. R., & Gray, R. D. (2003). Diversification and cumulative evolution in New Caledonian crow tool manufacture. *Proceedings of the National Academy of Sciences of the USA, 270*, 867–874.

Hunt, P. S., Holloway, J. L., & Scordalakes, E. M. (2001). Social interaction with an intoxicated sibling can result in increased intake of ethanol by periadolescent rats. *Developmental Psychobiology, 38*, 101–109.

Hurley, S., Clark, A., & Kiverstein, J. (2008). The Shared Circuits Model (SCM): How control, mirroring, and simulation can enable imitation, deliberation, and mindreading. *Behavioral and Brain Sciences, 31*(1), 1–58.

Huttenlocher, P. R., & Dabholkar, A. S. (1997). Regional differences in synaptogenesis in human cerebral cortex. *Journal of Comparative Neurology, 387*, 167–178.

Iacoboni, M. (2009). Imitation, empathy, and mirror neurons. *Annual Review of Psychology, 60*, 653–670.

Iacoboni, M., & Dapretto, M. (2006). The mirror neuron system and the consequences of its dysfunction. *Nature Reviews Neuroscience, 7*(12), 942–951.

Iacoboni, M., Molnar-Szakacs, I., Gallese, V., Buccino, G., Mazziotta, J. C., & Rizzolatti, G. (2005). Grasping the intentions of others with one's own mirror neuron system. *PLoS Biology, 3*, 529–535.

Iacoboni, M., Woods, R., Brass, M., Bekkering, H., Mazziotta, J. C., & Rizzolatti, G. (1999). Cortical mechanisms of human imitation. *Science, 286*, 2526–2528.

Ickes, W. (1993). Empathic accuracy. *Journal of Personality, 61*(4), 587–610.

Ickes, W., Gesn, P. R., & Graham, T. (2000). Gender differences in empathic accuracy: Differential ability or differential motivation? *Personal Relationships, 7*(1), 95–110.

Insel, T. R., & Harbaugh, C. R. (1989). Lesions of the hypothalamic paraventricular nucleus disrupt the initiation of maternal-behavior. *Physiology and Behavior, 45*(5), 1033–1041.

Insel, T. R., & Shapiro, L. E. (1992). Oxytocin receptor distribution reflects social-organization in monogamous and polygamous voles. *Proceedings of the National Academy of Sciences of the USA, 89*(13), 5981–5985.

Iriki, A. (2006). The neural origins and implications of imitation, mirror neurons and tool use. *Current Opinion in Neurobiology, 16*, 660–667.

Iriki, A., & Sakura, O. (2008). The neuroscience of primate intellectual evolution: Natural selection and passive and intentional niche construction. *Philosophical Transactions of the Royal Society of London, Series B, 363*, 2229–2241.

Iriki, A., Tanaka, M., & Iwamura, Y. (1996). Coding of modified body schema during tool use by macaque postcentral neurons. *NeuroReport, 7*, 2325–2330.

Isen, A. M., & Levin, P. F. (1972). Effect of feeling good on helping: Cookies and kindness. *Journal of Personality and Social Psychology, 21*(3), 384–388.

Ishibashi, H., Hihara, S., Takahashi, M., Heike, T., Yokota, T., & Iriki, A. (2002). Tool-use learning selectively induces expression of brain-derived neurotrophic factor, its receptor trkB, and neurotrophin 3 in the intraparietal cortex of monkeys. *Cognitive Brain Research, 14*, 3–9.

Ito, T. A., & Urland, G. R. (2003). Race and gender on the brain: Electrocortical measures of attention to the race and gender of multiply categorizable individuals. *Journal of Personality and Social Psychology, 85*(4), 616–626.

Jabbi, M., Swart, M., & Keysers, C. (2007). Empathy for positive and negative emotions in gustatory cortex. *NeuroImage, 34*, 1744–1753.

Jackson, P. L., Meltzoff, A. N., & Decety, J. (2005). How do we perceive the pain of others? A window into the neural processes involved in empathy. *NeuroImage, 24*(3), 771–779.

Jahoda, G. (1982). *Psychology and anthropology: A psychological perspective.* London: Academic Press.

James, W. (1884). What is an emotion? *Mind, 9,* 188–205.

Jankoviak, W. R., & Fischer, E. F. (1992). A cross-cultural perspective on romantic love. *Ethology, 31,* 149–155.

Jellema, T., Baker, C. I., Wicker, B., & Perrett, D. I. (2000). Neural representation for the perception of the intentionality of actions. *Brain and Cognition, 44*(2), 280–302.

Jellema, T., Maassen, G., & Perrett, D. I. (2004). Single cell integration of animate form, motion and location in the superior temporal cortex of the macaque monkey. *Cerebral Cortex, 14*(7), 781–790.

Jellema, T. & Perrett, D. I. (2005). Neural basis for the perception of goal-directed actions. In A. Easton and N. J. Emery (Eds.), *The cognitive neuroscience of social behavior.* Hove, UK: Psychology Press.

Jenkins, J. M., & Astington, J. W. (1996). Cognitive factors and family structure associated with theory of mind development in young children. *Developmental Psychobiology, 32,* 70–78.

Joffe, T. H. (1997). Social pressures have selected for an extended juvenile period in primates. *Journal of Human Evolution, 32,* 593–605.

Johnson, D. W., & Johnson, F. P. (1987). *Joining together: Group theory and group skills.* Englewood Cliffs, NJ: Prentice-Hall.

Johnson, J. G., Cohen, P., Smailes, E. M., Kasen, S., & Brook, J. S. (2002). Television viewing and aggressive behavior during adolescence and adulthood. *Science, 295*(5564), 2468–2471.

Johnson, M. H., Dziurawiec, S., Ellis, H. D., & Morton, J. (1991). Newborns' preferential tracking of face-like stimuli and its subsequent decline. *Cognition, 40,* 1–19.

Johnson, M. H., Griffin, R., Csibra, G., Halit, H., Farroni, T., de Haan, M., Tucker, L. A., Baron-Cohen, S., & Richards, J. (2005). The emergence of the social brain network: Evidence from typical and atypical development. *Developmental Psychopathology, 17,* 599–619.

Johnson-Laird, P. N., & Oatley, K. (1992). Basic emotions, rationality, and folk theory. *Cognition and Emotion, 6*(3–4), 201–223.

Johnston, L. (2002). Behavioral mimicry and stigmatization. *Social Cognition, 20*(1), 18–35.

Jorgensen, B. W., & Cervone, J. C. (1978). Affect enhancement in the pseudorecognition task. *Personality and Social Psychology Bulletin, 4,* 285–288.

Josephs, O., & Henson, R. N. A. (1999). Event-related functional magnetic resonance imaging: Modelling, inference and optimization. *Philosophical Transactions of the Royal Society of London, Series B, 354,* 1215–1228.

Kahana-Kalman, R., & Walker-Andrews, A. S. (2001). The role of person familiarity in young infants' perception of emotional expressions. *Child Development, 72*(2), 352–369.

Kalin, N. H., Shelton, S. E., & Barksdale, C. M. (1988). Opiate modulation of separation-induced distress in non-human primates. *Brain Research, 440*(2), 285–292.

Kanner, L. (1943). Autistic disturbances of affective contact. *Nervous Child, 2,* 217–250.

Kanwisher, N. (2000). Domain specificity in face perception. *Nature Neuroscience, 3,* 759–763.

Kanwisher, N., McDermott, J., & Chun, M. M. (1997). The fusiform face area: A module in human extrastriate cortex specialised for face perception. *Journal of Neuroscience, 17,* 4302–4311.

Kanwisher, N., & Yovel, G. (2006). The fusiform face area: A cortical region specialized for the perception of faces. *Philosophical Transactions of the Royal Society of London, Series B, 361*(1476), 2109–2128.

Kaplan, J. T., & Iacoboni, M. (2006). Getting a grip on other minds: Mirror neurons, intention understanding, and cognitive empathy. *Social Neuroscience, 1*(3–4), 175–183.

Kapogiannis, D., Barbey, A. K., Su, M., Zamboni, G., Krueger, F., & Grafman, J. (2009). Cognitive and neural foundations of religious belief. *Proceedings of the National Academy of Sciences of the USA, 106*(12), 4876–4881.

Kawai, M. (1965). Newly-acquired pre-cultural behavior of the natural troop of Japanese monkeys on Koshima Islet. *Primates, 6,* 1–30.

Kawamura, S. (1959). The process of sub-culture propagation among Japanese macaques. *Primates, 2,* 43–60.

Kaye, K., & Fogel, A. (1980). The temporal structure of face-to-face communication between mothers and infants. *Developmental Psychology, 16*(5), 454–464.

Kaye, K., & Wells, A. J. (1980). Mothers' jiggling and the burst–pause pattern in neonatal feeding. *Infant Behavior and Development, 3,* 29–46.

Keizer, K., Lindenberg, S., & Steg, L. (2008). The spreading of disorder. *Science, 322*(5908), 1681–1685.

Kelley, W. M., Macrae, C. N., Wyland, C. N., Caglar, S., Inati, S., & Heatherton, T. F. (2002). Finding the self? An event related fMRI study. *Journal of Cognitive Neuroscience, 14,* 785–794.

Kellogg, C. K., Awatramani, G. B., & Piekut, D. T. (1998). Adolescent development alters stressor-induced Fos immunoreactivity in rat brain. *Neuroscience, 83*(3), 681–689.

Kennett, J. (2002). Autism, empathy and moral agency. *Philosophical Quarterly, 52,* 340–357.

Kenward, B., Weir, A. A. S., Rutz, C., & Kacelnik, A. (2005). Tool manufacture by naive juvenile crows. *Nature, 433,* 121.

Kerns, J. G., Cohen, J. D., MacDonal, A. W., Cho, R. Y., Stenger, V. A., & Carter, C. S. (2004). Anterior cingulate conflict monitoring and adjustments in control. *Science, 303,* 1023–1026.

Kiehl, K. A. (2008). *Without morals: The cognitive neuroscience of criminal psychopaths*. Cambridge, MA: MIT Press.

Kiehl, K. A., Smith, A. M., Hare, R. D., Mendrek, A., Forster, B. B., Brink, J., & Liddle, P. F. (2001). Limbic abnormalities in affective processing by criminal psychopaths as revealed by functional magnetic resonance imaging. *Biological Psychiatry, 50*, 677–684.

Killgore, W. D. S., & Yurgelun-Todd, D. A. (2010). Cerebral correlates of amygdala responses during nonconscious perception of facial affect in adolescent and pre-adolescent children. *Cognitive Neuroscience, 1*, 33–43.

King, M., & Wilson, A. (1975). Evolution at two levels in humans and chimpanzees. *Science, 188*, 107–116.

King-Casas, B., Tomlin, D., Anen, C., Camerer, C. F., Quartz, S. R., & Montague, P. R. (2005). Getting to know you: Reputation and trust in a two-person economic exchange. *Science, 308*(5718), 78–83.

Kipps, C. M., Duggins, A. J., McCusker, E. A., & Calder, A. J. (2007). Disgust and happiness recognition correlate with anteroventral insula and amygdala volume respectively in preclinical Huntington's disease. *Journal of Cognitive Neuroscience, 19*, 1206–1217.

Kirsch, P., Esslinger, C., Chen, Q., Mier, D., Lis, S., Siddhanti, S., Gruppe, H., Mattay, V. S., Gallhofer, B., & Meyer-Lindenberg, A. (2005). Oxytocin modulates neural circuitry for social cognition and fear in humans. *Journal of Neuroscience, 25*(49), 11489–11493.

Klein, S. B., Loftus, J., & Kihlstrom, J. F. (1996). Self-knowledge of an amnesic patient: Toward a neuropsychology of personality and social psychology. *Journal of Experimental Psychology: General, 125*(3), 250–260.

Klein, S. B., Rozendal, K., & Cosmides, L. (2002). A social-cognitive neuroscience analysis of the self. *Social Cognition, 20*, 105–135.

Klinnert, M. D., Campos, J. J., & Source, J. (1983). Emotions as behavior regulators: Social referencing in infancy. In R. Plutchik & H. Kellerman (Eds.), *Emotions in early development*. New York: Academic Press.

Kluver, H., & Bucy, P. C. (1939). Preliminary analysis of functions of the temporal lobes in monkeys. *Archives of Neurology and Psychiatry, 42*, 979–1000.

Knoch, D., Pascual-Leone, A., Meyer, K., Treyer, V., & Fehr, E. (2006). Diminishing reciprocal fairness by disrupting the right prefrontal cortex. *Science, 314*(5800), 829–832.

Knoch, D., Schneider, F., Schunk, D., Hohmann, M., & Fehr, E. (2009). Disrupting the prefrontal cortex diminishes the human ability to build a good reputation. *Proceedings of the National Academy of Sciences of the USA, 106*(49), 20895–20899.

Knutson, B., Adams, C. M., Fong, G. W., & Hommer, D. (2001). Anticipation of increasing monetary reward selectively recruits nucleus accumbens [Electronic version]. *Journal of Neuroscience, 21*(16).

Knutson, K. M., Mah, L., Manly, C. F., & Grafman, J. (2007). Neural correlates of automatic beliefs about gender and race. *Human Brain Mapping, 28*(10), 915–930.

Knutson, K. M., Wood, J. N., Spampinato, M. V., & Grafman, J. (2006). Politics on the brain: An MRI investigation. *Social Neuroscience, 1*(1), 25–40.

Kobayashi, C., Glover, G. H., & Temple, E. (2006). Cultural and linguistic influence on neural bases of 'theory of mind': An fMRI study with Japanese bilinguals. *Brain and Language, 98*(2), 210–220.

Koenigs, M., & Tranel, D. (2007). Irrational economic decision-making after ventromedial prefrontal damage: Evidence from the Ultimatum Game. *Journal of Neuroscience, 27*, 951–956.

Koenigs, M., Young, L., Adolphs, R., Tranel, D., Cushman, F., Hauser, M., & Damasio, A. (2007). Damage to the prefrontal cortex increases utilitarian moral judgements. *Nature, 446*(7138), 908–911.

Kohlberg, L., Levine, C., & Hewer, A. (1983). Moral stages: A current formulation and response to critics. In J. A. Meacham (Ed.), *Contributions to human development*. Basel: Karger.

Kolb, B., & Whishaw, I. Q. (2002). *Fundamentals of human neuropsychology* (5th edition). New York: Worth/Freeman.

Koob, G. F. (1992). Dopamine, addiction and reward. *Seminars in the Neurosciences, 4*, 139–148.

Kosfeld, M., Heinrichs, M., Zak, P. J., Fischbacher, U., & Fehr, E. (2005). Oxytocin increases trust in humans. *Nature, 435*(7042), 673–676.

Kosslyn, S. M. (1999). If neuroimaging is the answer, what is the question? *Philosophical Transactions of the Royal Society of London, Series B, 354*, 1283–1294.

Kotelchuck, M., Zelazo, P. R., Akgan, J., & Spelke, E. (1975). Infant reaction to parental separations when left with familiar and unfamiliar adults. *Journal of Genetic Psychology, 126*(2), 255–262.

Kramer, R. S. S., Arend, I., & Ward, R. (2010). Perceived health from biological motion predicts voting behaviour. *Quarterly Journal of Experimental Psychology, 63*(4), 625–632.

Kringelbach, M. L. (2005). The human orbitofrontal cortex: Linking reward to hedonic experience. *Nature Reviews Neuroscience, 6*, 691–702.

Kringelbach, M. L., & Rolls, E. T. (2003). Neural correlates of rapid, context-dependent reversal learning in a simple model of human social interaction. *NeuroImage, 20*, 1371–1383.

Krueger, F., Barbey, A. K., & Grafman, J. (2009). The medial prefrontal cortex mediates social event knowledge. *Trends in Cognitive Sciences, 13*(3), 103–109.

Krueger, F., McCabe, K., Moll, J., Kriegeskorte, N., Zahn, R., Strenziok, M., Heinecke, A., & Grafman, J. (2007). Neural correlates of trust. *Proceedings of the National Academy of Sciences of the USA*, *104*(50), 20084–20089.

Kuhl, P. K. (2007). Is speech learning 'gated' by the social brain? *Developmental Science*, *10*(1), 110–120.

Kuhl, P. K., Coffey-Corina, S., Padden, D., & Dawson, G. (2005). Links between social and linguistic processing of speech in preschool children with autism: Behavioral and electrophysiological measures. *Developmental Science*, *8*(1), F1–F12.

Kuhl, P. K., Tsao, F. M., & Liu, H. M. (2003). Foreign-language experience in infancy: Effects of short-term exposure and social interaction on phonetic learning. *Proceedings of the National Academy of Sciences of the USA*, *100*, 9096–9101.

Kumsta, R., Kreppner, J., Rutter, M., Beckett, C., Castle, J., Stevens, S., & Sonuga-Barke, E. J. (2010). III. Deprivation-specific psychological patterns. *Monographs of the Society for Research in Child Development*, *75*(1), 48–78.

LaBar, K. S., Gatenby, J. C., Gore, J. C., Le Doux, J. E., & Phelps, E. A. (1998). Human amygdala activation during conditioned fear acquisition and extinction: A mixed-trial fMRI study. *Neuron*, *20*(5), 937–945.

Lakin, J. L., & Chartrand, T. L. (2003). Using nonconscious behavioral mimicry to create affiliation and rapport. *Psychological Science*, *14*, 334–339.

Lamm, C., Batson, C. D., & Decety, J. (2007). The neural substrate of human empathy: Effects of perspective-taking and cognitive appraisal. *Journal of Cognitive Neuroscience*, *19*(1), 42–58.

Lang, P. J., Bradley, M. M., & Cuthbert, B. N. (1990). Emotion, attention and the startle reflex. *Psychological Review*, *97*, 377–395.

Langlois, J. H., & Roggman, L. A. (1990). Attractive faces are only average. *Psychological Science*, *1*(2), 115–121.

Langton, S. R. H., & Bruce, V. (1999). Reflexive visual orienting in response to the social attention of others. *Visual Cognition*, *6*, 541–567.

Lavie, N. (1995). Perceptual load as a necessary condition for selective attention. *Journal of Experimental Psychology: Human Perception and Performance*, *21*, 451–468.

Lawrence, A. D., Calder, A. J., McGowan, S. V., & Grasby, P. M. (2002). Selective disruption of the recognition of facial expressions of anger. *NeuroReport*, *13*(6), 881–884.

Lawrence, J. H., & DeLuca, C. J. (1983). Myoelectric signal versus force relationship in different human muscles. *Journal of Applied Physiology*, *54*, 1653–1659.

Lazarus, R. (1984). On the primacy of cognition. *American Psychologist*, *39*, 124–126.

Le Bon, G. (1903). *The crowd: A study of the popular mind.* London: T. F. Unwin.

Le Doux, J. E. (1996). *The emotional brain.* New York: Simon & Schuster.

Le Doux, J. E. (2000). Emotion circuits in the brain. *Annual Review of Neuroscience*, *23*, 155–184.

Le Doux, J. E. (2002). *Synaptic self: How our brains become who we are.* New York: Viking.

Le Doux, J. E., Iwata, J., Cicchetti, P., & Reis, D. (1988). Differential projections of the central amygdaloid nucleus mediate autonomic and behavioral correlates of conditioned fear. *Journal of Neuroscience*, *8*, 2517–2529.

Lea, S. E. G., & Webley, P. (2006). Money as tool, money as drug: The biological psychology of a strong incentive. *Behavioral and Brain Sciences*, *29*, 161–209.

Leakey, R. E. (1994). *The origin of humankind.* London: Weidenfeld & Nicolson.

Leavens, D. A., Hopkins, W. D., & Bard, K. A. (2005). Understanding the point of chimpanzee pointing: Epigenesis and ecological validity. *Current Directions in Psychological Science*, *14*(4), 185–189.

Lee, D. (2008). Game theory and neural basis of social decision making. *Nature Neuroscience*, *11*(4), 404–409.

Leekam, S. R., & Perner, J. (1991). Does the autistic child have a metarepresentational deficit? *Cognition*, *40*, 203–218.

Lefebvre, L., & Bouchard, J. (2003). Social learning about food in birds. In D. M. Fragazsy & S. Perry (Eds.), *The biology of traditions: Models and evidence.* Cambridge, UK: Cambridge University Press.

Lemche, E., Giampietro, V. P., Surguladze, S. A., Amaro, E. J., Andrew, C. M., Williams, S. C. R., Brammer, M. J., Lawrence, N., Maier, M. A., Russell, T. A., Simmons, A., Ecker, C., Joraschky, P., & Phillips, M. L. (2006). Human attachment security is mediated by the amygdala: Evidence from combined fMRI and psychophysiological measures. *Human Brain Mapping*, *27*, 623–635.

Lenggenhager, B., Tadi, T., Metzinger, T., & Blanke, O. (2007). Video ergo sum: Manipulating bodily self-consciousness. *Science*, *317*, 1096–1099.

Leppanen, J. M., Moulson, M. C., Vogel-Farley, V. K., & Nelson, C. A. (2007). An ERP study of emotional face processing in the adult and infant brain. *Child Development*, *78*(1), 232–245.

Leslie, A. M. (1987). Pretence and representation: The origins of 'Theory of Mind'. *Psychological Review*, *94*, 412–426.

Leslie, A. M., Mallon, R., & Di Corcia, J. A. (2006). Transgressors, victims and cry babies: Is basic moral judgment spared in autism? *Social Neuroscience*, *1*, 270–283.

Lewis, M., & Brooks-Gunn, J. (1979). *Social cognition and the acquisition of self.* New York: Plenum Press.

Lewis, M., & Carmody, D. P. (2008). Self-representation and brain development. *Developmental Psychology, 44,* 1329–1334.

Libet, B. (1985). Unconscious cerebral initiative and the role of conscious will in voluntary action. *Behavioral and Brain Sciences, 8,* 529–566.

Libet, B., Gleason, C. A., Wright, E. W., & Pearl, D. K. (1983). Time of conscious intention to act in relation to onset of cerebral activity (readiness potential): The unconscious initiation of a freely voluntary act. *Brain, 102,* 623–642.

Lieberman, M. D., & Cunningham, W. (2009). Type I and Type II error concerns in fMRI research: Re-balancing the scale. *Social Cognitive and Affective Neuroscience, 4,* 423–428.

Lieberman, M. D., Eisenberger, N. I., Crockett, M. J., Tom, S. M., Pfeifer, J. H., & Way, B. M. (2007). Putting feelings into words: Affect labeling disrupts amygdala activity in response to affective stimuli. *Psychological Science, 18*(5), 421–428.

Lieberman, M. D., Hariri, A., Jarcho, J. M., Eisenberger, N. I., & Bookheimer, S. Y. (2005). An fMRI investigation of race-related amygdala activity in African-American and Caucasian-American individuals. *Nature Neuroscience, 8*(6), 720–722.

Light, S. N., Coan, J. A., Zahn-Waxler, C., Frye, C., Goldsmith, H. H., & Davidson, R. J. (2009). Empathy is associated with dynamic change in prefrontal brain electrical activity during positive emotion in children. *Child Development, 80*(4), 1210–1231.

Lin, Z. C., & Han, S. H. (2009). Self-construal priming modulates the scope of visual attention. *Quarterly Journal of Experimental Psychology, 62*(4), 802–813.

Lin, Z. C., Lin, Y., & Han, S. H. (2008). Self-construal priming modulates visual activity, underlying global/local perception. *Biological Psychology, 77*(1), 93–97.

Lipps, T. (1903). Einfuhlung, inner nachahnubg, und organempfindungen. *Archiv fuer die Gesamte Psychologie, 1,* 185–204.

Little, A. C., Burt, D. M., & Perrett, D. I. (2006). What is good is beautiful: Face preference reflects desired personality. *Personality and Individual Differences, 41*(6), 1107–1118.

Liu, Y., & Wang, Z. X. (2003). Nucleus accumbens oxytocin and dopamine interact to regulate pair bond formation in female prairie voles. *Neuroscience, 121*(3), 537–544.

Ljungberg, T., Apicella, P., & Schultz, W. (1992). Responses of monkey dopamine neurons during learning of behavioral reactions. *Journal of Neurophysiology, 67*(1), 145–163.

Lloyd, B., & Duveen, G. (1990). A semiotic analysis of the development of the social representation of gender. In G. Duveen & B. Lloyd (Eds.), *Social representation and the development of knowledge.* Cambridge, UK: Cambridge University Press.

Lloyd-Fox, S., Blasi, A., & Elwell, C. E. (2010). Illuminating the developing brain: The past, present and future of functional near-infrared spectroscopy. *Neuroscience and Biobehavioral Review, 34,* 269–284.

Lohmann, H., & Tomasello, M. (2003). The role of language in the development of false belief understanding: A training study. *Child Development, 74,* 1130–1144.

Lorenz, K. (1966). *On aggression.* New York: Harcourt Brace.

Lykken, D. T. (1957). A study of anxiety in the sociopathic personality. *Journal of Abnormal and Social Psychology, 55,* 6–10.

Macchi Cassia, V., Turati, C., & Simion, F. (2004). Can a nonspecific bias toward top-heavy patterns explain new-borns face preference? *Psychological Science, 15,* 379–383.

MacDonald, G., & Leary, M. R. (2005). Why does social exclusion hurt? The relationship between social and physical pain. *Psychological Bulletin, 103,* 202–223.

MacLean, P. D. (1949). Psychosomatic disease and the 'visceral brain': Recent developments bearing on the Papez theory of emotion. *Psychosomatic Medicine, 11,* 338–353.

MacLoed, C. M., & MacDonald, P. A. (2000). Interdimensional interference in the Stroop effect: Uncovering the cognitive and neural anatomy of attention. *Trends in Cognitive Sciences, 4,* 383–391.

Macmillan, M. B. (1986). A wonderful journey through skull and brains: The travels of Mr. Gage's tamping iron. *Brain and Cognition, 5,* 67–107.

Macrae, C. N., & Bodenhausen, G. V. (2000). Social cognition: Thinking categorically about others. *Annual Review of Psychology, 51,* 93–120.

Main, M., Kaplan, N., & Cassidy, J. (1985). Security in infancy, childhood and adulthood: A move to the level of representation. *Monographs of the Society for Research in Child Development, 50*(1–2).

Marazziti, D. (2009). Neurobiology and hormonal aspects of romantic relationships. In M. D. Haan & M. R. Gunnar (Eds.), *Handbook of developmental social neuroscience.* New York: Guilford Press.

Marazziti, D., Akiskal, H. S., Rossi, A., & Cassano, G. B. (1999). Alteration of the platelet serotonin transporter in romantic love. *Psychological Medicine, 29*(3), 741–745.

Marazziti, D., & Canale, D. (2004). Hormonal changes when falling in love. *Psychoneuroendocrinology, 29*(7), 931–936.

Marcar, V. L., Strassle, A. E., Loenneker, T., Schwarz, U., & Martin, E. (2004). The influence of cortical maturation on the BOLD response: An fMRI study of visual cortex in children. *Pediatric Research, 56,* 967–974.

Marcus, G. B. (1986). Stability and change in political attitudes: Observe, recall and 'explain'. *Political Behavior, 8,* 21–44.

Marcus-Newhall, A., Pedersen, W. C., Carlson, M., & Miller, N. (2000). Displaced aggression is alive and well: A meta-analytic review. *Journal of Personality and Social Psychology, 78*(4), 670–689.

Markus, H. R., & Kitayama, S. (1991). Culture and the self: Implications for cognition, emotion, and motivation. *Psychological Review, 98*(2), 224–253.

Markus, H. R., Uchida, Y., Omoregie, H., Townsend, S. S. M., & Kitayama, S. (2006). Going for the gold: Models of agency in Japanese and American contexts. *Psychological Science, 17*(2), 103–112.

Maslow, A. H. (1943). A theory of human motivation. *Psychological Review, 50*, 370–396.

Master, S. L., Eisenberger, N. I., Taylor, S. E., Naliboff, B. D., Shirinyan, D., & Lieberman, M. D. (2009). A picture's worth: Partner photographs reduce experimentally induced pain. *Psychological Science, 20*(11), 1316–1318.

Masuda, T., & Nisbett, R. E. (2006). Culture and change blindness. *Cognitive Science, 30*(2), 381–399.

Materna, S., Dicke, P. W., & Thier, P. (2008). The posterior superior temporal sulcus is involved in social communication not specific for the eyes. *Neuropsychologia, 46*(11), 2759–2765.

Matsumoto, D., Yoo, S. H., Fontaine, J., Anguas-Wong, A. M., Arriola, M., Ataca, B., Bond, M. H., Boratav, H. B., Breugelmans, S. M., Cabecinhas, R., Chae, J., Chin, W. H., Comunian, A. L., Degere, D. N., Djunaidi, A., et al. (2008). Mapping expressive differences around the world: The relationship between emotional display rules and individualism versus collectivism. *Journal of Cross-Cultural Psychology, 39*(1), 55–74.

Matthews, G., & Wells, A. (1999). The cognitive science of attention and emotion. In T. Dalgleish & M. J. Power (Eds.), *Handbook of cognition and emotion*. New York: Wiley.

Maynard Smith, J. (1982). *Evolution and the theory of games*. Cambridge, UK: Cambridge University Press.

Mazur, A., & Booth, A. (1998). Testosterone and dominance in men. *Behavioral and Brain Sciences, 21*(3), 353–397.

Mazur, A., Booth, A., & Dabbs, J. (1992). Testosterone and chess competition. *Social Psychology Quarterly, 55*, 70–77.

McCabe, K., Houser, D., Ryan, L., Smith, V., & Trouard, T. (2001a). A functional imaging study of cooperation in two-person reciprocal exchange. *Proceedings of the National Academy of Sciences of the USA, 98*, 11832–11835.

McCabe, K., Houser, D., Ryan, L., Smith, V., & Trouard, T. (2001b). A functional imaging study of cooperation in two-person reciprocal exchange. *Proceedings of the National Academy of Sciences of the USA, 98*, 11832–11835.

McClure, S., Lee, J., Tomlin, D., Cypert, K., Montague, L., & Montague, P. R. (2004). Neural correlates of behavioural preferences for culturally familiar drinks. *Neuron, 44*, 379–387.

McConahay, J. B. (1986). Modern racism, ambivalence, and the Modern Racism Scale. In J. F. Dovidio & S. L. Gaertner (Eds.), *Prejudice, discrimination and racism*. New York: Academic Press.

McConnell, A. R., & Leibold, J. M. (2001). Relations among the implicit association test, discriminatory behavior, and explicit measures of racial attitudes. *Journal of Experimental Social Psychology, 37*(5), 435–442.

McLoed, P., Dittrich, W., Driver, J., Perrett, D., & Zihl, J. (1996). Preserved and impaired detection of structure from motion by a 'motion-blind' patient. *Visual Cognition, 3*, 363–391.

McNeil, J. E., & Warrington, E. K. (1993). Prosopagnosia: A face-specific disorder. *Quarterly Journal of Experimental Psychology, 46A*, 1–10.

Meeren, H. K. M., van Heijnsbergen, C., & de Gelder, B. (2005). Rapid perceptual integration of facial expression and emotional body language. *Proceedings of the National Academy of Sciences of the USA, 102*(45), 16518–16523.

Mehler, J., Bertoncini, J., Barriere, M., & Jassikgerschenfeld, D. (1978). Infant recognition of mother's voice. *Perception, 7*(5), 491–497.

Meltzoff, A. N. (1995). Understanding the intentions of others: Re-enactment of intended acts by 18-month-old children. *Developmental Psychology, 31*, 838–850.

Meltzoff, A. N. (2007). 'Like me': A foundation for social cognition. *Developmental Science, 10*, 126–134.

Meltzoff, A. N., & Borton, R. W. (1979). Intermodal matching by human neonates. *Nature, 282*, 403–404.

Meltzoff, A. N., & Decety, J. (2003). What imitation tells us about social cognition: A rapprochement between developmental psychology and cognitive neuroscience. *Philosophical Transactions of the Royal Society of London, Series B, 358*(1431), 491–500.

Meltzoff, A. N., & Moore, M. K. (1977). Imitation of facial and manual gestures by human neonates. *Science, 198*, 75–78.

Meltzoff, A. N., & Moore, M. K. (1983). Newborn infants imitate adult facial gestures. *Child Development, 54*, 702–709.

Milgram, S. (1963). Behavioral study of obedience. *Journal of Abnormal and Social Psychology, 67*, 371–378.

Milgram, S. (1974). *Obedience to authority*. London: Tavistock.

Miller, E. K., & Cohen, J. D. (2001). An integrative theory of prefrontal cortex function. *Annual Review of Neuroscience, 24*, 167–202.

Mills, D., & Conboy, B. T. (2009). Early communicative development and the social brain. In M. D. Haan & M. R. Gunnar (Eds.), *Handbook of developmental social neuroscience*. New York: Guilford Press.

Mineka, S., & Cook, M. (1993). Mechanisms involved in the observational conditioning of fear. *Journal of Experimental Psychology: General, 122,* 23–38.

Mitchell, J. P. (2009). Social psychology as a natural kind. *Trends in Cognitive Sciences, 13,* 246–251.

Mitchell, J. P., Banaji, M. R., & Macrae, C. N. (2005a). General and specific contributions of the medial prefrontal cortex to knowledge about mental states. *NeuroImage, 28*(4), 757–762.

Mitchell, J. P., Banaji, M. R., & Macrae, C. N. (2005b). The link between social cognition and self-referential thought in the medial prefrontal cortex. *Journal of Cognitive Neuroscience, 17,* 1306–1315.

Mitchell, J. P., Heatherton, T. F., & Macrae, C. N. (2002). Distinct neural systems subserve person and object knowledge. *Proceedings of the National Academy of Sciences of the USA, 99*(23), 15238–15243.

Mitchell, J. P., Nosek, B. A., & Banaji, M. R. (2003). Contextual variations in implicit evaluation. *Journal of Experimental Psychology: General, 132*(3), 455–469.

Mitchell, P., & Ropar, D. (2004). Visuo-spatial abilities in autism: A review. *Infant and Child Development, 13*(3), 185–198.

Mitchell, R. W., & Anderson, J. R. (1993). Discrimination learning of scratching, but failure to obtain imitation and self-recognition in a long-tailed macaque. *Primates, 34,* 301–309.

Mithen, S. (2007). Did farming arise from a misapplication of social intelligence? *Philosophical Transactions of the Royal Society of London, Series B, 362,* 705–718.

Mogenson, G. J., Jones, D. L., & Yim, C. Y. (1980). From motivation to action: Functional interface between the limbic system and the motor system. *Progress in Neurobiology, 14,* 69–97.

Moll, J., de Oliveira-Souza, R., Eslinger, P. J., Bramati, I. E., Mourao-Miranda, J., Andreiuolo, P. A., & Pessoa, L. (2002). The neural correlates of moral sensitivity: A functional magnetic resonance imaging investigation of basic and moral emotions. *Journal of Neuroscience, 22*(7), 2730–2736.

Moll, J., de Oliveira-Souza, R., Moll, F. T., Ignacio, F. A., Bramati, I. E., Caparelli-Daquer, E. M., & Eslinger, P. J. (2005). The moral affiliations of disgust: A functional MRI study. *Cognitive and Behavioral Neurology, 18*(1), 68–78.

Moll, J., de Oliveira-Souza, R., Zahn, R., & Grafman, J. (2008). *The cognitive neuroscience of moral emotions.* Cambridge, MA: MIT Press.

Moll, J., Krueger, F., Zahn, R., Pardini, M., de Oliveira-Souza, R., & Grafman, J. (2006). Human fronto-mesolimbic networks guide decisions about charitable donation. *Proceedings of the National Academy of Sciences of the USA, 103*(42), 15623–15628.

Moll, J., & Schulkin, J. (2009). Social attachment and aversion in human moral cognition. *Neuroscience and Biobehavioral Reviews, 33*(3), 456–465.

Moll, J., Zahn, R., de Oliveira-Souza, R., Krueger, F., & Grafman, J. (2005). The neural basis of human moral cognition. *Nature Reviews Neuroscience, 6*(10), 799–809.

Moreland, R. L., & Beach, S. R. (1992). Exposure effects in the classroom: The development of affinity amongst students. *Journal of Experimental Social Psychology, 28,* 255–276.

Moreno, J. D. (2003). Neuroethics: An agenda for neuroscience and society. *Nature Reviews Neuroscience, 4*(2), 149–153.

Moretti, L., & Di Pellegrino, G. (2010). Disgust selectively modulates reciprocal fairness in economic interactions. *Emotion, 10,* 169–180.

Moro, V., Urgesi, C., Pernigo, S., Lanteri, P., Pazzaglia, M., & Aglioti, S. M. (2008). The neural basis of body form and body action agnosia. *Neuron, 60*(2), 235–246.

Morris, J., Friston, K. J., Buechel, C., Frith, C. D., Young, A. W., Calder, A. J., & Dolan, R. J. (1998). A neuromodulatory role for the human amygdala in processing emotional facial expressions. *Brain, 121,* 47–57.

Morris, J., Frith, C. D., Perrett, D., Rowland, D., Young, A. W., Calder, A. J., & Dolan, R. J. (1996). A differential neural response in the human amygdala to fearful and happy facial expressions. *Nature, 383,* 812–815.

Morris, J., Ohmann, A., & Dolan, R. (1999). A sub-cortical pathway to the right amygdala mediating 'unseen' fear. *Proceedings of the National Academy of Sciences of the USA, 96,* 1680–1685.

Morton, J., & Johnson, M. H. (1991). CONSPEC and CONLERN: A 2-process theory of infant face recognition. *Psychological Review, 98*(2), 164–181.

Moses, L. J., Baldwin, D. A., Rosicky, J. G., & Tidball, G. (2001). Evidence for referential understanding in the emotions domain at twelve and eighteen months. *Child Development, 72*(3), 718–735.

Mukamel, R., Ekstrom, A. D., Kaplan, J. T., Iacoboni, M., & Fried, I. (2010). Single-neuron responses in humans during execution and observation of actions. *Current Biology, 8,* 750–756.

Mummery, C. J., Patterson, K., Price, C. J., Ashburner, J., Frackowiak, R. S. J., & Hodges, J. R. (2000). A voxel-based morphometry study of semantic dementia: Relationship between temporal lobe atrophy and semantic memory. *Annals of Neurology, 47,* 36–45.

Mundy, P., Card, J., & Fox, N. (2000). EEG correlates of the development of infant joint attention skills. *Developmental Psychobiology, 36*(4), 325–338.

Murnighan, J. K., & Saxon, M. S. (1998). Ultimatum bargaining by children and adults. *Journal of Economic Psychology, 19*(4), 415–445.

Murray, E. A., & Baxter, M. G. (2006). Cognitive neuroscience and nonhuman primates: Lesion studies. In C. Senior, T. Russell, M. S. Gazzaniga (Eds.), *Methods in mind.* Cambridge, MA: MIT Press.

Neisser, U. (1988). Five kinds of self-knowledge. *Philosophical Psychology, 1*, 35–59.

Nelson, C. A., & DeHaan, M. (1996). Neural correlates of infants' visual responsiveness to facial expressions of emotion. *Developmental Psychobiology, 29*(7), 577–595.

Neumann, I. D., Toschi, N., Ohl, F., Torner, L., & Kromer, S. A. (2001). Maternal defence as an emotional stressor in female rats: Correlation of neuroendocrine and behavioural parameters and involvement of brain oxytocin. *European Journal of Neuroscience, 13*(5), 1016–1024.

Newcomb, T. M. (1961). *The acquaintance process.* New York: Holt, Rinehart & Winston.

Newman, J. P., & Lorenz, A. R. (2002). Response modulation and emotion processing: Implications for psychopathy and other dysregulatory psychopathology. In R. J. Davidson, J. Scherer, & H. H. Goldsmith (Eds.), *Handbook of affective sciences.* Oxford: Oxford University Press.

Nicholls, M. E. R., Ellis, B. E., Clement, J. G., & Yoshino, M. (2004). Detecting hemifacial asymmetries in emotional expression with three-dimensional computerized image analysis. *Proceedings of the Royal Society of London, Series B, 271*(1540), 663–668.

Nieder, A. (2005). Counting on neurons: The neurobiology of numerical competence. *Nature Reviews Neuroscience, 6*, 1–14.

Nisbett, R. E., & Cohen, D. (1996). *The culture of honor: The psychology of violence in the south.* Boulder, CO: Westview Press.

Nisbett, R. E., Peng, K. P., Choi, I., & Norenzayan, A. (2001). Culture and systems of thought: Holistic versus analytic cognition. *Psychological Review, 108*(2), 291–310.

Nisbett, R. E., & Wilson, T. D. (1977). Halo effect: Evidence for unconscious alteration of judgments. *Journal of Personality and Social Psychology, 35*(4), 250–256.

Norman, D. A., & Shallice, T. (1986). Attention to action. In R. J. Davidson, G. E. Schwartz, & D. Shapiro (Eds.), *Consciousness and self regulation.* New York: Plenum Press.

Nosek, B. A., Banaji, M. R., & Greenwald, A. G. (2002). Math = male, me = female, therefore math ≠ me. *Journal of Personality and Social Psychology, 83*(1), 44–59.

Nowak, M. (2006). Five rules for the evolution of cooperation. *Science, 314*(5805), 1560–1563.

Nowak, M., & Sigmund, K. (2005). Evolution of indirect reciprocity by image scoring. *Nature, 437*, 1291–1298.

Nunez, P. L. (1981). *Electric fields of the brain: The neurophysics of EEG.* London: Oxford University Press.

Nuttin, J. M. (1985). Narcissism beyond gestalt and awareness: The name letter effect. *European Journal of Social Psychology, 15*(3), 353–361.

O'Connell, R. G., Dockree, P. M., Bellgrove, M. A., Kelly, S. P., Hester, R., Garavan, H., Robertson, I. H., & Foxe, J. J. (2007). The role of cingulate cortex in the detection of errors with and without awareness: A high-density electrical mapping study. *European Journal of Neuroscience, 25*(8), 2571–2579.

O'Connor, M. F., Wellisch, D. K., Stanton, A. L., Eisenberger, N. I., Irwin, M. R., & Lieberman, M. D. (2008). Craving love? Enduring grief activates brain's reward center. *NeuroImage, 42*(2), 969–972.

O'Connor, T. G., Bredenkamp, D., & Rutter, M. (1999). Attachment disturbances and disorders in children exposed to early severe deprivation. *Infant Mental Health Journal, 20*(1), 10–29.

O'Connor, T. G., & Rutter, M. (2000). Attachment disorder behavior following early severe deprivation: Extension and longitudinal follow-up. *Journal of the American Academy of Child and Adolescent Psychiatry, 39*(6), 703–712.

O'Doherty, J., Kringelbach, M. L., Rolls, E. T., Hornak, J., & Andrews, C. (2001). Abstract reward and punishment representations in the human orbitofrontal cortex. *Nature Neuroscience, 4*, 95–102.

Oberman, L. M., Hubbard, E. M., McCleery, J. P., Altschuler, E. L., Ramachandran, V. S., & Pineda, J. A. (2005). EEG evidence for mirror neuron dysfunction in autism spectrum disorders. *Cognitive Brain Research, 24*, 190–198.

Oberman, L. M., & Ramachandran, V. S. (2007). The simulating social mind: The role of the mirror neuron system and simulation in the social and communicative deficits of autism spectrum disorders. *Psychological Bulletin, 133*(2), 310–327.

Oberman, L. M., Winkielman, P., & Ramachandran, V. S. (2007). Face to face: Blocking facial mimicry can selectively impair recognition of emotional expressions. *Social Neuroscience, 2*(3–4), 167–178.

Ochsner, K. N., Bunge, S. A., Gross, J. J., & Gabrieli, J. D. E. (2002). Rethinking feelings: An fMRI study of the cognitive regulation of emotion. *Journal of Cognitive Neuroscience, 14*(8), 1215–1229.

Ochsner, K. N., & Lieberman, M. D. (2001). The emergence of social cognitive neuroscience. *American Psychologist, 56*, 717–734.

Ochsner, K. N., Ray, R. D., Cooper, J. C., Robertson, E. R., Chopra, S., Gabrieli, J. D. E., & Gross, J. J. (2004). For better or for worse: Neural systems supporting the cognitive down- and up-regulation of negative emotion. *NeuroImage, 23*(2), 483–499.

Ogawa, S., Lee, T. M., Kay, A. R., & Tank, D. W. (1990). Brain magnetic resonance imaging with contrast dependent on blood oxygenation. *Proceedings of the National Academy of Sciences of the USA, 87*, 9862–9872.

Ohman, A., Flykt, A., & Esteves, F. (2001). Emotion drives attention: Detecting the snake in the grass. *Journal of Experimental Psychology: General, 130*, 466–478.

Ohman, A., & Soares, J. J. F. (1994). Unconscious anxiety: Phobic responses to masked stimuli. *Journal of Abnormal Psychology, 102*, 121–132.

Olds, J. (1956). Pleasure centers of the brain. *Scientific American, 195*, 105–116.

Olds, J., & Milner, P. (1954). Positive reinforcement produced by electrical stimulation of septal area and other regions of the rat brain. *Journal of Comparative and Physiological Psychology, 47*, 419–427.

Olson, K. R., & Spelke, E. S. (2008). Foundations of cooperation in young children. *Cognition, 108*(1), 222–231.

Olsson, A., & Phelps, E. A. (2004). Learned fear of 'unseen' faces after Pavlovian, observational, and instructed fear. *Psychological Science, 15*(12), 822–828.

Öngür, D., & Price, J. L. (2000). The organization of networks within the orbital and medial prefrontal cortex of rats, monkeys and humans. *Cerebral Cortex, 10*, 206–219.

Onishi, K. H., & Baillargeon, R. (2005). Do 15-month-old infants understand false beliefs? *Science, 308*, 255–258.

Oosterhof, N. N., & Todorov, A. (2008). The functional basis of face evaluation. *Proceedings of the National Academy of Sciences of the USA, 105*(32), 11087–11092.

Ortony, A., Clore, G. L., & Collins, A. (1988). *The cognitive structure of emotions.* New York: Cambridge University Press.

Ortony, A., & Turner, T. J. (1990). What's basic about basic emotions. *Psychological Review, 97*(3), 315–331.

Otten, L. J., & Rugg, M. D. (2005). Interpreting event-related brain potentials. In T. C. Handy (Ed.), *Event-related potentials: A methods handbook.* Cambridge, MA: MIT Press.

Ozonoff, S., Pennington, B. F., & Rogers, S. J. (1991). Executive function deficits in high-functioning autistic individuals: Relationship to theory of mind. *Journal of Child Psychology and Psychiatry, 32*, 1081–1105.

Paik, H., & Comstock, G. (1994). The effects of television violence on antisocial-behavior: A meta-analysis. *Communication Research, 21*(4), 516–546.

Panksepp, J. (1998). Affective neuroscience: The foundations of human and animal emotions. New York: Oxford University Press.

Panksepp, J. (2005). Why does separation distress hurt? Comment on MacDonald and Leary (2005). *Psychological Bulletin, 131*(2), 224–230.

Panksepp, J. (2007). Neurologizing the psychology of affects: How appraisal-based constructivism and basic emotion theory can coexist. *Perspectives on Psychological Science, 2*, 281–296.

Panksepp, J., Herman, B. H., Vilberg, T., Bishop, P., & DeEskinazi, F. G. (1980). Endogenous opioids and social behavior. *Neuroscience and Biobehavioral Reviews, 4*, 473–487.

Papez, J. W. (1937). A proposed mechanism of emotion. *Archives of Neurology and Psychiatry, 38*(4), 725–743.

Park, K. A., & Waters, E. (1989). Security of attachment and preschool friendships. *Child Development, 60*, 1076–1081.

Parkman, J. M., & Groen, G. (1971). Temporal aspects of simple additions and comparison. *Journal of Experimental Psychology, 92*, 437–438.

Pascalis, O., de Haan, M., Nelson, C. A., & de Schonen, S. (1998). Long-term recognition memory for faces assessed by visual paired comparison in 3- and 6-month-old infants. *Journal of Experimental Psychology: Learning, Memory and Cognition, 24*(1), 249–260.

Pascual-Leone, A., Bartres-Faz, D., & Keenan, J. P. (1999). Transcranial magnetic stimulation: Studying the brain–behaviour relationship by induction of 'virtual lesions'. *Philosophical Transactions of the Royal Society of London, Series B, 354*, 1229–1238.

Pascual-Leone, A., Houser, C., Reeves, K., Shotland, L. M., Grafman, J., Sato, S., Valls-Sole, J., Brasil-Neto, J. P., Wassermann, E. M., & Cohen, L. G. (1993). Safety of rapid-rate transcranial magnetic stimulation in normal volunteers. *Electroencephalogy and Clinical Neurophysiology, 89*, 120–130.

Paulus, M. P., Rogalsky, C., Simmons, A., Feinstein, J. S., & Stein, M. B. (2003). Increased activation in the right insula during risk-taking decision making is related to harm avoidance and neuroticism. *NeuroImage, 19*(4), 1439–1448.

Payne, B. K. (2001). Prejudice and perception: The role of automatic and controlled processes in misperceiving a weapon. *Journal of Personality and Social Psychology, 81*(2), 181–192.

Pedersen, C. A., Ascher, J. A., Monroe, Y. L., & Prange, A. J. (1982). Oxytocin induces maternal-behavior in virgin female rats. *Science, 216*(4546), 648–650.

Peelen, M. V., & Downing, P. E. (2005). Selectivity for the human body in the fusiform gyrus. *Journal of Neurophysiology, 93*(1), 603–608.

Peelen, M. V., & Downing, P. E. (2007). The neural basis of visual body perception. *Nature Reviews Neuroscience, 8*(8), 636–648.

Pelham, B. W., Mirenberg, M. C., & Jones, J. T. (2002). Why Susie sells seashells by the seashore: Implicit egotism and major life decisions. *Journal of Personality and Social Psychology, 82*(4), 469–487.

Pelphrey, K. A., Singerman, J. D., Allison, T., & McCarthy, G. (2003). Brain activation evoked by perception of gaze shifts: The influence of context. *Neuropsychologia, 41*(2), 156–170.

Pempek, T. A., Yermolayeva, Y. A., & Calvert, S. L. (2009). College students' social networking experiences on Facebook. *Journal of Applied Developmental Psychology, 30*(3), 227–238.

Penn, D. C., & Povinelli, D. J. (2007). On the lack of evidence that non-human animals possess anything remotely resembling a 'theory of mind'. *Philosophical Transactions of the Royal Society of London, Series B, 362*, 731–744.

Penton-Voak, I. S., Pound, N., Little, A. C., & Perrett, D. I. (2006). Personality judgments from natural and composite facial images: More evidence for a 'kernel of truth' in social perception. *Social Cognition, 24*(5), 607–640.

Perner, J., Aichhorn, M., Kronbichler, M., Staffen, W., & Ladurner, G. (2006). Thinking of mental and other representations: The roles of left and right temporo-parietal junction. *Social Neuroscience, 1*(3–4), 245–258.

Perner, J., Frith, U., Leslie, A. M., & Leekam, S. R. (1989). Exploration of the autistic child's theory of mind: Knowledge, belief and communication. *Child Development, 60*, 689–700.

Perner, J., & Ruffman, T. (1995). Episodic memory and autonoetic consciousness: Developmental evidence and a theory of childhood amnesia. *Journal of Experimental Child Psychology, 59*(3), 516–548.

Perner, J., & Ruffman, T. (2005). Infants' insight into the mind: How deep? *Science, 308*, 214–216.

Perner, J., & Wimmer, H. (1985). 'John thinks that Mary thinks that...': Attribution of second-order beliefs by 5- to 10-year-old children. *Journal of Experimental Child Psychology, 39*, 437–471.

Perrett, D., & Mistlin, A. (1990). Perception of facial characteristics by monkeys. In W. Stebbins & M. Berkley (Eds.), *Comparative perception. Volume 2. Complex signals.* New York: Wiley.

Perrett, D. I., Harries, M. H., Bevan, R., Thomas, S., Benson, P. J., Mistlin, A. J., Chitty, A. J., Hietanen, J. K., & Ortega, J. E. (1989). Frameworks of analysis for the neural representation of animate objects and actions. *Journal of Experimental Biology, 146*, 87–113.

Perrett, D. I., Hietanen, J. K., Oram, M. W., & Benson, P. J. (1992). Organisation and functions of cells responsive to faces in the temporal cortex. *Philosophical Transactions of the Royal Society of London, Series B, 335*, 23–30.

Perrett, D. I., May, K. A., & Yoshikawa, S. (1994). Facial shape and judgments of female attractiveness. *Nature, 368*(6468), 239–242.

Perrett, D. I., Smith, P., Potter, D., Mistlin, A., Head, A., Milner, A., & Jeeves, M. (1985). Visual cells in the temporal cortex sensitive to face view and gaze direction. *Proceedings of the Royal Society of London, Series B, 223*, 293–317.

Perry, S., Baker, M., Fedigan, L., Gros-Louis, J., Jack, K., MacKinnon, K. C., Manson, J. H., Panger, M., & Rose, L. (2003). Social conventions in white-face capuchins monkeys: Evidence for behavioral traditions in a neotropical primate. *Current Anthropology, 44*, 241–268.

Persaud, N., McLeod, P., & Cowey, A. (2007). Post-decision wagering objectively measures awareness. *Nature Neuroscience, 10*(2), 257–261.

Persinger, M. A. (1983). Religious and mystical experiences as artifacts of temporal lobe function: A general hypothesis. *Perceptual and Motor Skills, 57*, 1255–1262.

Peterson, C. C., & Siegal, M. (1995). Deafness, conversation and theory of mind. *Journal of Child Psychology and Psychiatry and Allied Disciplines, 36*(3), 459–474.

Petrinovich, L., & O'Neill, P. (1996). Influence of wording and framing effects on moral intuitions. *Ethology and Sociobiology, 17*, 145–171.

Pfaus, J. G., Damsma, G., Nomikos, G. G., Wenkstern, D. G., Blaha, C. D., Phillips, A. G., & Fibiger, H. C. (1990). Sexual-behavior enhances central dopamine transmission in the male-rat. *Brain Research, 530*(2), 345–348.

Phelps, E. A. (2006). Emotion and cognition: Insights from studies of the human amygdala. *Annual Review of Psychology, 57*, 27–53.

Phelps, E. A., O'Connor, K. J., Cunningham, W. A., Funayama, E. S., Gatenby, J. C., Gore, J. C., & Banaji, M. R. (2000). Performance on indirect measures of race evaluation predicts amygdala activation. *Journal of Cognitive Neuroscience, 12*, 729–738.

Phillips, M. L., Young, A. W., Senior, C., Brammer, M., Andrews, C., Calder, A. J., Bullmore, E. T., Perrett, D. I., Rowland, D., Williams, S. C. R., Gray, J. A., & David, A. S. (1997). A specific neural substrate for perceiving facial expressions of disgust. *Nature, 389*, 495–498.

Phillips, R. G., & Le Doux, J. E. (1992). Differential contribution of amygdala and hippocampus to cued and contextual fear conditioning. *Behavioral Neuroscience, 106*(2), 274–285.

Piaget, J. (1932). *The moral judgment of the child.* London: Routledge & Kegan Paul.

Piazza, M., Izard, V., Pinel, P., Le Bihan, D., & Dehaene, S. (2004). Tuning curves for approximate numerosity in the human intraparietal sulcus. *Neuron, 44*, 547–555.

Pica, P., Lemer, C., Izard, V., & Dehaene, S. (2004). Exact and approximate arithmetic in an Amazonian indigene group with a reduced number lexicon. *Science, 306*, 499–503.

Piliavin, J. A., Dovidio, J. F., Gaertner, S. L., & Clark, R. D. (1981). *Emergency intervention.* New York: Academic Press.

Pillemer, D. B., & White, S. H. (1989). Childhood events recalled by children and adults. In H. W. Reese (Ed.), *Advances in child development and behavior.* New York: Academic Press.

Pine, D. S., Cohen, P., Gurley, D., Brook, J., & Ma, Y. J. (1998). The risk for early-adulthood anxiety and depressive disorders in adolescents with anxiety and depressive disorders. *Archives of General Psychiatry, 55*(1), 56–64.

Pineda, J. A. (2005). The functional significance of mu rhythms: Translating 'seeing' and 'hearing' into 'doing'. *Brain Research Reviews, 50*(1), 57–68.

Pinker, S., & Bloom, P. (1990). Natural-language and natural-selection. *Behavioral and Brain Sciences, 13*, 707–726.

Pinker, S., & Jackendoff, R. (2005). The faculty of language: What's special about it? *Cognition*, *95*, 201–236.

Pitcher, D., Garrido, L., Walsh, V., & Duchaine, B. C. (2008). Transcranial magnetic stimulation disrupts the perception and embodiment of facial expressions. *Journal of Neuroscience*, *28*(36), 8929–8933.

Plant, E. A., & Devine, P. G. (1998). Internal and external motivation to respond without prejudice. *Journal of Personality and Social Psychology*, *75*(3), 811–832.

Plassmann, H., O'Doherty, J., Shiv, B., & Rangel, A. (2008). Marketing actions can modulate neural representations of experienced pleasantness. *Proceedings of the National Academy of Sciences of the USA*, *105*(3), 1050–1054.

Platek, S. M., Critton, S. R., Myers, T. E., & Gallup, G. G. (2003). Contagious yawning: The role of self-awareness and mental state attribution. *Cognitive Brain Research*, *17*(2), 223–227.

Plutchik, R. (1980). *Emotion: A psychoevolutionary synthesis.* New York: Harper & Row.

Poldrack, R. A. (2006). Can cognitive processes be inferred from neuroimaging data? *Trends in Cognitive Sciences*, *10*, 59–63.

Posner, M. I. (1978). *Chronometric explorations of mind.* Hillsdale, NJ: Lawrence Erlbaum Associates, Inc.

Povinelli, D. J., & Simon, B. B. (1998). Young children's understanding of briefly versus extremely delayed images of the self: Emergence of the autobiographical stance. *Developmental Psychology*, *34*(1), 188–194.

Premack, D., & Woodruff, G. (1978). Does the chimpanzee have a theory of mind? *Behavioral and Brain Sciences*, *1*, 515–526.

Preston, S. D., & de Waal, F. B. M. (2002). Empathy: Its ultimate and proximate bases. *Behavioral and Brain Sciences*, *25*(1), 1–72.

Provine, R. R. (1996). Contagious yawning and laughter: Significance for sensory feature detection, motor pattern generation, imitation, and the evolution of social behaviour. In C. M. Heyes & B. G. Galef (Eds.) *Social Learning in Animals: The roots of culture.* San Diego, CA, Academic Press, 179–208.

Purves, D. (1994). *Neural activity and the growth of the brain.* Cambridge, UK: Cambridge University Press.

Putman, P., Hermans, E., & van Honk, J. (2004). Emotional Stroop performance for masked angry faces: It's BAS, not BIS. *Emotion*, *4*(3), 305–311.

Quinn, P. C., Yahr, J., Kuhn, A., Slater, A. M., & Pascalis, O. (2002). Representation of the gender of human faces by infants: A preference for female. *Perception*, *31*(9), 1109–1121.

Quiroga, R. G., Reddy, L., Kreiman, G., Koch, C., & Fried, I. (2005). Invariant visual representation by single neurons in the human brain. *Nature*, *435*, 1102–1107.

Raafat, R. M., Chater, N., & Frith, C. (2009). Herding in humans. *Trends in Cognitive Sciences*, *13*(10), 420–428.

Rabbie, J. M., & Horwitz, M. (1969). Arousal of ingroup–outgroup bias by a chance win or loss. *Journal of Personality and Social Psychology*, *13*(3), 269–277.

Raichle, M. E. (1987). Circulatory and metabolic correlates of brain function in normal humans. In F. Plum & V. Mountcastle (Eds.), *Handbook of physiology: The nervous system.* Baltimore: Williams & Wilkins.

Ramachandran, V. S. (2000). *Mirror neurons and imitation learning as the driving force behind 'the great leap forward' in human evolution.* Available from: www.edge.org.

Ramachandran, V. S., & Oberman, L. M. (2006). Broken mirrors: A theory of autism. *Scientific American*, *295*(5), 62–69.

Ramamurthi, B. (1988). Stereotactic operation in behaviour disorders: Amygdalotomy and hypothalamotomy. *Acta Neurochirurgica, Supplementum (Wein)*, *44*, 152–157.

Rand, D. G., Dreber, A., Ellingsen, T., Fudenberg, D., & Nowak, M. A. (2009). Positive interactions promote public cooperation. *Science*, *325*(5945), 1272–1275.

Reader, S. M., & Laland, K. N. (2002). Social intelligence, innovation, and enhanced brain size in primates. *Proceedings of the National Academy of Sciences of the USA*, *99*, 4436–4441.

Reby, D., McComb, K., Cargnelutti, B., Darwin, C., Fitch, W. T., & Clutton-Brock, T. (2005). Red deer stags use formants as assessment cues during intrasexual agonistic interactions. *Proceedings of the Royal Society of London, Series B*, *272*, 941–947.

Reicher, S. (1984). Social influence in the crowd: Attitudinal and behavioral effects of de-individuation in conditions of high and low group salience. *British Journal of Social Psychology*, *23*, 341–350.

Reicher, S., & Haslam, S. A. (2006). Rethinking the psychology of tyranny: The BBC prison study. *British Journal of Social Psychology*, *45*, 1–40.

Reiss, A. L., Abrams, M. T., Singer, H. S., Ross, J. L., & Denckla, M. B. (1996). Brain development, gender and IQ in children: A volumetric imaging study. *Brain*, *119*, 1763–1774.

Reiss, D., & Marino, L. (2001). Mirror self-recognition in the bottlenose dolphin: A case of cognitive convergence. *Proceedings of the National Academy of Sciences of the USA*, *98*, 5937–5942.

Renfrew, C. (2007). *Prehistory: Making of the human mind.* London: Weidenfield & Nicolson.

Repa, J. C., Muller, J., Apergis, J., Desrochers, T. M., Zhou, Y., & Le Doux, J. E. (2001). Two different lateral amygdala cell populations contribute to the initiation and storage of memory. *Nature Neuroscience*, *4*(7), 724–731.

Rhodes, G. (2006). The evolutionary psychology of facial beauty. *Annual Review of Psychology*, *57*, 199–226.

Rhodes, G., Sumich, A., & Byatt, G. (1999). Are average facial configurations attractive only because of their symmetry? *Psychological Science*, *10*(1), 52–58.

Richardson, M. P., Strange, B. A., & Dolan, R. J. (2004). Encoding of emotional memories depends on amygdala and hippocampus and their interactions. *Nature Neuroscience, 7,* 278–285.

Ridley, M. (2003). *Nature via nurture.* London: Fourth Estate.

Rilling, J. K., Glenn, A. L., Jairam, M. R., Pagnoni, G., Goldsmith, D. R., Elfenbein, H. A., & Lilienfeld, S. O. (2007). Neural correlates of social cooperation and non-cooperation as a function of psychopathy. *Biological Psychiatry, 61*(11), 1260–1271.

Rilling, J. K., Goldsmith, D. R., Glenn, A. L., Jairam, M. R., Elfenbein, H. A., Dagenais, J. E., Murdock, C. D., & Pagnoni, G. (2008). The neural correlates of the affective response to unreciprocated cooperation. *Neuropsychologia, 46*(5), 1256–1266.

Rilling, J. K., Gutman, D. A., Zeh, T. R., Pagnoni, G., Berns, G. S., & Kilts, C. D. (2002). A neural basis for social cooperation. *Neuron, 35*(2), 395–405.

Rilling, J. K., Sanfey, A. G., Aronson, J. A., Nystrom, L. E., & Cohen, J. D. (2004). The neural correlates of theory of mind within interpersonal interactions. *NeuroImage, 22*(4), 1694–1703.

Rizzolatti, G. (2005). The mirror neuron system and imitation. In S. Hurley & N. Chater (Eds.), *Perspectives on imitation: From neuroscience to social science* (Vol. 1). Cambridge, MA: MIT Press.

Rizzolatti, G., & Craighero, L. (2004). The mirror-neuron system. *Annual Review of Neuroscience, 27,* 169–192.

Rizzolatti, G., & Fabbri-Destro, M. (2010). Mirror neurons: From discovery to autism. *Experimental Brain Research, 200*(3–4), 223–237.

Rizzolatti, G., Fadiga, L., Fogassi, L., & Gallese, V. (1996). Premotor cortex and the recognition of motor actions. *Cognitive Brain Research, 3,* 131–141.

Rizzolatti, G., Fogassi, L., & Gallese, V. (2002). Motor and cognitive functions of the ventral premotor cortex. *Current Opinion in Neurobiology, 12,* 149–154.

Rizzolatti, G., Fogassi, L., & Gallese, V. (2006). Mirrors in the mind. *Scientific American, Nov,* 30–37.

Robbins, T. W., Cador, M., Taylor, J. R., & Everitt, B. J. (1989). Limbic–striatal interactions in reward-related processes. *Neuroscience and Biobehavioral Reviews, 13*(2-3), 155–162.

Robertson, E. M., Theoret, H., & Pascual-Leone, A. (2003). Studies in cognition: The problems solved and created by transcranial magnetic stimulation. *Journal of Cognitive Neuroscience, 15,* 948–960.

Robins, L. N., Tipp, J., & Przybeck, T. (1991). Antisocial personality. In L. N. Robins & D. A. Reiger (Eds.), *Psychiatric disorders in America.* New York: Free Press.

Rolls, E. T. (1996). The orbitofrontal cortex. *Philosophical Transactions of the Royal Society of London, Series B, 351,* 1433–1444.

Rolls, E. T. (2005). *Emotion explained.* Oxford: Oxford University Press.

Rolls, E. T., Hornak, J., Wade, D., & McGrath, J. (1994). Emotion-related learning in patients with social and emotional changes associated with frontal damage. *Journal of Neurology, Neurosurgery and Psychiatry, 57,* 1518–1524.

Rolls, E. T., & Tovee, M. J. (1995). Sparseness of the neuronal representation of stimuli in the primate temporal visual cortex. *Journal of Neurophysiology, 73,* 713–726.

Rorden, C., & Karnath, H. O. (2004). Using human brain lesions to infer function: A relic from a past era in the fMRI age? *Nature Reviews Neuroscience, 5,* 813–819.

Ross, M., & Wilson, A. E. (2002). It feels like yesterday: Self-esteem, valence of personal past experiences, and judgments of subjective distance. *Journal of Personality and Social Psychology, 82*(5), 792–803.

Rosvold, H. E., Mirsky, A. F., & Pribram, K. H. (1954). Influence of amygdalectomy on social behaviour in monkeys. *Journal of Comparative Physiological Psychology, 47,* 173–178.

Rotshtein, P., Henson, R. N. A., Treves, A., Driver, J., & Dolan, R. J. (2005). Morphing Marilyn into Maggie dissociates physical and identity face representations in the brain. *Nature Neuroscience, 8,* 107–113.

Rousselet, G. A., Mace, M. J.-M., & Thorpe, M. F. (2004). Animal and human faces in natural scenes: How specific to human faces is the N170 ERP component? *Journal of Vision, 4,* 13–21.

Roy, P., Rutter, M., & Pickles, A. (2004). Institutional care: associations between overactivity and lack of selectivity in social relationships. *Journal of Child Psychology and Psychiatry, 45*(4), 866–873.

Rozin, P., Haidt, J., & McCauley, C. R. (1993). Disgust. In M. Lewis & J. M. Haviland (Eds.), *Handbook of emotions.* New York: Guilford Press.

Rubin, K. H., Bukowski, W., & Parker, J. G. (2006). Peer interactions, relationships, and groups. In W. Damon, R. M. Lerner, & N. Eisenberg (Eds.), *Handbook of child psychology: Vol. 3. Social, emotional, and personality development.* New York: Wiley.

Ruby, P., & Decety, J. (2004). How would you feel versus how do you think she would feel? A neuroimaging study of perspective-taking with social emotions. *Journal of Cognitive Neuroscience, 16*(6), 988–999.

Rudebeck, P. H., Buckley, M. J., Walton, M. E., & Rushworth, M. F. S. (2006). A role for the macaque anterior cingulate gyrus in social valuation. *Science, 313*(5791), 1310–1312.

Rudebeck, P. H., Walton, M. E., Smyth, A. N., Bannerman, D. M., & Rushworth, M. F. S. (2006). Separate neural pathways process different decision costs. *Nature Neuroscience, 9*(9), 1161–1168.

Rupniak, N. M. J., Carlson, E. C., Harrison, T., Oates, B., Seward, E., Owen, S., de Felipe, C., Hunt, S., & Wheeldon, A. (2000). Pharmacological blockade or genetic deletion of substance P (NK1) receptors attenuates neonatal vocalisation in guinea-pigs and mice. *Neuropharmacology, 39*, 1413–1421.

Rushworth, M. F. S., Behrens, T. E. J., Rudebeck, P. H., & Walton, M. E. (2007). Contrasting roles for cingulate and orbitofrontal cortex in decisions and social behaviour. *Trends in Cognitive Sciences, 11*(4), 168–176.

Russell, J. (1997). *Autism as an executive disorder*. Oxford: Oxford University Press.

Sabbagh, M. A., Bowman, L. C., Evraire, L. E., & Ito, J. M. B. (2009). Neurodevelopmental correlates of theory of mind in preschool children. *Child Development, 80*(4), 1147–1162.

Sagiv, N., & Bentin, S. (2001). Structural encoding of human and schematic faces: Holistic and part based processes. *Journal of Cognitive Neuroscience, 13*, 1–15.

Sahdra, B., & Ross, M. (2007). Group identification and historical memory. *Personality and Social Psychology Bulletin, 33*(3), 384–395.

Sai, F. Z. (2005). The role of the mother's voice in developing mother's face preference: Evidence for intermodal perception at birth. *Infant and Child Development, 14*(1), 29–50.

Said, C. P., Sebe, N., & Todorov, A. (2009). Structural resemblance to emotional expressions predicts evaluation of emotionally neutral faces. *Emotion, 9*(2), 260–264.

Sally, D., & Hill, E. (2006). The development of interpersonal strategy: Autism, theory-of-mind, cooperation and fairness. *Journal of Economic Psychology, 27*(1), 73–97.

Samson, D. (2009). Reading other people's mind: Insights from neuropsychology. *Journal of Neuropsychology, 3*, 3–16.

Samson, D., Apperly, I. A., Chiavarino, C., & Humphreys, G. W. (2004). Left temporoparietal junction is necessary for representing someone else's belief. *Nature Neurosecience, 7*, 499–500.

Sanfey, A., Rilling, J., Aaronson, J., Nystron, L., & Cohen, J. (2003). Probing the neural basis of economic decision-making: An fMRI investigation of the ultimatum game. *Science, 300*, 1755–1758.

Sato, A., & Yasuda, A. (2005). Illusion of sense of self-agency: Discrepancy between the predicted and actual sensory consequences of actions modulates the sense of self-agency, but not the sense of self-ownership. *Cognition, 94*(3), 241–255.

Saver, J. L., & Damasio, A. R. (1991). Preserved access and processing of social knowledge in a patient with acquired sociopathy due to ventromedial frontal damage. *Neuropsychologia, 29*, 1241–1249.

Saxe, R. (2006). Uniquely human social cognition. *Current Opinion in Neurobiology, 16*(2), 235–239.

Saxe, R., & Kanwisher, N. (2003). People thinking about thinking people: The role of the temporo-parietal junction in 'theory of mind'. *NeuroImage, 19*, 1835–1842.

Saxe, R., & Powell, L. J. (2006). It's the thought that counts: specific brain regions for one component of theory of mind. *Psychological Science, 17*, 692–699.

Saxe, R., & Wexler, A. (2005). Making sense of another mind: The role of the right temporo-parietal junction. *Neuropsychologia, 43*(10), 1391–1399.

Schacter, S., & Singer, J. E. (1962). Cognitive, social, and physiological determinants of emotional state. *Psychology Review, 69*, 379–399.

Schaffer, H. R. (1984). *The child's entry into the social world*. London: Academic Press.

Schaffer, H. R. (1996). *Social development*. Oxford: Blackwell.

Schaffer, H. R., & Emerson, P. E. (1964). The development of social attachments in infancy. *Monographs of the Society for Research in Child Development, 29*(3).

Scherer, K. R., Banse, R., & Wallbott, H. G. (2001). Emotion inferences from vocal expression correlate across languages and cultures. *Journal of Cross-Cultural Psychology, 32*, 76–92.

Schilhab, T. S. S. (2004). What mirror self-recognition in nonhumans can tell us about aspects of self. *Biology and Philosophy, 19*, 111–126.

Schjødta, U., Stødkilde-Jørgensenb, H., Geertza, A. W., & Roepstorff, A. (2008). Rewarding prayers. *Neuroscience Letters, 443*, 165–168.

Schneider, W., & Shiffrin, R. M. (1977). Controlled and automatic human information processing: I. Detection, search and attention. *Psychological Review, 84*, 1–66.

Scholz, J., Triantafyllou, C., Whitfield-Gabrieli, S., Brown, E. N., & Saxe, R. (2009). Distinct regions of right temporo-parietal junction are selective for theory of mind and exogenous attention. *PloS One, 4*(3), 7.

Schultz, W., Apicella, P., Scarnati, E., & Ljungberg, T. (1992). Neuronal-activity in monkey ventral striatum related to the expectation of reward. *Journal of Neuroscience, 12*(12), 4595–4610.

Schultz, W., Dayan, P., & Montague, P. R. (1997). A neural substrate of prediction and reward. *Science, 275*(5306), 1593–1599.

Scott, L. S., & Monesson, A. (2009). The origin of biases in face perception. *Psychological Science, 20*(6), 676–680.

Scott, S., Young, A. W., Calder, A. J., Hellawell, D. J., Aggleton, J. P., & Johnson, M. (1997). Auditory recognition of emotion after amygdalectomy: Impairment of fear and anger. *Nature, 385*, 254–227.

Scoville, W. B., & Milner, B. (1957). Loss of recent memory after bilateral hippocampal lesions. *Journal of Neurology, Neurosurgery and Psychiatry, 20*, 11–21.

Sebanz, N., & Shiffrar, M. (2009). Detecting deception in a bluffing body: The role of expertise. *Psychonomic Bulletin and Review, 16*(1), 170–175.

Sebastian, C., Viding, E., Williams, K. D., & Blakemore, S. J. (2010). Social brain development and the affective consequences of ostracism in adolescence. *Brain and Cognition, 72*(1), 134–145.

Senju, A., Southgate, V., White, S., & Frith, U. (2009). Mindblind eyes: An absence of spontaneous theory of mind in Asperger syndrome. *Science, 325*(5942), 883–885.

Sergent, J., & Signoret, J.-L. (1992). Varieties of functional deficits in prosopagnosia. *Cerebral Cortex, 2,* 375–388.

Serino, A., Bassolino, M., Farne, A., & Ladavas, E. (2007). Extended multisensory space in blind cane users. *Psychological Science, 18,* 642–648.

Seyfarth, R. M., & Cheney, D. L. (2002). What are big brains for? *Proceedings of the National Academy of Sciences of the USA, 99,* 4141–4142.

Shah, A., & Frith, U. (1983). Islet of ability in autistic-children: A research note. *Journal of Child Psychology and Psychiatry and Allied Disciplines, 24,* 613–620.

Shallice, T., & Burgess, P. (1996). The domain of supervisory process and temporal organization of behaviour. *Philosophical Transactions of the Royal Society of London, Series B, 351,* 1405–1412.

Shamay-Tsoory, S. G., Aharon-Peretz, J., & Perry, D. (2009). Two systems for empathy: A double dissociation between emotional and cognitive empathy in inferior frontal gyrus versus ventromedial prefrontal lesions. *Brain, 132,* 617–627.

Shamay-Tsoory, S. G., Tibi-Elhanany, Y., & Aharon-Peretz, J. (2006). The ventromedial prefrontal cortex is involved in understanding affective but not cognitive theory of mind stories. *Social Neuroscience, 1*(3–4), 149–166.

Shatz, M., Wellman, H. M., & Silber, S. (1983). The acquisition of mental verbs: A systematic investigation of the 1st reference to mental state. *Cognition, 14,* 301–321.

Shaver, P. R., Morgan, H. J., & Wu, S. (1996). Is love a 'basic' emotion? *Personal Relationships, 3,* 81–96.

Shellock, F. G. (2004). *Reference manual for magnetic resonance safety, implants and devices.* Los Angeles, CA: Biomedical Research Publishing.

Shepher, J. (1971). Mate selection amongst second generation Kibbutz adolescents and adults: Incest avoidance and negative imprinting. *Archives of Sexual Behavior, 1,* 293–307.

Sherman, P. W. (1977). Nepotism and evolution of alarm calls. *Science, 197*(4310), 1246–1253.

Shih, M., Pittinsky, T. L., & Ambady, N. (1999). Stereotype susceptibility: Identity salience and shifts in quantitative performance. *Psychological Science, 10*(1), 80–83.

Siegel, A., Roeling, T. A. P., Gregg, T. R., & Kruk, M. R. (1999). Neuropharmacology of brain-stimulation-evoked aggression. *Neuroscience and Biobehavioral Reviews, 23*(3), 359–389.

Sigman, M., Mundy, P., Ungerer, J., & Sherman, T. (1986). Social Interactions of autistic, mentally retarded, and normal children and their caregivers. *Journal of Child Psychology and Psychiatry, 27,* 647–656.

Simner, M. L. (1971). Newborns' response to the cry of another infant. *Developmental Psychology, 5,* 136–150.

Simons, D. J., & Chabris, C. F. (1999). Gorillas in our midst: Sustained inattentional blindness for dynamic events. *Perception, 28,* 1059–1074.

Simpson, J. A. (1990). Influence of attachment styles on romantic relationships. *Journal of Personality and Social Psychology, 59,* 971–980.

Singer, T., Critchley, H. D., & Preuschoff, K. (2009). A common role of insula in feelings, empathy and uncertainty. *Trends in Cognitive Sciences, 13*(8), 334–340.

Singer, T., Kiebel, S. J., Winston, J. S., Dolan, R. J., & Frith, C. D. (2004a). Brain responses to the acquired moral status of faces. *Neuron, 41*(4), 653–662.

Singer, T., Seymour, B., O'Doherty, J., Kaube, H., Dolan, R. J., & Frith, C. D. (2004b). Empathy for pain involves the affective but not the sensory components of pain. *Science, 303,* 1157–1162.

Singer, T., Seymour, B., O'Doherty, J. P., Stephan, K. E., Dolan, R. J., & Frith, C. D. (2006). Empathic neural responses are modulated by the perceived fairness of others. *Nature, 439,* 466–469.

Singh, L., Morgan, J. L., & Best, C. T. (2002). Infants' listening preferences: Baby talk or happy talk? *Infancy, 3*(3), 365–394.

Small, D. M., Gregory, M. D., Mak, Y. E., Gitelman, D., Mesulam, M. M., & Parrish, T. (2003). Dissociation of neural representation of intensity and affective valuation in human gustation. *Neuron, 39*(4), 701–711.

Small, D. M., Zatorre, R. J., Dagher, A., Evans, A. C., & Jones-Gotman, M. (2001). Changes in brain activity related to eating chocolate: From pleasure to aversion. *Brain, 124,* 1720–1733.

Smeltzer, M. D., Curtis, J. T., Aragona, B. J., & Wang, Z. X. (2006). Dopamine, oxytocin, and vasopressin receptor binding in the medial prefrontal cortex of monogamous and promiscuous voles. *Neuroscience Letters, 394*(2), 146–151.

Smetana, J. G. (1981). Preschool children's conceptions of moral and social rules. *Child Development, 52*(4), 1333–1336.

Smetana, J. G. (1985). Preschool children's conceptions of transgressions: Effects of varying moral and conventional domain-related attributes. *Developmental Psychology, 21*(1), 18–29.

Smith, C. A., & Lazarus, R. S. (1990). Emotion and adaptation. In L. A. Pervin (Ed.), *Handbook of personality: Theory and research.* New York: Guilford Press.

Smith, M. C., Smith, M. K., & Ellgring, H. (1996). Spontaneous and posed facial expression in Parkinson's disease. *Journal of the International Neuropsychological Society, 2*, 383–391.

Sodian, B., & Frith, U. (1992). Deception and sabotage in autistic, retarded and normal children. *Journal of Child Psychology and Psychiatry, 33*, 591–605.

Soken, N. H., & Pick, A. D. (1999). Infants' perception of dynamic affective expressions: Do infants distinguish specific expressions? *Child Development, 70*(6), 1275–1282.

Southgate, V., & Hamilton, A. F. C. (2008). Unbroken mirrors: Challenging a theory of autism. *Trends in Cognitive Sciences, 12*, 225–229.

Southgate, V., Johnson, M. H., El Karoui, I., & Csibra, G. (2010). Motor system activation reveals infants' on-line prediction of others' goals. *Psychological Science, 21*, 355–359.

Sowell, E. R., Thompson, P. M., Holmes, C. J., Batth, R., Jernigan, T. L., & Toga, A. W. (1999). Localizing age-related changes in brain structure between childhood and adolescence using statistical parametric mapping. *NeuroImage, 9*(6), 587–597.

Spangler, G., & Grossmann, K. E. (1993). Biobehavioral organization in securely and insecurely attached infants. *Child Development, 64*(5), 1439–1450.

Spielberger, C. D., Jacobs, G., Russell, S., & Crane, R. S. (1983). Assessment of anger: The state–trait anger scale. In J. N. Butcher & C. D. Spielberger (Eds.), *Advances in personality assessment* (*Vol. 2*). Hillsdale, NJ: Lawrence Erlbaum Associates, Inc.

Spinella, M. (2005). Prefrontal substrates of empathy: Psychometric evidence in a community sample. *Biological Psychology, 70*(3), 175–181.

Sprengelmeyer, R., Young, A. W., Calder, A. J., Karnat, A., Lange, H., Homberg, V., Perrett, D., & Rowland, D. (1996). Loss of disgust: Perception of faces and emotions in Huntington's disease. *Brain, 119*, 1647–1665.

Sprengelmeyer, R., Young, A. W., Sprengelmeyer, A., Calder, A. J., Rowland, D., Perrett, D., Homberg, V., & Lange, H. (1997). Recognition of facial expression: Selective impairment of specific emotions in Huntington's disease. *Cognitive Neuropsychology, 14*, 839–879.

Squire, L. R. (1992). Memory and the hippocampus: A synthesis from findings with rats, monkeys and humans. *Psychological Review, 99*, 195–231.

Stanton, S. J., Beehner, J. C., Saini, E. K., Kuhn, C. M., & LaBar, K. S. (2009). Dominance, politics, and physiology: Voters' testosterone changes on the night of the 2008 United States Presidential Election. *PLoS One, 4*, e7543.

Stanton, S. J., Wirth, M. M., Waugh, C. E., & Schultheiss, O. C. (2009). Endogenous testosterone levels are associated with amygdala and ventromedial prefrontal cortex responses to anger faces in men but not women. *Biological Psychology, 81*, 118–122.

Sternberg, R. J. (1986). A triangular theory of love. *Psychological Review, 93*(2), 119–135.

Sternberg, R. J. (1988). *The triangle of love.* New York: Basic Books.

Stewart, L., Battelli, L., Walsh, V., & Cowey, A. (1999). Motion perception and perceptual learning studied by magnetic stimulation. *Electroencephalography and Clinical Neurophysiology, 3*, 334–350.

Stillwell, A. M., Baumeister, R. F., & Del Priore, R. E. (2008). We're all victims here: Toward a psychology of revenge. *Basic and Applied Social Psychology, 30*(3), 253–263.

Stone, V. E., & Gerrans, P. (2006). What's domain-specific about theory of mind? *Social Neuroscience, 1*, 309–319.

Striano, T., Henning, A., & Stahl, D. (2005). Sensitivity to social contingencies between 1 and 3 months of age. *Developmental Science, 8*, 509–519.

Stroop, J. R. (1935). Studies of interference in serial verbal reactions. *Journal of Experimental Psychology: General, 106*, 404–426.

Stuss, D. T., & Benson, D. F. (1986). *The frontal lobes.* New York: Raven Press.

Stuss, D. T., Floden, D., Alexander, M. P., Levine, B., & Katz, D. (2001). Stroop performance in focal lesion patients: Dissociation of processes and frontal lobe lesion location. *Neuropsychologia, 39*, 771–786.

Stuss, D. T., Gallup, G. G., & Alexander, M. P. (2001). The frontal lobes are necessary for 'theory of mind'. *Brain, 124*, 279–286.

Suhler, C. L., & Churchland, P. S. (2009). Control: Conscious and otherwise. *Trends in Cognitive Sciences, 13*(8), 341–347.

Sutton, J., Smith, P. K., & Swettenham, J. (1999). Social cognition and bullying: Social inadequacy or skilled manipulation? *British Journal of Developmental Psychology, 17*, 435–450.

Swami, V., Furnham, A., Amin, R., Chaudhri, J., Joshi, K., Jundi, S., Miller, R., Mirza-Begum, J., Begum, F. N., Sheth, P., & Tovee, M. J. (2008). Lonelier, lazier, and teased: The stigmatizing effect of body size. *Journal of Social Psychology, 148*(5), 577–593.

Swingler, M. M., Sweet, M. A., & Carver, L. J. (2010). Brain–behavior correlations: Relationships between mother–stranger face processing and infants' behavioral responses to a separation from mother. *Developmental Psychology, 46*(3), 669–680.

Tager-Flusberg, H. (1992). Autistic children's talk about psychological states: Deficits in the early acquisition of a theory of mind. *Child Development, 63*, 161–172.

Tajfel, H., & Turner, J. C. (1986). The social identity theory of intergroup behavior. In S. Worchel & W. Austin (Eds.), *Psychology of intergroup relations.* Chicago: Nelson Hall.

Takahashi, H., Yahata, N., Koeda, M., Matsuda, T., Asai, K., & Okubo, Y. (2004). Brain activation associated with evaluative processes of guilt and embarrassment: An fMRI study. *Neuroimage, 23*(3), 967–974.

Talairach, J., & Tournoux, P. (1988). *A co-planar stereotactic atlas of the human brain.* Stuttgart: Thieme Verlag.

Tallis, F. (2005). *Love sick: Love as a mental illness.* London: de Capo Press.

Tamietto, M., & De Gelder, B. (2010). Neural bases of the non-conscious perception of emotional signals. *Nature Reviews Neuroscience, 11*, 697–709.

Tang, Y., Zhang, W., Chen, K., Feng, S., Ji, Y., Shen, J., Reiman, E. M., & Liu, Y. (2006). Arithmetic processing in the brain shaped by cultures. *Proceedings of the National Academy of Sciences of the USA, 103*, 10775–10780.

Tankersley, D., Stowe, C. J., & Huettel, S. A. (2007). Altruism is associated with an increased neural response to agency. *Nature Neuroscience, 10*(2), 150–151.

Taylor, J. C., Wiggett, A. J., & Downing, P. E. (2007). Functional MRI analysis of body and body part representations in the extrastriate and fusiform body areas. *Journal of Neurophysiology, 98*(3), 1626–1633.

Taylor, S. E., Gonzaga, G. C., Klein, L. C., Hu, P. F., Greendale, G. A., & Seeman, T. E. (2006). Relation of oxytocin to psychological stress responses and hypothalamic-pituitary-adrenocortical axis activity in older women. *Psychosomatic Medicine, 68*(2), 238–245.

Taylor, S. E., Saphire-Bernstein, S., & Seeman, T. E. (2010). Are plasma oxytocin in women and plasma vasopressin in men biomarkers of distressed pair-bond relationships? *Psychological Science, 21*, 3–7.

Theoret, H., Halligan, E., Kobayashi, M., Fregni, F., Tager-Flusberg, H., & Pascual-Leone, A. (2005). Impaired motor facilitation during action observation in individuals with autism spectrum disorder. *Current Biology, 15*(3), R84–R85.

Thomas, K. M., & Casey, B. J. (2003). Methods for imaging the developing brain. In M. De Haan & M. H. Johnson (Eds.), *The cognitive neuroscience of development.* New York: Psychology Press.

Thomas, K. M., & Nelson, C. A. (1996). Age related changes in the electrophysiological response to visual stimulus novelty: A topographical approach. *Electroencephalography and Clinical Neurophysiology, 98*, 294–308.

Thompson, P. M., Schwartz, C., Lin, R. T., Khan, A. A., & Toga, A. W. (1996). Three-dimensional statistical analysis of sulcal variability in the human brain. *Journal of Neuroscience, 16*, 4261–4274.

Thomson, J. J. (1976). Killing, letting die, and the trolley problem. *Monist, 59*, 204–207.

Thomson, J. J. (1986). *Rights, restitution and risk: Essays in moral theory.* Cambridge, MA: Harvard University Press.

Tiihonen, J., Rossi, R., Laakso, M. P., Hodgins, S., Testa, C., Perez, J., Repo-Tiihonen, E., Vaurio, O., Soininen, H., Aronen, H. J., Kononen, M., Thompson, P. A., & Frisoni, G. B. (2008). Brain anatomy of persistent violent offenders: More rather than less. *Psychiatry Research: Neuroimaging, 163*(3), 201–212.

Tinbergen, N. (1951). *The study of instinct.* London: Oxford University Press.

Titchener, E. B. (1909). *Lectures on the experimental psychology of the thought processes.* New York: Macmillan.

Tizard, B., & Rees, J. (1975). Effect of early institutional rearing on behavior problems and affectional relationships of 4-year-old children. *Journal of Child Psychology and Psychiatry and Allied Disciplines, 16*(1), 61–73.

Todorov, A., & Duchaine, B. (2008). Reading trustworthiness in faces without recognizing faces. *Cognitive Neuropsychology, 25*(3), 395–410.

Todorov, A., & Engell, A. D. (2008). The role of the amygdala in implicit evaluation of emotionally neutral faces. *Social Cognitive and Affective Neuroscience, 3*(4), 303–312.

Todorov, A., Mandisodza, A. N., Goren, A., & Hall, C. C. (2005). Inferences of competence from faces predict election outcomes. *Science, 308*(5728), 1623–1626.

Todorov, A., Said, C. P., Engell, A. D., & Oosterhof, N. N. (2008). Understanding evaluation of faces on social dimensions. *Trends in Cognitive Sciences, 12*(12), 455–460.

Tomasello, M. (1999). *The cultural origins of human cognition.* Boston, MA: Harvard University Press.

Tomasello, M. (2009). *Why we cooperate.* Cambridge, MA: Bradford Books.

Tomasello, M., Hare, B., Lehmann, H., & Call, J. (2007). Reliance on head versus eyes in the gaze following of great apes and human infants: The cooperative eye hypothesis. *Journal of Human Evolution, 52*(3), 314–320.

Tomasello, M., & Todd, J. (1983). Joint attention and lexical acquisition style. *First Language, 4*, 197–212.

Toth, N. (1985). Archeological evidence for preferential right-handedness in the Lower Pleistocene, and its possible implications. *Journal of Human Evolution, 14*, 607–614.

Tranel, D., Bechara, A., & Denburg, N. L. (2002). Asymmetric functional roles of right and left ventromedial prefrontal cortices in social conduct, decision-making, and emotional processing. *Cortex, 38*(4), 589–612.

Tranel, D., & Damasio, H. (1995). Neuroanatomical correlates of electrodermal skin conductance responses. *Psychophysiology, 31*, 427–438.

Tranel, D., Damasio, A. R., & Damasio, H. (1988). Intact recognition of facial expression, gender and age in patients with impaired recognition of face identity. *Neurology, 38*, 690–696.

Tranel, D., Damasio, H., & Damasio, A. R. (1995). Double dissociation between overt and covert face recognition. *Journal of Cognitive Neuroscience, 7*, 425–432.

Tranel, D., Fowles, D. C., & Damasio, A. R. (1985). Electrodermal discrimination of familiar and unfamiliar faces: A methodology. *Psychophysiology, 22*(4), 403–408.

Trevarthen, C., & Hubley, P. (1978). Secondary intersubjectivity. In A. Lock (Ed.), *Action, gesture and symbol*. London: Academic Press.

Trivers, R. L. (1971). Evolution of reciprocal altruism. *Quarterly Review of Biology, 46*(1), 35–57.

Trivers, R. L., & Hare, H. (1975). Haplodiploidy and evolution of social insects. *Science, 191*(4224), 249–263.

Tseng, K. Y., & O'Donnell, P. (2005). Post-pubertal emergence of prefrontal cortical up states induced by D-1-NMDA co-activation. *Cerebral Cortex, 15*(1), 49–57.

Tulving, E. (1983). *Elements of episodic memory*. Oxford: Oxford University Press.

Turati, C., Cassia, V. M., Simion, F., & Leo, I. (2006). Newborns' face recognition: Role of inner and outer facial features. *Child Development, 77*(2), 297–311.

Turiel, E. (1983). *The development of social knowledge: Morality and convention*. Cambridge, UK: Cambridge University Press.

Tybur, J. M., Lieberman, D., & Griskevicius, V. (2009). Microbes, mating, and morality: Individual differences in three functional domains of disgust. *Journal of Personality and Social Psychology, 97*(1), 103–122.

Tzourio-Mazoyer, N., De Schonen, S., Crivello, F., Reutter, B., Aujard, Y., & Mazoyer, B. (2002). Neural correlates of woman face processing by 2-month-old infants. *NeuroImage, 15*(2), 454–461.

Uchino, B. N., Cacioppo, J. T., & Kiecolt-Glaser, J. K. (1996). The relationship between social support and physiological processes: A review with emphasis on underlying mechanisms and implications for health. *Psychological Bulletin, 119*(3), 488–531.

Uddin, L. Q., Iacoboni, M., Lange, C., & Keenan, J. P. (2007). The self and social cognition: The role of cortical midline structures and mirror neurons. *Trends in Cognitive Sciences, 11*, 153–157.

Ulrich, R. E., & Azrin, N. H. (1962). Reflexive fighting in response to aversive stimulation. *Journal of the Experimental Analysis of Behavior, 5*(4), 511–520.

Umilta, M. A., Escola, L., Intskirveli, I., Grammont, F., Rochat, M., Caruana, F., Jezzini, A., Gallese, V., & Rizzolatti, G. (2008). When pliers become fingers in the monkey motor system. *Proceedings of the National Academy of Sciences of the USA, 105*(6), 2209–2213.

Umilta, M. A., Kohler, E., Gallese, V., Fogassi, L., Fadiga, L., Keysers, C., & Rizzolatti, G. (2001). I know what you are doing: A neurophysiological study. *Neuron, 25*, 287–295.

Ungerleider, L. G., & Mishkin, M. (1982). Two cortical systems. In D. J. Ingle, M. A. Goodale, & R. J. W. Mansfield (Eds.), *Analysis of visual behaviour*. Cambridge, MA: MIT Press.

Urgesi, C., Berlucchi, G., & Aglioti, S. M. (2004). Magnetic stimulation of extrastriate body area impairs visual processing of nonfacial body parts. *Current Biology, 14*(23), 2130–2134.

Vaish, A., Carpenter, M., & Tomasello, M. (2009). Sympathy through affective perspective taking and its relation to prosocial behavior in toddlers. *Developmental Psychology, 45*(2), 534–543.

Valentine, T., Darling, S., & Donnelly, M. (2004). Why are average faces attractive? The effect of view and averageness on the attractiveness of female faces. *Psychonomic Bulletin and Review, 11*(3), 482–487.

van Baaren, R., Holland, R. W., Kawakami, K., & van Knippenberg, A. (2004). Mimicry and prosocial behavior. *Psychological Science, 15*(1), 71–74.

van Baaren, R., Janssen, L., Chartrand, T. L., & Dijksterhuis, A. (2009). Where is the love? The social aspects of mimicry. *Philosophical Transactions of the Royal Society of London, Series B, 364*(1528), 2381–2389.

van Elst, L. T., Woermann, F. G., Lemieux, L., Thompson, P. J., & Trimble, M. R. (2000). Affective aggression in patients with temporal lobe epilepsy: A quantitative MRI study of the amygdala. *Brain, 123*, 234–243.

van Erp, A. M. M., & Miczek, K. A. (2000). Aggressive behavior, increased accumbal dopamine, and decreased cortical serotonin in rats. *Journal of Neuroscience, 20*(24), 9320–9325.

Van Hoesen, G. W., Morecraft, R. J., & Vogt, B. A. (1993). Connections of the monkey cingulate cortex. In B. A. Vogt & M. Gabriel (Eds.), *The neurobiology of the cingulate cortex and limbic thalamus: A comprehensive handbook*. Boston, MA: Birkhauser.

van Honk, J., & Schutter, D. L. G. (2007). Vigilant and avoidant responses to angry facial expressions: Dominance and submission motives. In E. Harmon-Jones & P. Winkielman (Eds.), *Social neuroscience: Integrating biological and psychological explanations of social behavior*. New York: Guilford Press.

van Honk, J., Tuiten, A., Hermans, E., Putman, P., Koppeschaar, H., Thijssen, J., Verbaten, R., & van Doornen, L. (2001a). A single administration of testosterone induces cardiac accelerative responses to angry faces in healthy young women. *Behavioral Neuroscience, 115*(1), 238–242.

van Honk, J., Tuiten, A., van den Hout, M., Putman, P., de Haan, E., & Stam, H. (2001b). Selective attention to unmasked and masked threatening words: Relationships to trait anger and anxiety. *Personality and Individual Differences, 30*(4), 711–720.

Van Ijzendoorn, M. H., & Kroonenberg, P. M. (1988). Cross-cultural patterns of attachment: A meta-analysis of the Strange Situation. *Child Development, 59*(1), 147–156.

van't Wout, M., Kahn, R. S., Sanfey, A. G., & Aleman, A. (2006). Affective state and decision-making in the Ultimatum Game. *Experimental Brain Research, 169*, 564–568.

Vanman, E. J., Paul, B. Y., Ito, T. A., & Miller, N. (1997). The modern face of prejudice and structural features that moderate the effect of cooperation on affect. *Journal of Personality and Social Psychology, 73*(5), 941–959.

Varlinskaya, E. I., & Spear, L. P. (2002). Acute effects of ethanol on social behavior of adolescent and adult rats: Role of familiarity of the test situation. *Alcoholism: Clinical and Experimental Research, 26*(10), 1502–1511.

Veit, R., Flor, H., Erb, M., Hermann, C., Lotze, M., Grodd, W., & Birbaumer, N. (2002). Brain circuits involved in emotional learning in antisocial behavior and social phobia in humans. *Neuroscience Letters, 328*(3), 233–236.

Voorn, P., Vanderschuren, L., Groenewegen, H. J., Robbins, T. W., & Pennartz, C. M. A. (2004). Putting a spin on the dorsal–ventral divide of the striatum. *Trends in Neurosciences, 27*(8), 468–474.

Vritcka, P., Anderson, F., Grandjean, D., Sander, D., & Vuilleumier, P. (2008). Individual differences in attachment style modulates human amygdala and striatum activity during social appraisal. *PLoS One, 3*, e2868.

Vuilleumier, P. (2005). How brains beware: Neural mechanisms of emotional attention. *Trends in Cognitive Sciences, 9*, 585–594.

Vul, E., Harris, C., Winkielman, P., & Pashler, H. (2009). Puzzlingly high correlations in fMRI studies of emotion, personality, and social cognition. *Perspectives on Psychological Science, 4*(3), 274–290.

Wagner, A. D., Schacter, D. L., Rotte, M., Koutstaal, W., Maril, A., Dale, A. M., Rosen, B. R., & Buckner, R. I. (1998). Building memories: Remembering and forgetting of verbal experiences as predicted by brain activity. *Science, 281*, 1188–1191.

Wahba, A., & Bridgewell, L. (1976). Maslow reconsidered: A review of research on the need hierarchy theory. *Organizational Behavior and Human Performance, 15*, 212–240.

Walker, A. E. (1940). A cytoarchitectural study of the prefrontal area of the macaque monkey. *Journal of Comparative Neurology, 73*, 59–86.

Walsh, V., & Cowey, A. (1998). Magnetic stimulation studies of visual cognition. *Trends in Cognitive Sciences, 2*, 103–110.

Walster, E., Aronson, V., Abrahams, D., & Rottman, L. (1966). Importance of physical attractiveness in dating behavior. *Journal of Personality and Social Psychology, 4*, 508–516.

Walton, M. E., Bannerman, D. M., Alterescu, K., & Rushworth, M. F. S. (2003). Functional specialization within medial frontal cortex of the anterior cingulate for evaluating effort-related decisions. *Journal of Neuroscience, 23*(16), 6475–6479.

Wang, A. T., Lee, S. S., Sigman, M., & Dapretto, M. (2006). Neural basis of irony comprehension in children with autism: The role of prosody and context. *Brain, 129*, 932–943.

Ward, J. (2010). *The student's guide to cognitive neuroscience*. Hove, UK: Psychology Press.

Warneken, F., Hare, B., Melis, A. P., Hanus, D., & Tomasello, M. (2007). Spontaneous altruism by chimpanzees and young children. *Plos Biology, 5*(7), 1414–1420.

Warneken, F., & Tomasello, M. (2006). Altruistic helping in human infants and young chimpanzees. *Science, 311*(5765), 1301–1303.

Warneken, F., & Tomasello, M. (2008). Extrinsic rewards undermine altruistic tendencies in 20-month-olds. *Developmental Psychology, 44*(6), 1785–1788.

Wassermann, E. M. (1996). Risk and safety of transcranial magnetic stimulation: Report and suggested guidelines from the International Workshop on the Safety of Repetitive Transcranial Magnetic Stimulation, June 5-7. *Electroencephalogy and Clinical Neurophysiology, 108*, 1–16.

Wassermann, E. M., Cohen, L. G., Flitman, S. S., Chen, R., & Hallett, M. (1996). Seizures in healthy people with repeated 'safe' trains of transcranial magnetic stimulation. *Lancet, 347*, 825–826.

Waters, E., Merrick, S., Treboux, D., Crowell, J., & Albersheim, L. (2000). Attachment security in infancy and early adulthood: A twenty-year longitudinal study. *Child Development, 71*(3), 684–689.

Way, B. M., Taylor, S. E., & Eisenberger, N. I. (2009). Variation in the mu-opioid receptor gene (OPRM1) is associated with dispositional and neural sensitivity to social rejection. *Proceedings of the National Academy of Sciences of the USA, 106*, 15079–15084.

Wegner, D. M. (2002). *The illusion of conscious will*. Cambridge, MA: MIT Press.

Weiskrantz, L. (1956). Behavioral changes associated with ablations of the amygdaloid complex in monkeys. *Journal of Comparative Physiological Psychology, 49*, 381–391.

Wellman, H. (2002). Understanding the psychological world: Developing a theory of mind. In U. Goswami (Ed.), *Childhood cognitive development*. Oxford: Blackwell.

Wellman, H., & Lagattuta, K. H. (2000). Developing understandings of mind. In S. Baron-Cohen, H. Tager-Flusberg, & D. Cohen (Eds.), *Understanding other minds: Perspectives from developmental cognitive neuroscience*. Oxford: Oxford University Press.

Westermarck, E. (1891). *The history of human marriage*. London: Macmillan.

White, S., Burgess, P. W., & Hill, E. L. (2009). Impairments on 'open-ended' executive function tests in autism. *Autism Research*, 2(3), 138–147.

White, S., Hill, E., Winston, J., & Frith, U. (2006). An islet of social ability in Asperger syndrome: Judging social attributes from faces. *Brain and Cognition*, 61(1), 69–77.

White, S., O'Reilly, H., & Frith, U. (2009). Big heads, small details and autism. *Neuropsychologia*, 47(5), 1274–1281.

Whiten, A., & Byrne, R. W. (1988). The Machiavellian intelligence hypothesis. In R. W. Byrne & A. Whiten (Eds.), *Maciavellian intelligence: Social complexity and the evolution of intellect in monkeys, apes and humans*. Oxford: Oxford University Press.

Whiten, A., Horner, V., & de Waal, F. B. M. (2005). Conformity to cultural norms of tool use in chimpanzee. *Nature*, 437, 737–740.

Whiten, A., & van Schaik, C. P. (2007). The evolution of animal 'cultures' and social intelligence. *Philosophical Transactions of the Royal Society of London, Series B*, 362, 603–620.

Wicker, B., Keysers, C., Plailly, J., Royet, J. P., Gallese, V., & Rizzolatti, G. (2003). Both of us disgusted in my insula: The common neural basis of seeing and feeling disgust. *Neuron*, 40, 655–664.

Wilkinson, G. S. (1988). Reciprocal altruism in bats and other mammals. *Ethology and Sociobiology*, 9(2–4), 85–100.

Wilkinson, R., & Pickett, K. (2009). *The spirit level: Why equality is better for everyone*. London: Penguin.

Will, J. A., Self, P. A., & Datan, N. (1976). Maternal-behavior and perceived sex of infant. *American Journal of Orthopsychiatry*, 46(1), 135–139.

Williams, G. C. (1966). *Adaptation and natural selection: A critique of some current evolutionary thought*. Princeton: Princeton University Press.

Williams, J. H. G., Whiten, A., & Singh, T. (2004). A systematic review of action imitation in autistic spectrum disorder. *Journal of Autism and Developmental Disorders*, 34(3), 285–299.

Williams, J. R., Catania, K. C., & Carter, C. S. (1992). Development of partner preferences in female prairie voles (Microtus ochrogaster): The role of social and sexual experience. *Hormones and Behavior*, 26(3), 339–349.

Willingham, D. T., & Dunn, E. W. (2003). What neuroimaging and brain localization can do, cannot do, and should not do for social psychology. *Journal of Personality and Social Psychology*, 85(4), 662–671.

Wilson, R. S., Krueger, K. R., Arnold, S. E., Schneider, J. A., Kelly, J. F., Barnes, L. L., Tang, Y. X., & Bennett, D. A. (2007). Loneliness and risk of Alzheimer disease. *Archives of General Psychiatry*, 64(2), 234–240.

Wimmer, H., & Perner, J. (1983). Beliefs about beliefs: Representation and the constraining function of wrong beliefs in youg children's understanding of deception. *Cognition*, 13, 103–128.

Winston, J. S., Gottfried, J. A., Kilner, J. M., & Dolan, R. J. (2005). Integrated neural representations of odor intensity and affective valence in human amygdala. *Journal of Neuroscience*, 25(39), 8903–8907.

Winston, J. S., O'Doherty, J., & Dolan, R. J. (2003). Common and distinct neural responses during direct and incidental processing of multiple facial emotions. *NeuroImage*, 20, 84–97.

Winston, J. S., Strange, B. A., O'Doherty, J., & Dolan, R. J. (2002). Automatic and intentional brain responses during evaluation of trustworthiness of faces. *Nature Neuroscience*, 5(3), 277–283.

Wohlschlager, A., Gattis, M., & Bekkering, H. (2003). Action generation and action perception in imitation: An instance of the ideomotor principle. *Philosophical Transactions of the Royal Society of London, Series B*, 358, 501–515.

Wolf, A. P. (1995). *Sexual attraction and childhood associations*. Stanford, CA: Stanford University Press.

Wolpert, D. M., Ghahramani, Z., & Jordan, M. I. (1995). An internal model for sensorimotor integration. *Science*, 269, 1880–1882.

Wommack, J. C., Liu, Y., & Zuoxin, W. (2009). Animal models of romantic relationships. In M. D. Haan & M. R. Gunnar (Eds.), *Handbook of developmental social neuroscience*. New York: Guilford Press.

Young, A. W., Hellawell, D. J., Van de Wal, C., & Johnson, M. (1996). Facial expression processing after amygdalectomy. *Neuropsychologia*, 34, 31–39.

Younger, J., Aron, A., Parke, S., Chatterjee, N., & Mackey, S. (2010). Viewing pictures of a romantic partner reduces experimental pain: Involvement of neural reward systems. *PloS One*, 5(10), 7.

Yovel, G., & Kanwisher, N. (2005). The neural basis of the behavioral face-inversion effect. *Current Biology*, 15(24), 2256–2262.

Yovel, G., Tambini, A., & Brandman, T. (2008). The asymmetry of the fusiform face area is a stable individual characteristic that underlies the left-visual-field superiority for faces. *Neuropsychologia*, 46(13), 3061–3068.

Zahavi, A. (1975). Mate selection: A selection for a handicap. *Journal of Theoretical Biology*, 53, 205–214.

Zahavi, A. (1995). Altruism as a handicap: The limitations of kin selection and reciprocity. *Journal of Avian Biology*, 26(1), 1–3.

Zahn, R., Moll, J., Krueger, F., Huey, E. D., Garrido, G., & Grafman, J. (2007). Social concepts are represented in the superior anterior temporal cortex. *Proceedings of the National Academy of Sciences of the USA*, 104, 6430–6435.

Zajonc, R. B. (1980). Feeling and thinking: Preferences need no inferences. *American Psychologist*, *35*(2), 151–175.

Zak, P. J., Kurzban, R., Ahmadi, S., Swerdloff, R. S., Park, J., Efremidze, L., Redwine, K., Morgan, K., & Matzner, W. (2009). Testosterone administration decreases generosity in the Ultimatum Game. *PLoS One*, *4*, e8330.

Zaki, J., Weber, J., Bolger, N., & Ochsner, K. (2009). The neural bases of empathic accuracy. *Proceedings of the National Academy of Sciences of the USA*, *106*(27), 11382–11387.

Zangl, R., & Mills, D. L. (2007). Increased brain activity to infant-directed speech in 6- and 13-month-old infants. *Infancy*, *11*(1), 31–62.

Zebrowitz, L. A., Andreoletti, C., Collins, M. A., Lee, S. Y., & Blumenthal, J. (1998). Bright, bad, babyfaced boys: Appearance stereotypes do not always yield self-fulfilling prophecy effects. *Journal of Personality and Social Psychology*, *75*(5), 1300–1320.

Zebrowitz, L. A., & McDonald, S. M. (1991). The impact of litigants baby-facedness and attractiveness on adjudications in small claims courts. *Law and Human Behavior*, *15*(6), 603–623.

Zhu, Y., Zhang, L., Fan, J., & Han, S. H. (2007). Neural basis of cultural influence on self-representation. *NeuroImage*, *34*(3), 1310–1316.

Zihl, J., Von Cramon, D., & Mai, N. (1983). Selective disturbance of movement vision after bilateral brain damage. *Brain*, *106*, 313–340.

Zimbardo, P. G. (1972). *The Stanford Prison Experiment: A simulation study of the psychology of imprisonment*. Palo Alto, CA: Stanford University.

Zimbardo, P. G., Maslach, C., & Haney, C. (1999). Reflections on the Stanford Prison Experiment: Genesis, transformations, consequences. In T. Blass (Ed.), *Obedience to authority: Current perspectives on the Milgram paradigm*. Mahwah, NJ: Lawrence Erlbaum Associates, Inc.

Author index

Aaronson, J., 71, 170, 171
Abrahams, D., 181
Abrams, M. T., 263
Ackerman, P., 162
Adam, E. K., 198
Adams, C. M., 97
Adams, J. E., 196
Adams, J. M., 181
Adams, R. B., 116, 118
Adolphs, R., 5, 41, 42, 83, 84, 85, 86, 108, 109, 125, 237, 238, 244
Aggleton, J. P., 83, 85
Aglioti, S. M., 112, 131
Aguirre, G. K., 35
Ahad, P., 115, 259
Aharon-Peretz, J., 136
Ahmadi, S., 246
Aichhorn, M., 145
Ainsworth, M. D. S., 185, 186, 187
Akalis, S., 138
Akgan, J., 184
Akhtar, N., 265
Akiskal, H. S., 183
Akitsuki, Y., 251
Albersheim, L., 188
Aleman, A., 171
Alexander, G. E., 97
Alexander, M. P., 94, 143
Alleva, E., 257
Alley, T. R., 120
Allison, T., 29, 106, 116
Allport, G. W., 4, 214
Aloe, L., 257
Alpert, N. M., 244
Alterescu, K., 95
Altschuler, E. L., 151, 152
Alvard, M., 175
Amaral, D. G., 76
Amaro, E. J., 188
Ambady, N., 118, 213
American Psychiatric Association, 146, 246, 266
Amin, R., 125
Amodio, D. M., 143, 144, 217, 218, 219, 220, 231, 279
Anders, S., 86
Anderson, A. K., 87
Anderson, A. W., 104, 105, 106
Anderson, C. A., 239, 243
Anderson, F., 188, 189
Anderson, J. R., 61, 129
Anderson, S. W., 93, 237, 248, 251
Andreiuolo, P. A., 231
Andreoletti, C., 124
Andrew, C. M., 188

Andrews, C., 87, 92
Anen, C., 3, 174, 180
Anguas-Wong, A. M., 107
Antoun, N., 42, 87
Apergis, J., 83
Apicella, P., 97
Apperly, I. A., 141, 142, 144, 272, 279, 280
Aprile, T., 131
Aragona, B. J., 191
Arend, I., 125
Arevalo, J. M., 198
Armony, J. L., 231, 232
Arnold, S. E., 179, 198
Aron, A., 162, 183, 213
Aron, E. N., 162, 213
Aronen, H. J., 248
Aronson, J. A., 144
Aronson, V., 181
Arriola, M., 107
Arzy, S., 268
Asai, K., 231, 232
Asch, S. E., 164, 221
Ascher, J. A., 190
Ashburner, J., 143
Ashwin, C., 150
Ashwin, E., 150
Asperger, H., 146
Astington, J. W., 275
Ataca, B., 107
Atkinson, L., 189
Attwell, D., 34
Auger, A. P., 190
Aujard, Y., 261
Auyeung, B., 150
Avenanti, A., 131
Awatramani, G. B., 278
Axelrod, R., 166, 167, 168
Aziz-Zadeh, L., 134
Azrin, N. H., 241

Back, K., 181
Baddeley, A., 252
Baillargeon, R., 148, 270, 271
Baird, A. A., 118
Baird, G., 119, 146, 270
Baker, C. I., 106, 107, 119
Baker, M., 57
Baker, S. C., 142
Baldwin, D. A., 262
Bale, T. L., 190
Ballantyne, A. O., 42
Banaji, M. R., 8, 141, 143, 144, 210, 213, 214, 215, 216, 217, 218
Bandura, A., 135, 230, 242, 243
Banerjee, K., 234

Bannerman, D. M., 95, 166
Banse, R., 115
Bar-Haim, Y., 260
Barbaranelli, C., 230
Barbey, A. K., 144, 162, 210, 223
Barch, D. M., 94
Bard, K. A., 61, 118
Barenboim, C., 275
Bargh, J. A., 16, 114, 132, 133, 253
Barker, A. T., 44
Barksdale, C. M., 196
Barnes, L. L., 179, 198
Baron, R. A., 123
Baron-Cohen, S., 116, 117, 118, 119, 132, 136, 139, 141, 146, 147, 148, 149, 150, 233, 262, 270
Barr, A., 175, 176
Barraclough, N. E., 106, 107
Barrett, C., 175, 176
Barrientos, S., 162
Barriere, M., 259
Bartels, A., 182
Barton, J. S. J., 104
Bartres-Faz, D., 45, 46
Bassolino, M., 65
Batelli, L., 44
Bateson, M., 160
Batson, C. D., 130, 136, 161, 162, 163
Batson, J. G., 162
Batth, R., 277
Baumeister, R. F., 196, 198, 241, 253, 254
Baumgartner, T., 192
Baxter, M. G., 43, 85, 89
Bayliss, A. P., 116, 117
Bayly, M. J., 162
Beach, S. R., 181
Beardsall, L., 275
Beauregard, M., 223
Bechara, A., 83, 87, 93, 196, 237, 248, 250, 251
Beckett, C., 193
Bedny, M., 144, 272
Beehner, J. C., 246
Beer, J. S., 219
Begum, F. N., 125
Behen, M. E., 194, 257
Behrens, T. E. J., 94
Behrmann, M., 153
Bekkering, H., 60, 61, 66, 265
Belin, P., 115, 259
Bellgrove, M. A., 219, 220
Belsky, J., 187
Bender, D. B., 103
Bennett, D. A., 179, 198
Benson, D. F., 252

Benson, P. J., 66, 116
Bentin, S., 29
Benuzzi, F., 66
Berg, J., 173
Berkowitz, L., 240, 241
Berlucchi, G., 112
Bernhardt, P. C., 246
Berns, G. S., 168, 169
Berntson, G. G., 4
Berry, D. S., 124
Berry, J. D., 198
Berry Mendes, W., 23
Berscheid, E., 121
Berthoz, S., 231, 232
Bertoncini, J., 259
Best, C. T., 267
Bevan, R., 66
Birbaumer, N., 247, 248
Birch, K., 162
Bird, C. M., 143, 237
Bird, G., 148
Bisarya, D., 149
Bishop, P., 5, 196
Blackhart, G. C., 196
Blackmore, S., 56
Blaha, C. D., 97, 191
Blair, J. R. J., 90, 91, 93, 95, 232, 233, 234,
  236, 237, 247
Blair, K., 93, 95
Blair, R. J. R., 148, 231, 232
Blairy, S., 23
Blake, R., 106, 112
Blakemore, S. J., 206, 272, 277, 279, 280
Blanke, O., 67, 206, 207, 268
Blasi, A., 263
Blehar, M. C., 185, 186, 187
Bliss, E., 245
Bloch, M., 222
Blood, A. J., 93
Bloom, P., 53
Bluemke, M., 215
Blumenthal, J., 124, 277
Bodamer, J., 103
Bodenhausen, G. V., 213
Boduroglu, A., 211
Bojeson, S., 91
Bolger, D. J., 62
Bolger, N., 132
Bolhuis, J. J., 184
Bolyanatz, A., 175, 176
Bond, M. H., 107
Bookheimer, S. Y., 90, 152, 218
Booth, A., 242, 245, 246
Boratav, H. B., 107
Borg, J. S., 235
Boria, S., 152
Borton, R. W., 17
Bos, P. A., 246
Bottini, G., 143
Botvinick, M., 94
Botvinick, M. M., 94
Bouchard, J., 57

Bourgeois, P., 136
Bowlby, J., 185
Bowles, S., 175
Bowman, L. C., 271
Boyd, R., 175
Boyer, P., 204, 222
Bradley, M. M., 23
Bramati, I. E., 87, 231
Bramham, J., 93, 136, 244, 249, 250
Brammer, M., 87
Brammer, M. J., 106, 188
Branchi, I., 257
Brandman, T., 105
Brandt, J. R., 162
Brass, M., 66
Braver, T. S., 94
Bredenkamp, D., 193
Brennan, K. A., 187
Breugelmans, S. M., 107
Bridgewell, L., 89
Brindley, R., 148
Brindley, R. M., 153
Brink, J., 248
Brook, J., 280
Brook, J. S., 243
Brooks, R., 262
Brooks-Gunn, J., 268
Broome, M. R., 182
Brophy, M., 142
Brown, E. N., 145
Brown, H. R., 92, 237
Brown, J., 274, 275
Brown, M., 118
Brown, R., 214
Brown, S. L., 162, 163
Brown, S. M., 94
Brownlow, S., 124
Bruce, V., 15, 102, 103, 104, 108, 110,
  111, 116
Brugger, P., 67, 206
Brunswick, N., 142
Buccino, G., 66
Buchanan, T. W., 42, 85, 86
Buck, A., 174
Buckley, M. J., 95
Buckley, T., 162
Buckner, R. I., 37
Bucy, P. C., 81, 244
Buechel, C., 85
Bufalari, I., 131
Bukowski, H., 157
Bulbulia, J., 222
Bullmore, E. T., 87
Bullock, P. R., 93, 136, 244, 249, 250
Bunge, S. A., 90, 188
Burgess, P., 252
Burgess, P. W., 148
Burnham.,D., 267
Burrows, L., 253
Burt, D. M., 122
Bush, G., 93, 94, 143
Bushman, B. J., 239, 243

Bushnell, I. W. R., 261
Buss, D. M., 121, 122
Buttelmann, D., 61
Byatt, G., 121
Byrne, R. W., 16, 50, 52

Cabecinhas, R., 107
Cabeza, R., 86
Cacioppo, J. T., 4, 23, 25, 98, 138, 179, 192,
  197, 198
Cador, M., 97
Caglar, S., 210, 212
Cahill, L., 72, 83
Calder, A. J., 41, 42, 83, 85, 87, 108, 244
Call, J., 49, 61, 116
Callaghan, T., 274
Caltran, G., 264
Calvert, G. A., 65, 106
Calvert, S. L., 179
Camerer, C., 175
Camerer, C. F., 3, 96, 97, 174, 180
Campbell, C., 252, 253
Campbell, R., 103, 116, 117, 118, 146, 147
Campos, J. J., 89, 262
Canale, D., 183
Canessa, N., 66
Canli, T., 86
Cannon, W. B., 75
Caparelli-Daquer, E. M., 87
Capgras, J., 110
Caprara, G. V., 230
Card, J., 262
Cardenas, J., 175, 176
Cardinal, R. N., 96
Cargnelutti, B., 54
Carlson, E. C., 196
Carlson, M., 242
Carmichael, S., 76
Carmody, D. P., 268
Carpenter, M., 61, 265, 272
Carr, L., 133, 134
Carr, T. S., 245
Carre, J. M., 123
Carroll, N., 142
Carter, C. S., 8, 94, 188, 190, 191, 192
Caruana, F., 133
Carver, L. J., 187
Casey, B. J., 263, 278
Caspi, A., 251
Cassano, G. B., 183
Cassia, V. M., 261
Cassidy, J., 188
Casteli, F., 143, 237
Castellanos, F. X., 277
Castelli, F., 138, 143, 147
Castle, J., 193
Catania, K. C., 190
Cattaneo, L., 152
Cavada, C., 77
Cervone, J. C., 181
Chabris, C. F., 18
Chae, J., 107

Chakrabarti, B., 150
Chaminade, T., 162
Chan, A. W., 111
Chance, M., 115
Chandler, J., 214, 215
Chandler, S., 146
Chapman, H. A., 87
Charman, T., 119, 146, 270
Chartrand, T. L., 16, 114, 132, 133, 136, 161
Chater, N., 221
Chatterjee, N., 183
Chaudhri, J., 125
Cheesman, J., 18
Chen, I. Y., 150
Chen, K., 62
Chen, M., 253
Chen, Q., 191
Chen, R., 46
Cheney, D. L., 51
Cheng, Y., 150
Cheng, Y. W., 136, 150
Chersi, F., 66, 133
Chiavarino, C., 144
Chin, W. H., 107
Chisholm, K., 194
Chittka, L., 52
Chitty, A. J., 66
Cho, M. M., 191
Cho, R. Y., 94
Chochon, F., 62
Choi, I., 149
Chomsky, N., 53, 54
Chopra, S., 90
Chou, K. H., 150
Choudhury, S., 279
Chugani, D. C., 194, 257
Chugani, H. T., 194, 257
Chun, M. M., 105
Churchland, P. S., 14, 253
Cialdini, R. B., 162, 163
Ciaramelli, E., 237
Cicchetti, P., 5, 85
Cima, M., 237, 247
Cipolotti, L., 77, 85, 91, 237
Civai, C., 171
Clark, A., 63, 64, 130
Clark, F., 233, 247
Clark, R. D., 161
Clark, S., 213
Clasen, L. S., 277
Claux, M. L., 274
Clement, J. G., 73
Clements, W. A., 270, 271
Clore, G. L., 79, 80
Clutton-Brock, T., 54
Coan, J. A., 273, 274
Coffey-Corina, S., 267
Cohen, D., 238, 239
Cohen, J., 71, 170, 171
Cohen, J. D., 8, 94, 144, 232, 233, 252
Cohen, L., 62
Cohen, L. G., 46

Cohen, P., 243, 280
Cohen, S., 94
Cole, S. W., 198
Coles, M. G. H., 94, 219
Collins, A., 79
Collins, D., 38
Collins, M. A., 124
Company, T., 77
Comstock, G., 243
Comunian, A. L., 107
Conboy, B. T., 267
Conley, M., 85
Connor, R. C., 55
Conway, M. A., 207, 208
Cook, M., 90
Cooke, D. F., 64
Cooper, G., 108, 109
Cooper, J. C., 90
Corcoran, R., 143
Coricelli, G., 93
Corkin, S., 209
Corp, N., 16, 52
Corradi-Dell'Acqua, C., 171
Cosmides, L., 6, 209
Cossu, G., 152
Costa, P. T., 18, 122
Couppis, M. H., 244
Covert, A. E., 219, 220
Cowey, A., 18, 44, 116
Cox, A., 119, 270
Craig, A. D., 87
Craig, I. W., 251
Craighero, L., 6, 137, 162
Crane, R. S., 241
Creswell, C. S., 150
Critchley, H. D., 23, 77, 85, 87, 93, 94
Critton, S. R., 129
Crivello, F., 261
Crockett, M. J., 90
Crone, E. A., 273, 279
Cronin, H., 245
Cross, P., 116, 118
Crowell, J., 188
Crutcher, M. D., 97
Cruz-Rizzolo, R. J., 77
Csibra, G., 116, 262, 265, 267, 270
Cunningham, W., 19
Cunningham, W. A., 8, 86, 217, 218, 219
Curtin, G., 219, 220
Curtis, J. T., 191
Cushman, F., 229, 237, 238, 244
Custance, D. M., 61
Cuthbert, B. N., 23
Cypert, K., 93

Dabbs, J., 246
Dabbs, J. M., 10, 245, 246, 251
Dabholkar, A. S., 258, 277
Dagenais, J. E., 168
Dagher, A., 92
Dale, A. M., 37, 94
Damasio, A., 41, 83, 237, 238, 244

Damasio, A. R., 4, 5, 42, 75, 83, 85, 87, 92, 93, 103, 108, 109, 110, 125, 166, 196, 206, 248, 249, 237
Damasio, H., 5, 23, 41, 77, 83, 87, 92, 93, 103, 108, 109, 110, 196, 237, 248, 249, 251
Damsma, G., 97, 191
D'Andrea, I., 257
Dapretto, M., 151, 152, 278
Darley, J. M., 8, 232, 233
Darling, S., 121
Darwin, C., 54, 58, 72, 79
Dasgupta, N., 216
Datan, N., 275
David, A. S., 87
Davidoff, J., 149
Davidson, R. J., 273, 274
Davies, M. S., 152
Davis, A. M., 190
Davis, M., 18, 132, 218
Dawkins, R., 56, 158
Dawson, G., 267
Dayan, P., 97
de Felipe, C., 196
de Gelder, B., 113, 114
De Gelder, B., 74
de Haan, E., 241
de Haan, M., 261, 262
de Oliveira-Souza, R., 87, 162, 163, 164, 231
de Quervain, D. J. F., 174
De Renzi, E., 103
De Schonen, S., 261
de Veer, M. W., 111
de Vignemont, F., 134, 233
de Waal, F. B. M., 56, 130, 161
De Waal, F. B. M., 159
DeBruine, L. M., 124
DeCasper, A. J., 259
Decety, J., 98, 131, 136, 137, 142, 144, 150, 162, 197, 198, 206, 251, 265, 270, 272
DeEskinazi, F. G., 5, 196
Degere, D. N., 107
DeHaan, M. 262
Dehaene, S., 62, 63, 94, 219
Dehaene-Lambertz, G., 62
Del Priore, R. E., 241
Delgado, M. R., 175
Della Sala, S., 252
DeLuca, C. J., 23
den Ouden, H., 279
Denburg, N. L., 250
Denckla, M. B., 263
Dennett, D. C., 60, 139, 140, 203
Deouell, L. Y., 29
DePauw, K. W., 110
Desmond, J. E., 86
D'Esposito, M., 35
Desrochers, T. M., 83
Devine, P. G. 216, 218, 219, 220, 236
Devos, T., 215
DeVries, A. C., 8, 188, 190, 191, 192

DeVries, M. B., 192
DeWall, C. N., 253, 254
Dewar, B. K., 77, 85
Di Corcia, J. A., 234
Di Fausto, V., 257
di Pellegrino, G., 66, 171, 237
Di Russo, F., 131
Diamond, J., 56
Dicke, P. W., 119
Dickhaut, J., 173
Diener, E., 221
Dijksterhuis, A., 61, 136
Dimberg, U., 23, 73, 109
Dimitrov, M., 249
Dinstein, I., 153
Dion, K., 121
Dittes, J. E., 182
Dittrich, W., 113
Djunaidi, A., 107
Dockree, P. M., 219, 220
Dodds, C. M., 111
Dolan, R., 84, 147
Dolan, R. J., 23, 77, 81, 83, 84, 85, 86, 87,
   93, 97, 105, 124, 134, 135, 136, 142, 164,
   165, 169, 170 173, 231, 232
Dolcos, F., 86
Dollard, J., 241
Donchin, E., 94, 219
Dondi, M., 264
Done, D. J., 206
Donnelly, M., 106, 112, 121
Doob, L., 241
Doris, J. M., 253
Dorsa, D. M., 190
Dougherty, D. D., 244
Dovidio, J. F., 161
Downing, P. E., 111, 112
Dreber, A., 174
Drevets, W., 191
Drevets, W. C., 93, 95
Drew, A., 119, 270
Driver, J., 105, 113
Dubeau, M. C., 133, 134
Duchaine, B. C., 103, 109, 125
Duggins, A. J., 87
Dumontheil, I., 272, 279, 280
Dunbar, R. I. M., 8, 50, 52, 53, 54
Duncan, B. D., 162
Dunn, E. W., 213
Dunn, J., 142, 274, 275
Dunton, B. C., 215
Duveen, G., 275
Dyck, J. L., 162
Dziurawiec, S., 17, 259, 260

Eastman, N., 252, 253
Eberhardt, J. L., 261
Ecker, C., 188
Eder, R. A., 275
Efremidze, L., 246
Ehlers, S., 147
Ehlert, U., 192
Ehrenkranz, J., 245

Eippert, F., 86
Eisenberg, N., 137, 273
Eisenberger, N. I., 90, 182, 183, 195,
   196, 218
Ekman, P., 74, 78, 79, 107, 180, 231
Ekstrom, A. D., 7
El Karoui, I., 265
Elfenbein, H. A., 168, 248
Elias, M., 246
Ellgring, H., 114
Ellingsen, T., 174
Elliot, A. J., 216, 218
Elliott, R., 23
Ellis, B. E., 73
Ellis, H., 110
Ellis, H. D., 17, 259, 260
Ellsworth, P., 74, 79
Elmehed, K., 23, 109
Elwell, C. E., 263
Emerson, P. E., 184, 185
Engel, A. K., 26
Engell, A. D., 124, 125, 232
Ensminger, J., 175, 176
Epley, N., 138
Erb, M., 247, 248
Erickson, K., 191
Ernst, M., 278
Escola, L., 133
Eslinger, P. J., 5, 87, 166, 231
Esslinger, C., 191
Esteves, F., 84
Etcoff, N. L., 41, 42, 83
Evans, A. C., 38, 92, 277
Everitt, B. J., 96, 97
Evraire, L. E., 271

Fabbri-Destro, M., 151, 152
Fabes, R. A., 137, 273
Fadiga, L., 66
Fagot, J., 149
Falk, A., 174
Fan, J., 212
Farah, M. J., 108, 203, 204, 252, 261
Farne, A., 65
Farnham, S. D., 214, 217
Farrer, C., 137
Farroni, T., 116, 262
Fazio, R. H., 215
Fedigan, L., 57
Fehr, E., 157, 171, 172, 173, 174, 175, 191
Feinstein, J. S., 87
Feldman Barrett, L., 88
Fellows, L. K., 108
Feng, S., 62
Fernandez, G., 246
Ferrari, P. F., 66, 67, 133, 272
Ferstl, E. C., 143, 144
Festinger, L., 181
Feys, J., 214
Fibiger, H, C., 97, 191
Fiedler, K., 215
Fielden, J. A., 246
Fifer, W. P., 259

Fine, C., 148
Finger, E. C., 232
Fink, G. R., 273, 274
Fiore, M., 257
Fischbacher, U., 172, 173, 174, 191
Fischer, E. F., 181
Fischer, H., 217, 218
Fischman, A. J., 244
Fiske, S. T., 143
Fitch, W. T., 54
Fleming, A. S., 189, 191
Fletcher, P., 147
Fletcher, P. C., 142
Flitman, S. S., 46
Floden, D., 94
Flor, H., 247, 248
Flory, J. D., 94
Flykt, A., 84
Fodor, J. A., 5, 141, 146
Fogassi, L., 65, 66, 67, 133, 272
Fogel, A., 262
Fong, G. W., 97
Fontaine, J., 107
Fonteneau, E., 149
Forster, B. B., 248
Forster, P., 59
Fortenbach, V. A., 162
Fowles, D. C., 109
Fox, C. J., 104
Fox, N., 262
Foxe, J. J., 219, 220
Frackowiak, R., 147
Frackowiak, R. S. J., 142, 143
Frady, R. L., 245
Fraedrich, E. M., 187
Fraley, R. C., 187
Frank, R., 249
Frank, R. H., 175
Fredrikson, M., 91, 222
Freeston, I. L., 44
Fregni, F., 152
Freud, S., 221
Fridlund, A. J., 23, 25
Fried, I., 7, 26, 27, 102
Friederici, A. D., 259
Friesen, W. V., 74, 78, 79
Frischen, A., 116, 117
Frisoni, G. B., 248
Friston, K. J., 85
Frith, C., 138, 143, 147, 221, 279
Frith, C. D., 5, 6, 7, 67, 83, 85, 97, 134, 135,
   136, 137, 138, 142, 143, 144, 148, 164,
   165, 169, 170, 173, 195, 206, 231, 272,
   279
Frith, U., 5, 67, 125, 138, 139, 142, 143,
   144, 146, 147, 148, 149, 153, 233, 237,
   270, 272
Fritsch, G. T., 75
Frye, C., 273, 274
Fudenberg, D., 174
Fujii, N., 67
Fultz, J., 162
Funayama, E. S., 8, 217, 218

Furnham, A., 125
Furst, C., 91
Fusar-Poli, P., 182
Fuster, J. M., 252

Gabrieli, J. D. E., 86, 90, 261
Gachter, S., 157, 173, 174
Gaertner, S. L., 161
Gaffan, D., 81
Gaillard, W. D., 263
Gailliot, M. T., 253, 254
Galaburda, A., 249
Galef, B. G., 59
Gallagher, H. L., 142
Gallagher, M., 85
Gallagher, S., 130, 203, 204
Gallegos, D. R., 15
Gallese, V., 65, 66, 87, 130, 133, 141, 206
Gallhofer, B., 191
Gallup, G. G., 129, 143
Gallup, G. G. J., 111
Galvan, A., 278
Gamer, M., 171
Ganel, T., 108
Garavan, H., 219, 220
Gardner, M., 280
Garnica, O., 267
Garrido, G., 143
Garrido, L., 109
Gatenby, J. C., 8, 83, 217, 218
Gattis, M., 60
Gauthier, I., 104, 105, 106
Gawronski, B., 216
Gazzola, V., 134
Geertza, A. W., 222
Gehring, W. J., 94, 219
Gergely, G., 60, 61, 265, 267, 270
Gerrans, P., 142
Gesierich, B., 66, 133
Gesn, P. R., 132
Getz, L. L., 8, 189, 190
Getz, S., 278
Ghahramani, Z., 204
Ghahremani, D. G., 261
Giampietro, V. P., 188
Gianaros, P. J., 94
Giedd, J. N., 277
Gil-White, F., 175
Gilbert, S. J., 148
Gillath, O., 188
Gillberg, C., 147
Gillihan, S. J., 108, 203, 204
Gintis, H., 175
Gitelman, D., 86
Gitelman, D. R., 116
Gleason, C. A., 205
Glenn, A. L., 168, 248
Glover, G. H., 274
Gobbini, M. I., 15, 104, 106, 108, 110
Gobrogge, K. L., 191
Gogtay, N., 277
Golarai, G., 261
Goldberg, E., 148, 252

Goldman, A., 130, 141
Goldman-Rakic, P. S., 252
Goldsmith, D. R., 168, 248
Goldsmith, H. H., 273, 274
Gonsalkorale, K., 219
Gonzaga, G. C., 193
Gonzalez, A., 189
Goodale, M. A., 108
Goodenough, O. R., 254
Gopnik, M., 141
Gordon, H. L., 118
Gore, J. C., 8, 83, 104, 105, 106, 217, 218
Goren, A., 101, 102
Gorenstein, E. E., 247
Goshen-Gottstein, Y., 108
Goss, B., 94, 219
Gosselin, F., 42, 85
Gosselin, N., 85
Gottfried, J. A., 86
Gould, S. J., 6, 53
Grabowski, T. J., 87, 196, 249
Grafman, J., 92, 143, 144, 162, 163, 164, 173, 210, 215, 219, 223, 231, 237, 249
Graham, J., 235, 236
Graham, M. D., 191
Graham, T., 132
Grammont, F., 133
Grandin, T., 234
Grandjean, D., 188, 189
Grandon, C. B., 263
Granqvist, P., 222
Grant, J., 117, 118, 146, 147
Grasby, P. M., 244
Gray, J. A., 87
Gray, R. D., 57
Graziano, M. S., 64
Greendale, G. A., 193
Greene, J. D., 8, 229, 232, 233
Greenstein, D., 277
Greenwald, A. G., 214, 215, 216, 217
Gregg, T. R., 5, 244
Gregory, M. D., 86
Greimel, E., 273, 274
Greve, D., 94
Griffin, R., 262
Griffin, T. M., 214, 215
Griffitt, C., 162
Griffitt, C. A., 162
Grill-Spector, K., 261
Griskevicius, V., 87, 230
Grodd, W., 247, 248
Groen, G., 15
Groenewegen, H. J., 96
Gros-Louis, J., 57
Gross, C. G., 103
Gross, J., 86
Gross, J. J., 90
Grossman, E., 106, 112
Grossman, K. E., 187
Grossmann, K. E., 185, 191
Grossmann, T., 259
Gruppe, H., 191
Gschwendner, T., 216

Gunnar, M. R., 194
Gupta, U., 181
Gurley, D., 280
Guroglu, B., 279
Gurunathan, N., 149
Gurven, M., 175, 176
Gutchess, A. H., 211
Guth, W., 166, 170
Guthrie, S., 138, 222
Gutman, D. A., 168, 169
Gwako, E., 175, 176

Hackett, G., 150, 268
Hadjikhani, N., 113, 151
Hagenfeldt, A., 222
Haggard, P., 205, 213
Haidt, J., 79, 90, 229, 230, 235, 236
Halit, H., 262
Hall, C. C., 101, 102
Hall, J., 96
Hallett, M., 46
Halligan, E., 152
Hamann, S., 86
Hamilton, A. F. C., 153
Hamilton, A. F. D., 153
Hamilton, W. D., 158, 160, 166, 167, 168
Han, J. S., 85
Han, S. H., 149, 211, 212
Haney, C., 201
Hansen, P. C., 106
Hanus, D., 272
Happe, F., 138, 142, 143, 147, 149
Harbaugh, C. R., 190
Hardin, M., 278
Hare, B., 49, 116, 118, 247, 272
Hare, H., 158
Hare, R. D., 236, 247, 248
Hare, T. A., 96, 97
Hariri, A., 218
Hariri, A. R., 90, 94
Harlow, H. F., 185
Harlow, J. M., 248
Harmon,-Jones, E., 218, 219, 220, 240
Harpur, T. J., 247
Harries, M. H., 66
Harris, C., 19
Harris, J. R., 275, 276
Harris, L. T., 143
Harrison, T., 196
Hart, A. J., 217, 218
Hart, S. D., 247
Hartley, S. L., 219, 220
Haslam, S. A., 201
Hassin, R., 101
Hatfield, T., 85
Hauser, M., 229, 237, 238, 244
Hauser, M. D., 8, 9, 54, 229, 230, 232, 234, 237, 247
Hawkley, L. C., 197, 198
Hawley, P. H., 240, 241, 242
Haxby, J., 106
Haxby, J. V., 15, 104, 106, 108, 110
Hayashi, K. M., 277

Haynes, J. D., 39
Hazan, C., 187
Head, A., 116
Healy, S. D., 52
Heatherton, T. F., 143, 210, 212
Heberlein, A. S., 108
Heeger, D. J., 153
Heider, F., 138
Heike, T., 67
Hein, G., 163
Heinecke, A., 173
Heinrichs, M., 191, 192
Hellawell, D. J., 83, 85
Henderson, L. M., 266
Henning, A., 264
Henrich, J., 175, 176
Henrich, N., 175, 176
Henrich, N. S., 175
Henson, R. N. A., 37, 105
Herman, B. H., 5, 196
Herman, D. H., 277
Herman, L. H., 55
Hermann, C., 247, 248
Hermans, E., 241, 245, 246
Hermans, E. J., 246
Hernandez-Lloreda, M. V., 49
Herpertz-Dahlmann, B., 273, 274
Herrmann, E., 49
Hess, U., 23, 136
Hesselink, J., 42
Hester, R., 219, 220
Hewer, A., 229
Heyes, C., 153
Heyes, C. M., 59
Heywood, C., 116
Hichwa, R. D., 87, 196
Hietanen, J. K., 66, 116
Hihara, S., 67
Hill, E., 125
Hill, E. L., 146, 148, 176
Hill, J., 132
Hill, K., 175
Hill, R. A., 52
Hitzig, E., 75
Hodes, R. M., 260
Hodges, J. R., 41, 42, 83, 143
Hodgins, S., 248
Hoffman, E., 106
Hoffman, E. A., 15, 104, 106, 108, 110
Hofmann, W., 216
Hoge, R. D., 35
Hogg, M. A., 9
Hohmann, M., 172
Holland, P., 85
Holland, R., 133, 161
Holloway, J. L., 281
Holmes, C. J., 277
Holmes, J., 94
Holmes, N. P., 65
Holmes, W. G., 158
Homberg, V., 87
Hommer, D., 97

Hooker, C. I., 116
Hoorens, V., 214
Hopkins, W. D., 118
Horenstin, J. A., 94
Hornak, J., 92, 93, 108, 136, 244, 249, 250
Horner, V., 56, 61
Horwitz, M., 214
Hosobuchi, Y., 196
Houser, D., 142, 173
Hsu, Y. Y., 136
Hu, P. F., 193
Hubbard, E. M., 151, 152
Huber, D., 191
Hubley, P., 265
Huebner, B., 234
Huettel, S. A., 162, 163
Huey, E. D., 143
Hughes, C., 148
Humphrey, N. K., 6, 50
Humphreys, G., 142
Humphreys, G. W., 144
Hung, D., 136, 150
Hunt, G. R., 57
Hunt, P. S., 281
Hunt, S., 196
Hurley, S., 130
Hursti, T., 91
Husain, M., 143, 237
Hussain, S., 142
Huttenlocher, P. R., 258, 277

Iacoboni, M., 7, 66, 111, 133, 134, 151, 152
Iadecola, C., 34
Iaria, G., 104
Ichinose, S., 67
Ickes, W., 132
Ignacio, F. A., 87
Inati, S., 210, 212
Insel, T. R., 190
Intskirveli, I., 133
Iriki, A., 64, 65, 67, 68
Irwin, M. R., 196
Isen, A. M., 253
Ishibashi, H., 67
Itakura, S., 274
Ito, J. M. B., 271
Ito, T. A., 23, 215, 220
Iversen, S. D., 106
Iwamura, Y., 65
Iwata, J., 5, 85
Izard, V., 62, 63

Jabbi, M., 132
Jack, A. I. 204
Jack, K., 57
Jackendoff, R., 54
Jackson, J. R., 215
Jackson, P. J., 137, 142
Jackson, P. L., 131, 137
Jacobs, G., 241
Jahoda, G., 221
Jairam, M. R., 168, 248

Jalinous, R., 44
James, W., 74
Jankoviak, W. R., 181
Janssen, L., 136
Jarcho, J. M., 196, 218
Jassikgerschenfeld, D., 259
Jeeves, M., 116
Jeffries, N. O., 277
Jellema, T., 113, 119
Jenike, M. A., 94
Jenkins, J. M., 275
Jernigan, T. L., 277
Jezzini, A., 133
Ji, Y., 62
Jia, X. X., 191
Jiang, Y. H., 111
Joffe, T. H., 53, 269
Joffily, M., 93
Johansson, M., 147
Johnsen, E., 85
Johnson, D. W., 202
Johnson, F. P., 202
Johnson, J. G., 243
Johnson, M., 83, 85
Johnson, M. H., 17, 116, 259, 260, 262, 265
Johnson, M. K., 86
Johnson-Laird, P. N., 79
Johnston, L., 136
Jones, D. L., 97
Jones, J. T., 214
Jones, L., 233, 247
Jones, M., 93, 95
Jones, W. H., 181
Jones-Gotman, M., 92
Joraschky, P., 188
Jordan, M. I., 204
Jorgensen, B. W., 181
Joseph, R. M., 151
Josephs, O., 37, 77, 85
Joshi, K., 125
Juhasz, C., 194, 257
Jundi, S., 125

Kacelnik, A., 57
Kahana-Kalman, R., 262
Kahn, R. S., 171
Kalin, N. H., 196
Kalogeras, J., 213
Kamel, N., 232
Kang, E. J., 86
Kanner, L., 146
Kanwisher, N., 6, 104, 105, 111, 145
Kaplan, J. T., 7, 134
Kaplan, N., 188
Kapogiannis, D., 223
Karbon, M., 137, 273
Karmiloff-Smith, A., 117, 118, 146, 147
Karnat, A., 87
Karnath, H. O. 42
Kasen, S., 243
Katz, D., 94
Kaube, H., 134, 195

Kawai, M., 60
Kawakami, K., 133, 161
Kawamura, S., 60
Kay, A. R., 35
Kaye, K., 262, 265
Keane, J., 42, 87, 244
Keenan, J. P., 45, 46, 111
Keizer, K., 253
Kelley, H. H., 182
Kelley, W. M., 210, 212
Kellogg, C. K., 278
Kelly, J. F., 179, 198
Kelly, S. P., 219, 220
Kennedy, C. H., 244
Kennett, J., 234
Kenward, B., 57
Kerns, J. G., 94
Keysers, C., 66, 87, 132, 134
Khan, A. A., 38
Kiebel, S. J., 169, 170, 173
Kiecolt-Glaser, J. K., 179, 192
Kiehl, K. A., 235, 248
Kihlstrom, J. F., 209
Killgore, W. D. S., 278
Kilner, J. M. 86
Kilts, C. D., 168, 169
Kim, D. A., 87
King, M., 49
King-Casas, B., 3, 174, 180
Kipps, C. M., 87
Kiraly, I., 60, 61, 265
Kirsch, P., 191
Kirschbaum, C., 192
Kitamura, C., 267
Kitayama, S., 211
Kiverstein, J., 130
Kleck, R. E., 116, 118
Klein, L. C., 193
Klein, S. B., 209
Klinnert, M. D., 89, 262
Kluver, H., 81, 244
Knickmeyer, R., 150
Knoch, D., 171, 172
Knowles, M. L., 196
Knutson, B., 97
Knutson, K. M., 215, 219
Kobayashi, C., 274
Kobayashi, M., 152
Koch, C., 26, 27, 102
Koch, S. P., 259
Koeda, M., 231, 232
Koenigs, M., 176, 237, 238, 244
Kohlberg, L., 229
Kohler, E., 66
Kolb, B., 29
Konig, P., 26
Kononen, M., 248
Konrad, K., 273, 274
Koob, G. F., 97, 280
Koppeschaar, H., 245, 246
Kosfeld, M., 191
Kosslyn, S. M., 41

Kotelchuck, M., 184
Koutstaal, W., 37
Kramer, R. S. S., 125
Kreiman, G., 26, 27, 102
Kreppner, J., 193
Kriegeskorte, N., 173
Kringelbach, M. L., 92
Kromer, S. A., 192
Kronbichler, M., 145
Kroonenberg, P. M., 184, 186
Krueger, F., 143, 144, 162, 163, 164, 173, 210, 223, 231
Krueger, K. R., 179, 198
Kruk, M. R., 5, 244
Kubota, J. T., 220
Kudielka, B. M., 198
Kuhl, P. K., 266, 267
Kuhn, A., 260
Kuhn, C. M., 246
Kumsta, R., 193
Kurzban, R., 246

Laakso, M. P., 248
LaBar, K. S., 83, 86, 246
Ladavas, E., 65, 237
Ladurner, G., 145
Lafaille, P., 115, 259
Lagattuta, K. H., 142
Lagravinese, G., 66
Lahey, B. B., 251
Lai, C., 262
Lakatos, K., 187
Lakin, J. L., 161
Laland, K. N., 52, 53
Lamm, C., 136, 206
Lamy, D., 260
Landis, T., 67, 103, 116, 206, 268
Lang, P. J., 23
Lange, C., 111
Lange, H., 87
Langlois, J. H., 121
Langton, S. R. H., 116
Lanteri, P., 112
Larhammar, D., 222
Larsson, M., 222
Lasko, M., 244
Lavie, N., 18
Lawrence, A. D., 87, 244
Lawrence, J. H., 23
Lawrence, N., 188
Lazarus, R., 136
Lazarus, R. S., 79
Le, H., 216
Le Bihan, D., 62
Le Bon, G., 221
Le Doux, J. E., 5, 74, 76, 81, 83, 84, 85, 89, 134, 208, 209, 217
Lea, S. E. G., 63
Leakey, R. E., 58
Leary, M. R., 194
Leavens, D. A., 118
Lee, D., 166

Lee, G. P., 248
Lee, J., 93
Lee, P. L., 150
Lee, S. S., 278
Lee, S. Y., 124
Lee, T. M., 35
Leekam, S. R., 147
Lefebvre, L., 57
Lehericy, S., 62
Lehmann, H., 116
Leibold, J. M., 216
Lemche, E., 188
Lemer, C., 62, 63
Lemieux, L., 244
Lenggenhager, B., 206, 207
Lenzi, G. L., 133, 134
Leo, I., 261
Leppanen, J. M., 262
Leslie, A., 139, 270
Leslie, A. M., 139, 141, 147, 234, 270
Lesorogol, C., 175, 176
Levin, P. F., 253
Levine, B., 94
Levine, C., 229
Lewis, B. P., 162, 163
Lewis, M., 268
Lewis, M. B., 110
Libet, B., 205, 213
Liddle, P. F., 248
Lieberman, D., 87, 230, 235
Lieberman, M. D., 5, 19, 90, 182, 183, 195, 196, 218
Light, S. N., 273, 274
Lilienfeld, S. O., 248
Lillard, A., 274
Lims, K. E., 136
Lin, C. P., 136, 150
Lin, R. T., 38
Lin, Y., 211
Lin, Z. C., 149, 211
Linchitz, R., 196
Lindenberg, S., 253
Lipps, T., 130
Lis, S., 191
Little, A. C., 122, 123
Little, T. D., 241
Liu, G. T., 261
Liu, H., 277
Liu, H. L., 136
Liu, H. M., 266
Liu, Y., 62, 189, 190, 191
Ljungberg, T., 97
Lloyd, B., 275
Lloyd-Fox, S., 263
Loenneker, T., 263
Loftus, J., 209
Lohmann, H., 275
Lombardo, M. V., 219
Lorenz, A. R., 247
Lorenz, K., 240
Lotze, M., 247, 248
Loucas, T., 146

Lowenberg, K., 232
Luce, C., 162, 163
Lui, F., 66
Lumsden, J., 148
Lusk, L., 277
Lutter, C., 246
Luttrell, L. M., 159
Luu, P., 93, 94, 143
Lykken, D. T., 247

Ma, Y. J., 280
Maassen, G., 113
Macchi Cassia V., 9, 260
MacDonald, A. M., 94
MacDonald, G., 194
MacDonald, P. A., 94
Mace, M. J.-M., 29
Mackey, S., 183
MacKinnon, K. C., 57
Macklin, M. L., 244
MacLean, P. D., 76
MacLoed, C. M., 94
Macmillan, M. B., 248, 249
Macrae, C. N., 141, 143, 144, 210, 212, 213
Mah, L., 219
Mai, N., 112
Maier, M. A., 188
Main, M., 188
Mak, Y. E., 86
Malik, O., 143, 237
Mallon, R., 234
Mandisodza, A. N., 101, 102
Manes, F., 42, 87, 244
Manly, C. F., 219
Mann, J., 55
Mansfield, E. M., 262
Manson, J. H., 57
Manuck, S. B., 94
Marazziti, D., 183
Marcar, V. L., 263
Marcus, G. B., 207
Marcus-Newhall, A., 242
Maril, A., 37
Marino, L., 55
Markus, H. R., 211
Marlowe, F. W., 175, 176
Marsh, A. A., 93, 95, 232
Martin, E., 263
Martin, J., 251
Masi, C. M., 198
Maslach, C., 201
Maslow, A. H., 88
Massaccesi, S., 262
Master, S. L., 182, 183
Masuda, T., 211
Maszk, P., 137
Materna, S., 119
Mathias, C. J., 23, 77, 85
Matsuda, T., 231, 232
Matsumoto, D., 107
Matsuzawa, T., 129
Mattay, V. S., 191

Matthews, G., 91
Matthews, K. A., 94
Matzner, W., 246
May, K. A., 121
Maynard Smith, J., 166
Mazoyer, B., 261
Mazur, A., 242, 245, 246
Mazziota, J. C., 66, 90, 133, 134
McCabe, K., 142, 173
McCarthy, G., 29, 106, 116
McCarthy, M. M., 190
McCarthy, P. M., 162
McCauley, C. R., 90
McClay, J., 251
McCleery, J. P., 151, 152
McClure, S., 93
McComb, K., 54
McConahay, J. B., 214
McConnell, A. R., 216
McCormick, C. M., 123
McCrae, R. R., 18, 122
McCusker, E. A., 87
McDermott, J., 105
McDonald, S. M., 124
McDuffie, A., 266
McElreath, R., 175, 176
McGaugh, J. L., 72, 83
McGhee, D. E., 215
McGowan, S. V., 244
McGrath, J., 92
McInerney, S. C., 217, 218
McLeod, P., 18
McLoed, P., 113
McMaster, M. R., 162
McNeil, J. E., 106
Meeren, H. K. M., 113
Mehler, J., 259
Meldrum, D., 146
Melis, A. P., 272
Mellot, D. S., 214, 217
Meltzoff, A. N., 17, 131, 262, 264, 265, 270, 272
Mendrek, A., 248
Menon, E., 262
Merikle, P. M., 18
Merrick, S., 188
Messner, C., 215
Mesulam, M. M., 86, 116
Metzinger, T., 206, 207
Meyer, D. E., 94, 219
Meyer, K., 171, 172
Meyer-Lindenberg, A., 191
Michalska, K. J., 251
Michel, C. M., 67, 206, 268
Miczek, K. A., 88, 244
Mier, D., 191
Mikulincer, M., 188
Milgram, S., 164, 165
Mill, J., 251
Miller, E. K., 252
Miller, N., 23, 215, 241, 242
Miller, R., 125

Mills, D., 267
Mills, D. L., 267
Milner, A., 116
Milner, B., 76, 209
Milner, P., 95
Mineka, S., 90
Mirenberg, M. C., 214
Mirsky, A. F., 82, 244
Mirza-Begum, J., 125
Mishkin, M., 104
Mistlin, A., 115, 116
Mistlin, A. J., 66
Mitchell, D. G. V., 232
Mitchell, J. P., 6, 141, 143, 144, 210, 213
Mitchell, P., 149
Mitchell, R. W., 61
Mithen, S., 59
Moffitt, T. E., 251
Mogenson, G. J., 97
Mohr, C., 67, 206, 268
Moll, C. K. E., 26
Moll, F. T., 87
Moll, J., 87, 143, 162, 163, 164, 173, 229, 231
Molnar-Szakacs, I., 66
Mondillo, K., 93, 95
Monesson, A., 261
Monroe, Y. L., 190
Montague, L., 93
Montague, P. R., 3, 93, 97, 174, 180
Monteleone, G., 98, 197, 198
Moon, S. Y., 104
Moore, M. K., 17, 264, 270, 272
Morecraft, R. J., 77
Moreland, R. L., 181
Morelli, S. A., 232
Moreno, J. D., 252
Moretti, L., 171
Morgan, H. J., 180
Morgan, J. L., 267
Morgan, K., 246
Morgan, V., 106, 112
Morison, S. J., 194
Moro, V., 112
Morris, E. G., 93, 136, 244, 249, 250
Morris, J., 83, 84, 85
Morris, R., 10, 245, 251
Morton, J., 17, 93, 95, 259, 260
Moses, L. J., 262
Moulson, M. C., 262
Mourao-Miranda, J., 231
Mowrer, O., 241
Muccioli, M., 237
Mukamel, R., 7
Muller, J., 83
Mullin, J. T., 261
Mummery, C. J., 143
Mundy, P., 117, 262
Murdock, C. D., 168
Murnighan, J. K., 273
Murphy, B., 137
Murray, E. A., 43, 85, 89

Muzik, O., 194, 257
Myers, T. E., 129
Myowa-Yamakoshi, M., 129

Nagell, K., 265
Nagy, F., 194, 257
Nakayama, K., 103
Naliboff, B. D., 182, 183, 196
Neelin, P., 38
Neighbor, G., 106, 112
Neisser, U., 203, 204
Nelson, B. C., 196
Nelson, C. A., 261, 262, 263, 264
Nettle, D., 160, 161
Neuberg, S. L., 162, 163
Neumann, I. D., 192
Newcomb, T. M., 182
Newman, J. P., 247
Nicholls, M. E. R., 73
Nieder, A., 62
Nisbett, R. E., 121, 149, 211, 238, 239
Niven, J., 52
Noll, D., 94
Nomikos, G. G., 97, 191
Norenzayan, A., 149
Norman, D. A., 252
Norris, C. J., 98, 197, 198
Northoff, G., 211
Nosek, B. A., 213, 214, 215, 217, 235, 236
Notoya, T., 67
Nowak, M., 160
Nowak, M. A., 174
Nugent, T. F., 277
Nunez, P. L., 28
Nusbaum, H., 98, 197, 198
Nuss, C. K., 198
Nuttin, J. M., 214
Nystrom, L. E., 8, 144, 232, 233
Nystron, L., 71, 170, 171

Oates, B., 196
Oatley, K., 79
Obayashi, S., 67
Oberecker, R., 259
Oberman, L. M., 109, 151, 152
Oboyle, C., 137
Ochsner, K., 132
Ochsner, K. N., 5, 90
O'Connell, R. G., 219, 220
O'Connor, K. J., 8, 217, 218
O'Connor, M. F., 196
O'Connor, T. G., 193, 194
Odden, H., 274
O'Doherty, J., 77, 84, 85, 92, 93, 96, 97,
    124, 134, 135, 136, 164, 165, 195, 244,
    249, 250
O'Doherty, J. P., 93
O'Donnell, P., 278
Ogawa, S., 35
Ohl, F., 192
Ohman, A., 84, 87
Ohmann, A., 84

Ojemann, G. A., 26
Ojima, H., 67
Okubo, Y., 231, 232
Olds, J., 95
Olson, K. R., 273
Olsson, A., 83
Omoregie, H., 211
O'Neill, P., 228
Öngür, D., 91
Onishi, K. H., 148, 270, 271
Oosterhof, N. N., 123, 124
Oram, M. W., 106, 107, 116
O'Reilly, H., 149
Orr, S. P., 244
Ortega, J. E., 66
Ortony, A., 79, 80
Ossewaarde, L., 246
Otten, L. J., 28
Owen, S., 196
Ozonoff, S., 148

Padden, D., 267
Padon, A. A., 108
Pagnoni, G., 168, 169, 248
Paik, H., 243
Paller, K. A., 116
Panger, M., 57
Panksepp, J., 5, 88, 196
Papez, J. W., 76
Paquette, V., 223
Pardini, M., 162, 163, 164
Park, D. C., 211
Park, J., 246
Park, K. A., 187
Parke, S., 183
Parker, J. G., 157
Parkinson, J. A., 96
Parkman, J. M., 15
Parrish, T., 86
Parrish, T. B., 116
Parvizi, J., 87, 196
Pascalis, O., 260, 261
Pascual-Leone, A., 45, 46, 67, 144, 152,
    171, 172, 206, 272
Pashler, H., 19
Pastorelli, C., 230
Patrick, C. J., 247
Patteri, I., 66
Patterson, K., 143
Paukner, A., 67, 272
Paul, B. Y., 23, 215
Paulescu, E., 143
Paulus, M. P., 87
Paus, T., 277
Payne, B. K., 219, 220
Pazzaglia, M., 112
Pearl, D. K., 205
Pedersen, C. A., 190
Pedersen, W. C., 242
Peelen, M. V., 111, 112
Pelham, B. W., 214
Pelphrey, K. A., 116

Pempek, T. A., 179
Peng, K. P., 149
Penn, D. C., 59
Pennartz, C. M. A., 96
Pennington, B. F., 148
Penton-Voak, I. S., 122, 123
Peretz, I., 85
Perez, E., 29
Perez, J., 248
Perfetti, C. A., 62
Perner, J., 139, 140, 145, 146, 147, 269, 270,
    271, 272
Pernigo, S., 112
Perrett, D., 83, 85, 87, 113, 115
Perrett, D. I., 41, 42, 66, 83, 87, 106, 107,
    113, 116, 119, 121, 122, 123
Perry, D., 136
Perry, S., 57
Persaud, N., 18
Persinger, M. A., 222
Pessoa, L., 231
Peters, T., 38
Peterson, C., 91
Peterson, C. C., 142
Petrinovich, L., 228
Petterson, M., 73
Pfaus, J. G., 97, 191
Pfeifer, J. H., 90, 152
Phelps, E. A., 8, 81, 83, 136, 175 217, 218
Phillips, A. G., 97, 191
Phillips, M. L., 87, 188
Phillips, R. G., 83
Phipps, M., 249
Piaget, J., 229
Piazza, M., 62
Pica, P., 62, 63
Pick, A. D., 262
Pickens, D., 106, 112
Pickett, K., 242
Pickles, A., 146, 194
Piefke, M., 273, 274
Piekut, D. T., 278
Pike, B., 115, 259
Pike, G. B., 35
Piliavin, J. A., 161
Pillemer, D. B., 269
Pine, D. C., 93, 95
Pine, D. S., 278, 280
Pineda, J. A., 151, 152
Pinel, P., 62
Pinker, S., 53, 54
Pitcher, D., 109
Pitkanen, A., 76
Pitman, R. K., 244
Pittinsky, T. L., 213
Plailly, J., 87
Plant, E. A., 219, 236
Plassmann, H., 93
Platek, S. M., 129
Pleydell-Pearce, C. W., 207, 208
Plumb, I., 132
Plutchik, R., 79

Poldrack, R. A., 8
Polkey, C. E., 93, 136, 244, 249, 250
Ponto, L. L. B., 87, 196
Porro, C. A., 66
Posner, M. I., 14, 93, 94, 143, 219
Potter, D., 116
Poulton, R., 251
Pound, N., 122, 123
Povinelli, D. J., 59, 269
Powell, A. L., 162
Powell, L. J., 145
Prange, A. J., 190
Premack, D., 139
Preston, S. D., 130
Preuschoff, K., 87, 163
Pribram, K. H., 82, 244
Price, C. J., 143
Price, J. L., 76, 91
Price, R., 106, 112
Pridgen, A., 92, 237
Prins, B., 72
Provine, 60
Przybeck, T., 247
Puce, A., 29, 106
Purves, D., 258
Putman, P., 241, 245, 246

Quartz, S. R., 3, 174, 180
Quayle, A. H., 110
Quinn, G. E., 261
Quinn, P. C., 260
Quiroga, R. G., 26, 27, 102

Raafat, R. M., 221
Rabbie, J. M., 214
Rabinowitz, C., 261
Raichle, M. E., 34
Ramachandran, V. S., 6, 109, 151, 152
Ramamurthi, B., 244
Ramsey, N. F., 246
Rand, D. G., 174
Rangel, A., 93, 96, 97
Rapoport, J. L., 277
Raste, Y., 132
Rauch, S. L., 244
Rausch, S. L., 217, 218
Ravine, M., 187
Ray, R. D., 90
Raye, C. L., 86
Read, A., 55
Reader, S. M., 52, 53
Reber, P. J., 116
Reboul-Lachaux, J., 110
Reby, D., 54
Reddy, L., 26, 27, 102
Redwine, K., 246
Rees, G., 39
Rees, J., 194
Rees, S. L., 191
Regard, M., 103, 116
Reicher, S., 201, 221
Reiman, E. M., 62

Reinoso-Suarez, F., 77
Reis, D., 5, 85
Reiss, A., 261
Reiss, A. L., 263
Reiss, D., 55
Renfrew, C., 59
Repa, J. C., 83
Repo-Tiihonen, E., 248
Reutter, B., 261
Rhodes, G., 121
Riad, J. K., 245
Richards, J., 262
Richardson, D. R., 123
Richardson, M. P., 81
Richler, J., 149
Ridley, M., 9
Rilling, J., 71, 170, 171
Rilling, J. K., 144, 168, 169, 248
Rivaud, S., 62
Rizzolatti, G., 6, 65, 66, 67, 87, 133, 137, 151, 152, 162
Robbins, P., 204
Robbins, T. W., 96, 97, 148
Roberts, G., 160, 161
Robertson, E. M., 46
Robertson, E. R., 90
Robertson, I. H., 219, 220
Robins, L. N., 247
Rocha-Miranda, C. E., 103
Rochat, M., 133
Rochat, P., 274
Rockland, C., 83
Rodkin, P. C., 241
Roeling, T. A. P., 5, 244
Roepstorff, A., 222
Rogalsky, C., 87
Rogers, S. J., 148
Roggman, L. A., 121
Rolls, E. T., 72, 79, 88, 89, 92, 93, 96, 102, 108, 136, 231, 244, 249, 250
Roozendaal, B., 83
Ropar, D., 149
Rorden, C., 42
Rose, L., 57
Rose, R. M., 198
Rosen, B. R., 37, 94
Rosicky, J. G., 262
Ross, D., 243
Ross, J. L., 263
Ross, L. L., 94
Ross, M., 208
Ross, S. A., 243
Rossi, A., 183
Rossi, R., 248
Rosvold, H. E., 82, 244
Rotshtein, P., 87, 105
Rotte, M., 37
Rottman, L., 181
Rousselet, G. A., 29
Rowe, C., 52
Rowland, D., 41, 42, 83, 85, 87
Roy, P., 194

Royet, J. P., 87
Rozendal, K., 209
Rozin, P., 90
Rozzi, S., 66, 133
Rubin, K. H., 157
Ruby, P., 137, 144, 162
Rudebeck, P. H., 94, 95, 166
Rudman, L. A., 214, 217
Ruffman, T., 147, 269, 271
Rugg, M. D., 28
Ruggiero, A., 67, 272
Rumiati, R. I., 171
Rupniak, N. M. J., 196
Rushworth, M. F. S., 94, 95, 166
Russell, J., 148
Russell, S., 241
Russell, T. A., 188
Rutherford, M., 148
Rutter, M., 193, 194
Rutz, C., 57
Ryan, L., 142, 173

Sabbagh, M. A., 271
Sagiv, N., 29
Sahdra, B., 208
Sai, F., 261
Sai, F. Z., 261
Said, C. P., 120, 124
Saini, E. K., 246
Sakura, O., 64, 67, 68
Salazar, A. M., 92, 237
Sally, D., 176
Salmi, P., 91
Samson, D., 141, 142, 144, 148
Sander, D., 188, 189
Sanfey, A., 71, 170, 171
Sanfey, A. G., 144, 171
Santelli, E., 152
Saphire-Bernstein, S., 192
Sato, A., 205
Saver, J. L., 237
Saxe, R., 6, 143, 144, 145, 162, 231, 268, 279
Saxe, R. R., 144, 272
Saxon, M. S., 273
Scarnati, E., 97
Scarpa, P., 143
Schacter, D. L., 37
Schacter, S., 74, 181
Schaffer, H. R., 184, 185, 265
Schelthammer, M., 174
Schenone, P., 143
Scherer, K. R., 115
Schilhab, T. S. S., 55, 111
Schjødta, U., 222
Schmitt, M., 216
Schmittberger, R., 166, 170
Schneider, F., 172
Schneider, J. A., 179, 198
Schneider, W., 62, 252
Schnyder, U., 174
Scholz, J., 145

Schuder, M., 194
Schulkin, J., 191, 229
Schulte-Ruther, M., 273, 274
Schultheiss, O. C., 245
Schultz, W., 96, 97
Schunk, D., 172
Schutter, D. L. G., 240
Schwab, K., 92, 237
Schwartz, C., 38
Schwartz, J. L. K., 215
Schwarz, U., 263
Schwarze, B., 166, 170
Schyns, P., 42, 85
Scordalakes, E. M., 281
Scott, A. A., 152
Scott, L. S., 261
Scott, S., 83, 85
Scoville, W. B., 76, 209
Sears, R., 241
Sebanz, N., 101
Sebastian, C., 277, 280
Sebe, N., 120
Seeck, M., 67, 206, 268
Seeman, T. E., 192, 193
Sejnowski, T. J., 14
Self, P. A., 275
Sellars, C., 247
Senior, C., 87
Senju, A., 147
Sergent, J., 106
Serino, A., 65
Seward, E., 196
Seyfarth, R. M. 51
Seymour, B., 97, 134, 135, 136, 164,
    165, 195
Shah, A., 149
Shallice, T., 77, 85, 252
Shamay-Tsoory, S. G., 136
Shapiro, L. E., 190
Shatz, M., 270
Shaver, P., 187
Shaver, P. R., 180, 188
Shaw, L. L., 162
Sheard, M. H., 245
Shellock, F. G., 40
Shelton, S. E., 196
Shen, J., 62
Shepher, J., 235
Sherman, J. W., 219
Sherman, P. W., 158
Sherman, T., 117
Sheth, P., 125
Shiffrar, M., 101
Shiffrin, R. M., 252
Shih, M., 213
Shin, L. M., 217, 218, 244
Shirinyan, D., 182, 183
Shiv, B., 93
Shuman, M., 111
Siddhanti, S., 191
Siegal, M., 142
Siegel, A., 5, 244

Sigman, M., 117, 152, 278
Sigmund, K., 160
Signoret, J.-L., 106
Silani, G., 163
Silber, S., 270
Simion, F., 9, 260, 261, 264
Simion, G., 116, 262
Simmel, M., 138
Simmons, A., 87, 188
Simner, M. L., 264
Simon, B. B., 269
Simonoff, E., 146
Simons, D. J., 18
Simpson, J. A., 187
Singer, H. S., 263
Singer, J. E., 74
Singer, T., 87, 97, 134, 135, 136, 163, 164,
    165, 169, 170, 173, 195
Singer, W., 27
Singerman, J. D., 116
Singh, L., 267
Singh, P., 181
Singh, S., 274
Singh, T., 153
Sinigaglia, C., 152
Sirigu, A., 93
Skudlarski, P., 104, 105, 106
Skuse, D. H., 150
Slater, A. M., 260
Slomkowski, C., 274
Smailes, E. M., 243
Small, D. M., 86, 92
Smeltzer, M. D., 191
Smetana, J. G., 236
Smith, A. M., 248
Smith, C. A., 79
Smith, M., 137, 233, 247
Smith, M. C., 114
Smith, M. K., 114
Smith, P., 116
Smith, P. K., 251
Smith, V., 142, 173
Smollan, D., 162, 213
Smyth, A. N., 95, 166
Snyder, J., 151
Soares, J. J. F., 84
Sodian, B., 147
Soininen, H., 248
Soken, N. H., 262
Sommerville, R. B., 8, 232, 233
Sonuga-Barke, E. J., 193
Sorensen, N., 214, 215
Source, J., 89, 262
Southgate, V., 147, 153, 265
Sowell, E. R., 277
Spampinato, M. V., 215
Spangler, G., 185, 187, 191
Sparaci, L., 152
Spear, L. P., 278, 280, 281
Spelke, E., 184
Spelke, E. S., 273
Spence, C., 65

Spence, M. J., 259
Spielberger, C. D., 241
Spilkin, A. M., 42
Spinella, M., 137
Spinelli, L., 206, 268
Sprengelmeyer, A., 87
Sprengelmeyer, P., 162
Sprengelmeyer, R., 87
Squire, L. R., 209
Staffen, W., 145
Stahl, D., 264
Stallen, M., 219
Stam, H., 241
Stanton, A. L., 196
Stanton, S. J., 245, 246
Steg, L., 253
Stein, M. B., 87
Steinberg, L., 280
Steineck, G., 91
Steiner, M., 191
Stenger, V. A., 94
Stephan, K., 97, 134, 135, 136, 164, 165
Sternberg, R. J., 180, 181
Sterzi, R., 143
Stevens, S., 193
Stewart, L., 44
Stillman, T. F., 253, 254
Stillwell, A. M., 241
Stødkilde-Jørgensenb, H., 222
Stone, V., 148
Stone, V. E., 142
Stoop, R., 191
Stowe, C. J., 162, 163
Strange, B. A., 81, 124
Strassle, A. E., 263
Strenziok, M., 173
Striano, T., 259, 264
Strickland, I., 247
Stroop, J. R., 94
Stuss, D. T., 94, 143, 252
Su, M., 223
Suh, K., 137
Suhler, C. L., 253
Sumich, A., 121
Sung, C. Y., 198
Suomi, S. J., 67, 272
Surguladze, S. A., 188
Susskind, J. M., 87
Sutton, J., 251
Swami, V., 125
Swart, M., 132
Sweet, M. A., 187
Swerdloff, R. S., 246
Swettenham, J., 119, 251, 270
Swingler, M. M., 187
Switzer, G., 273

Tadi, T., 206, 207
Tager-Flusberg, H., 147, 151, 152
Tajfel, H., 211
Takahashi, H., 231, 232
Takahashi, M., 67

Talaraich, J., 38
Tallis, F., 183
Tambini, A., 105
Tamietto, M., 74
Tanaka, M., 65, 67
Tang, Y., 62
Tang, Y. X., 179, 198
Tank, D. W., 35
Tankersley, D., 162, 163
Tapanya, S., 274
Tarr, M. J., 104, 105, 106
Tavassoli, T., 150
Taylor, A., 251
Taylor, C. S. R., 64
Taylor, J. L. C, 112
Taylor, J. R., 97
Taylor, K., 150
Taylor, S. E., 182, 183, 192, 193, 196
Taymans, S. E., 192
Tejedor, J., 77
Temple, E., 274
Terburg, D., 246
Terracciano, A., 122
Tesla, C., 274
Testa, C., 248
Theoret, H., 46, 152
Thier, P., 119
Thijssen, J., 245, 246
Thomas, C., 153
Thomas, K. M., 263, 264
Thomas, S., 66
Thompson, P. A., 248
Thompson, P. J., 244
Thompson, P. M., 38, 277
Thomson, J. J., 227
Thorpe, M. F., 29
Thunberg, M., 23, 109
Thut, G., 67, 206, 268
Tibi-Elhanany, Y., 136
Tidball, G., 262
Tiihonen, J., 248
Tinbergen, N., 184
Tipp, J., 247
Tipper, S. P., 116, 117
Titchener, E. B., 130
Tizard, B., 194
Todd, J., 266
Todorov, A., 101, 102, 120, 123, 124, 125
Toga, A. W., 38, 277
Tom, S. M., 90
Tomasello, M., 49, 57, 59, 60, 61, 116, 118, 157, 222, 265, 266, 272, 275
Tomlin, D., 3, 93, 174, 180
Tonnaer, F., 237, 247
Torner, L., 192
Toschi, N., 192
Toth, N., 59
Tournoux, P., 38
Tovee, M. J., 102, 125
Townsend, S. S., 211
Tracer, D., 175, 176

Tranel, D., 5, 15, 23, 41, 42, 77, 83, 85, 86, 92, 103, 108, 109, 110, 125, 176, 237, 238, 244, 249, 250, 251
Trauner, D., 42
Treboux, D., 188
Trevarthan, C., 265
Treves, A., 105
Treyer, V., 171, 172, 174
Triantafyllou, C., 145
Trimble, M. R., 244
Trivers, R. L., 158, 159
Trope, Y., 101
Trouard, T., 142, 173
Troyer, D., 273
Tsao, P. K., 266
Tseng, K. Y., 278
Tucker, D. M., 94, 219
Tucker, L. A., 262
Tuiten, A., 241, 245, 246
Tulving, E., 209
Turati, C., 9, 260, 261
Turiel, E., 227
Turner, J. C., 211
Turner, T. J., 79
Twenge, J. M., 198
Tybur, J. M., 87, 230
Tzeng, O. J. L., 150
Tzourio-Mazoyer, N., 261

Uchida, Y., 211
Uchino, B. N., 179, 192
Uddin, L., 111
Ulrich, R. E., 241
Umilta, M. A., 66, 133
Unge, P., 222
Ungerer, J., 117
Ungerleider, L. G., 104
Urgesi, C., 112
Urland, G. R., 220

Vaish, A., 272
Vaituzis, A. C., 277
Valentine, T., 121
Valind, S., 222
Valyear, K. F., 108
van Baaren, R., 133, 136, 161
Van de Wal, C., 83
Van den Bos, R., 111
van den Bos, W., 279
van den Hout, M., 241
van Doornen, L., 245, 246
van Elst, L. T., 244
van Erp, A. M. M., 88, 244
van Heijnsbergen, C., 113
Van Hoesen, G. W., 77
van Honk, J., 240, 241, 245, 246
Van Ijzendoorn, M. H., 184, 186
van Knippenberg, A., 133, 161
van Schaik, C. P., 50, 56, 57
Vance, S. L., 219
Vanderschuren, L., 96
Vanman, E. J., 23, 215

van't Wout, M., 171
Varlinskaya, E. I., 280, 281
Varney, L. L., 162
Vaughan, G. M., 9
Vaurio, O., 248
Veinante, P., 191
Veit, R., 86, 247
Verbaten, R., 245, 246
Viding, E., 277, 280
Viet, R., 248
Vilberg, T., 5, 196
Visalberghi, E., 67, 272
Vodel-Farley, V. K., 262
Vogt, B. A., 77, 94
Vollmer-Conna, U., 267
Von Cramon, D., 112
von Cramon, D. Y., 143, 144
Voorn, P., 96
Vritcka, P., 188, 189
Vuilleumier, P., 72, 188, 189
Vul, E., 19
Vythilingam, M., 93, 95

Wade, D., 92, 108, 136
Wagner, A. D., 37
Wahba, A., 89
Wahl, S., 185, 186, 187
Walker, A. E., 91
Walker, J., 117, 118, 146, 147
Walker-Andrews, A. S., 262
Wallbott, H. G., 115
Waller, N. G., 187
Walsh, V., 44, 109
Walster, E., 121, 181
Walton, M. E., 94, 95, 166
Wang, A. T., 278
Wang, Z. X., 191
Ward, J., 13
Ward, R., 125
Warden, D., 92, 237
Warneken, F., 272
Warrington, E. K., 106
Wassermann, E. M., 46, 264
Waters, E., 185, 186, 187, 188
Waugh, C. E., 245
Way, B. M., 90, 196
Waytz, A., 138
Weber, J., 132
Weber, M., 72
Webley, P., 63
Wegner, D. M., 205
Weinberger, N. M., 83
Weir, A. A. S., 57
Weiskopf, N., 86
Weiskrantz, L., 81
Wellisch, D. K., 196
Wellman, H., 141, 142, 270
Wellman, H. M., 270
Wells, A., 91
Wells, A. J., 265
Wells, R., 55
Welsh, R. C., 211

Wendelken, C., 188
Wenkstern, D. G., 97, 191
Westenberg, P. M., 273
Westermarck, E., 235
Wexler, A., 145, 162
Whalen, P. J., 217, 218
Wheeldon, A., 196
Wheelwright, S., 119, 132, 136, 148, 149, 270
Whishaw, I. Q., 29
White, S., 125, 147, 148, 149
White, S. H., 269
Whiten, A., 50, 56, 57, 61, 153
Whitfield-Gabrieli, S., 145, 261
Wicker, B., 87, 119
Wiens, S., 87
Wiggett, A. J., 112
Wilkinson, G. S., 158, 159
Wilkinson, R., 242
Will, J. A., 275
Williams, C. J., 215
Williams, G. C., 159
Williams, J. H. G., 153
Williams, J. R., 190, 191
Williams, K. D., 195, 277, 280
Williams, S. C. R., 87, 188
Williamson, C., 118, 247
Willingham, J. R., 213
Wilson, A., 49
Wilson, A. E., 208
Wilson, R. S., 179, 198
Wilson, T. D., 121
Wimmer, H., 139, 140, 146, 272
Winkielman, P., 19, 109

Winston, J., 125
Winston, J. S., 84, 86, 124, 169, 170, 173
Wirth, M. M., 245
Woermann, F. G., 244
Wohlschlager, A., 60
Wolf, A. P., 235
Wolpert, D. M., 204, 206
Wommack, J. C., 189, 190
Wood, J. N., 215
Woodruff, G., 139
Woods, R., 66
Wright, E. W., 205
Wu, S., 180
Wyland, C. N., 210, 212

Xiao, D., 106, 107
Xu, B., 263

Yahata, N., 231, 232
Yahr, J., 260
Yale, M. E., 266
Yang, C. Y., 150
Yasuda, A., 205
Yermolayeva, Y. A., 179
Yim, C. Y., 97
Yoder, P. J., 266
Yokota, T., 67
Yoo, S. H., 107
Yoshikawa, S., 121
Yoshino, M., 73
Young, A., 41, 42, 83
Young, A. W., 15, 83, 85, 87, 102, 103, 104, 108, 110, 111

Young, L., 229, 237, 238, 244
Youngblade, L., 274
Younger, J., 183
Yovel, G., 104, 105
Yurgelun-Todd, D. A., 278

Zahavi, A., 159
Zahn, R., 143, 162, 163, 164, 173, 231
Zahn, T. P., 249
Zahn-Waxler, C., 273, 274
Zajonc, R. B., 18
Zak, P. J., 191, 246
Zaki, J., 132
Zamboni, G., 223
Zangl, R., 267
Zanini, S., 77, 85
Zarahn, E., 35
Zatorre, R. J., 92, 93, 115, 259
Zebrowitz, L. A., 124
Zeh, T. R., 168, 169
Zeki, S., 182
Zelazo, P. R., 184
Zhang, L., 212
Zhang, W., 62
Zhao, Z., 86
Zhou, Y., 83
Zhu, Y., 212
Ziegler, S., 247
Zihl, J., 112, 113
Zijdenbos, A., 277
Ziker, J., 175, 176
Zimbardo, P. G., 201
Ziv, T., 260
Zuoxin, W., 189, 190

Subject index

Page numbers in **bold** indicate key term definitions.

Accuracy measurement, 15
Acquired sociopathy, 249–251
Action potential, **20**, 20–21, 22
Adolescence, **277**, 277–281
Affective empathy, 136
Agency, **204**, 204–206, 252
Aggression, 9, 10, **239**, 239–251
  biological basis, 243–246
  displaced, **242**
  faces, 123–124
  frustration–aggression model, 241
  instrumental, **236**
  pathological aspects, 246–251
  reactive, **236**
  social function, 240–242
  social learning, 242–243
Alcohol use, 280–281
Altruism, **157**, 158–164
  empathy-altruism model, **161**, 161–162
  evolutionary biological approaches, 158–160
  humans, 160–164
Altruistic punishment, **157**
Amnesics, 209–210
Amygdala, **81**
  adolescence, 278
  aggression, 243–244
  architecture and functions, 76–77
  attachment, 188
  cognitive control of emotions, 90
  fear response, 33, 81–86
  morality, 231, 232
  prejudice, 217–218
  psychopathy, 248
  reward learning, 85, 89
  trustworthiness judgment, 125
  unreciprocated cooperation, 169
Anger, 87–88, **230**, 240–241, 243–246;
    see also Aggression
Animal lesion studies, 43–44
Anterior, 30
Anterior cingulate cortex
  architecture and functions, 77
  cognitive and affective evaluation of responses, 93–95
  fear response, 85
  grief, 197
  pain response, 195
  response conflict, 219
  skin conductance response, 23
  social exclusion, 196
Anthropomorphism, **138**
Antisocial behavior, **227**

Antisocial personality disorder (ASPD), **246**, 246–247, 249–251
Anxiety disorders, 74
Aphasia, **142**
Appeasement behaviors, **240**
Arranged marriages, 181
Asperger's syndrome, 125, **146**, 150
Aspiration, 43
Assimilation, 212
Attachment, 180, **184**, 184–194
  adult relationships, 187–189
  hormones, 189–193
  individual differences in mother–child attachment, 185–187
  insecure/anxious, **186**
  insecure/avoidant, **186**
  institutionalization, 193–194
  neurotransmitters, 189–193
  securely attached, **186**
Attraction, 181–182
Attribution, **140**
Autism, 145–153, **146**
  broken mirror theory, **151**, 151–153
  diagnosis, 146
  empathy, 233–234
  executive dysfunction, 148
  extreme male brain, 149–150
  eye contact, 117–118
  imitation, 153
  mind-blindness, 146–148
  moral reasoning, 234
  motherese recognition, 267
  pretend play, 270
  proto-declarative pointing, 119
  savant skills, 148
  spectrum disorder, 146
  systemizing, 150
  weak central coherence, 148–149
Autobiographical knowledge, 207
Automatic behavior, 252
Autonomic nervous system (ANS), **22**, 24
Axon, **20**

Baby-faced appearance, 123–124
Basal ganglia, **32**, 32–33, 87, 96, 114
Basic emotions, **78**, 78–79, 88
Beauty and goodness link, 121–122
Behavior measurement, 14–19
Big Five, **122**, 122–123
Biological anthropology, **58**
Biological motion, **112**, 112–113, 125
Bipedalism, 58–59
Blank slate, **8**
Blind-person's cane, 65

Blind scoring, **17**
Block design, **36**, 36–37
Bobo doll, 242–243
Bodily response measurement, 22–25
Body language, 113–114
Body mass index (BMI), 125
Body perception, 111–115
Body scheme, 270
BOLD response, **35**
Bottlenose dolphins, 55
Brain
  adolescence, 277–278
  damage, 40–42
  embryonic and fetal development, 258–259
  emotional brain, 76–78
  organization and structure, 30–33
  size, 51–53
  terms of reference, 30
Breathing measures, 22
Broken mirror theory, **151**, 151–153
Bullying, 251

Cannon–Bard theory, **75**, 75–76
Capgras syndrome, **110**
Capital punishment, 238–239
Caudal, 30
Caudate nucleus, 32, 96, 174–175
Cell body, 20
Central gray, 196, **244**
Central nervous system, 22, 24, 31
Cerebral cortex, 30, 31
Cerebrovascular accidents (CVAs), **40**
Charity donations, 162, 163
Child development, 269–277
  adapting cognitive neuroscience methods, 263–264
  cultural influences, 274
  empathy, 272–274
  family influences, 274–275
  gender awareness, 275–276
  peer influences, 275
  prosocial behavior, 272–274
  self-awareness, 268–269
  self-recognition, 268–269
  theory of mind, 270–272
Childhood amnesia, **269**
Cingulate gyrus, 33
Cognition, measuring, 14–19
Cognitive appraisal in morality, 231
Cognitive empathy, 136
Cognitive model of face perception, 102–103
Cognitive psychology, **4**

Collectivist culture, **211**
Competition, 157
Compliance, **164**, 164–165
Complicated grief, 197
Conditioned taste aversion, 91
Conduct disorder, 251
Conspecifics, **101**
Contagion, **60**
Contamination, 87
Control processes, 219
Controlled behavior, 252–254
Conventional norms, **227**, 235–237
Cooperation, 157–158
Cortisol, 191
Crime of passion, 252
Criminality, 227
Crowds, 221
Cultural skills, 64–68
Culture, 9, **50**
    evolutionary origins, 55–61
    fairness norms, 175
    influence on development, 274
    social identity, 211–212
Culture of honor, 238
Culture pyramid, 56–57
Cumulative cultures, 57
CVAs, **40**
Cyberball game, 195–196

Death penalty, 238–239
Deception, 52
Dehumanization, **230**
Deindividuation, **221**
Delusions of control, **206**
Dendrites, **20**
Deoxyhemoglobin, 35
Development
    communication, 265–267
    cultural influences, 274
    embryonic and fetal brain development,
        258–259
    emotion recognition, 262
    empathy, 272–274
    face recognition, 259–262
    family influences, 274–275
    gaze following, 262
    gender awareness, 275–276
    imitation, 264–265
    interactions with others, 262
    joint attention, 262
    language learning, 266–267
    peer influences, 275
    prosocial behavior, 272–274
    self-awareness, 268–269
    self-recognition, 268–269
    social referencing, 262
    theory of mind, 270–272, 279
    turn-taking, **265**
    voice recognition, 259
Diencephalon, **33**
Differentiation, 212
Disgust, **86**, 86–87

Displaced aggression, 242
Display rules, **107**
Dissociations, **41**, 41–42
'Do-as-I-do' game, 61
Dolphins, 55
Domain specificity, **5**, 141–142
Dopamine system, 87, 97, 191, 244, 278
Dorsal, 30
Dorsal striatum, 174–175
Dorsal visual stream, **104**
Double dissociation, **42**
Drug-taking, 97, 280–281
Dual-task interference, 45
Dyadic interactions, **262**

Ego, **74**
Electroencephalography (EEG), 13, 27–28,
    28, 263
Electromyography (EMG), 22, **23**, 25
Electrophysiological methods, 25–29
Embedded figures, 149
Embodiment, **204**, 206
Emotional contagion, 264–265
Emotions, **71**, 71–72
    as appraisals, 79, 80
    basic emotions, **78**, 78–79, 88
    Cannon–Bard theory, **75**, 75–76
    categories, 78–88
    emotional body language perception,
        113–114
    emotional brain, 76–78
    evolutionary theory, 72–74
    facial emotion perception, 108–110
    Freudian theory, 74
    frontal lobes, 90
    infant recognition, 262
    James–Lange theory, **74**, 74–75
    limbic brain hypothesis, 76
    moral emotions, **79**, 230–231
    Papez circuit, **76**
    right brain hypothesis, 73
Empathizing, 149
Empathy, **130**
    affective, 136
    autism, 233–234
    cognitive, 136
    development, 272–274
    simulation theory, 130–139
Empathy-altruism model, **161**, 161–162
Empathy Quotient (EQ), 132
Error-related negativity, 94, 219–220
Event-based knowledge, 207
Event-related design, **36**, 36–37
Event-related potentials (ERPs), 13, 14,
    28, 28–29
    error-related negativity, 94, 219–220
    infant and children, 263, 264
Evolutionary theory
    altruism, 158–160
    emotions, 72–74
Executive functions, **148**
Experimentally-induced lesions, 43–44

Expressions, **71**
Extended cognition, **63**
External validity, 19
Extinction, **92**
Extrastriate body area (EBA), **111**, 111–112
Extreme male brain, 149–150
Eye gaze detection, 115–118, 262
Eyeblink startle response, **23**

Face perception, 15, 101–111
    aggressive faces, 123–124
    baby-faced appearance, 123–124
    beauty and goodness link, 121–122
    cognitive model of Bruce and Young,
        102–103
    emotion perception, 108–110
    event-related potentials, 28–29
    infants, 259–262
    neural basis, 103–107
    self-recognition, 111
    trustworthiness, 125
Face recognition unit, **102**
Facial expressions, primary reinforcers,
    89–91
Facial width-to-height ratio, 123
Factor analysis, **19**
Fairness norms, 175
False belief, **131**, 139, 270–271
    autism, 146–147
False discovery rate (FDR), **39**
False photograph tasks, 145
Family influences on development,
    274–275
Familywise error (FWE), **39**
Fear, 81–86
Fire, 59
First-order intentionality, **140**, 141
Fight-or-flight response, **243**
Footbridge Dilemma, 228, 232–233
Forward model, **204**
Free loaders/riders, **157**
Free will, 252
Freudian theory of emotions, 74
Friends, number of, 8
Frontal lobe
    cognitive control of emotions, 90
    morality, 237
    skin conductance response, 23
Frustration–aggression model, 241
Functional imaging, 33–40, 41
Functional magnetic resonance imaging
    (fMRI), 13, 33, 34
    data analysis, 38–39
    experimental design, 36–38
    infants and children, 263
    physiological underpinnings, 34–36
    safety issues, 40
    spatial and temporal resolution, 14, 35–36
Functional specialization, 41
Fusiform body area (FBA), **112**
Fusiform face area (FFA), **104**, 105–106,
    261–262

Gage, Phineas, 248–249
Galvanic skin response (GSR), 23
Game theory, **166**, 166–176
Gaze cueing, **116**
Gaze detection, 115–118, 262
Gender awareness, 275–276
Genocide, 135
Glia, **30**
Glioma, 41
Globus pallidus, 33, 96
Gray matter, **30**, 277
Greebles, 106
Grief, **196**, 196–197
Grooming behavior, 53, 54
Groups, definition, 202
Gyri, **31**

Habit, **96**
Habituation, **16**, 16–17
Halo effect, **121**
Heart rate measures, 22
Hedonic value, **71**
Helping behavior, 158–164
Hemodynamic methods, **33**, 33–40
Hemodynamic response function
    (HRF), **35**
Herding, **221**
Hierarchy of needs, **88**, 88–89
Hippocampus, 33
HM, 209
Homo erectus, 59
Homo sapiens, 59
Hormones
    aggression, 244–246
    attachment 189–193
    love, 183
Huntington's disease, 87, 97
Hurricane names, 214, 215
Hypothalamic-pituitary-adrenal (HPA)
    axis, **191**, 191–193, 198
Hypothalamus, **33**, 75

Id, **74**
Identity, 202–213
Imitation, **59**
    autism, 153
    empathy and, 132–134, 136
    infants, 264–265
    mirror neurons, 65–66
    social learning, 59, 60–61
    theory of mind, 270, 272
Immaturity, length of, 53
Implicit Association Test (IAT), **215**,
    215–216, 218, 219
Imprinting, **184**
Incest, 234–235
Independent self, **211**
Indirect reciprocity, **160**
Individualist culture, **211**
Infant development, 259–268
    adapting cognitive neuroscience
        methods, 263–264

communication, 265–267
emotion recognition, 262
face recognition, 259–262
gaze following, 262
imitation, 264–265
interactions with others, 262
joint attention, 262
language learning, 266–267
preferential looking technique, **16**,
    16–17
self-awareness, 268–269
self-recognition, 268–269
social referencing, 262
turn-taking, **265**
voice recognition, 259
Infant-directed speech, **267**
Inferior, 30
Inferior parietal lobe, 232
Ingroup favoritism, 214
Innovation, 52
Insecure/anxious, **186**
Insecure/avoidant, **186**
Institutionalization, 193–194
Instrumental aggression, **236**
Insula, 32, **86**
    architecture and functions, 77
    disgust, 86–87
    grief, 197
    morality, 231
    social exclusion, 196
    unfairness, 171
    unreciprocated cooperation, 169
Intentional stance, **60**, **140**
Intentions, 133
Inter-dependent self, **211**
Inter-observer reliability, **17**
Inter-rater reliability, **17**
Interactions, **157**
Interoception, **206**
Interpersonal Reactivity Index (IRI), 132
Interviews, 18–19
Intraparietal sulcus, 62, 67, 116
Intuition, 252
Intuitive moral grammar, 230, 233
Invasiveness, **13**, 13–14
Iowa Gambling Task, **92**, 92–93
Irony comprehension, 278–279

James–Lange theory, **74**, 74–75
Joint attention, **115**, 115–120, 262
'Joy of giving' effect, 162, 163

Kin selection, **158**
Kluver–Bucy syndrome, **81**, 244

Language
    evolution, 53–54
    learning, 266–267
    theory of mind development, 142
Larynx, **54**
Lateral, 31
Lateral geniculate body, 33

Law, 227
Lesion studies
    experimentally-induced lesions,
        43–44
    naturally occurring lesions, 14, 40–42
    reversible lesions, 43, 45
Levels of explanation, 7–10
Lie detector, 22
'Like me' hypothesis, 265
Limbic brain hypothesis, 76
Limbic–motor interface, 97
Limbic system, **33**
Local enhancement, **60**
Loneliness, 179, **197**, 197–198
Love, **180**, 180–183

Machiavellian intelligence
    hypothesis, 50
Magnetoencephalography
    (MEG), 13, 14
Mark test, 111, 269
Masking, **17**
Matching hypothesis, **181**
Material symbols, 62–63
Maturation time, 53
Measurement
    behavior and cognition, 14–19
    bodily responses, 22–25
Medial, 31
Medial geniculate body, 33
Meme, **56**
Memory
    self and, 207–210
    working memory, 252
Meningioma, 41
Mental chronometry, **14**
Mental states, **129**
    autism, 146–148
    reasoning about, 139–145
Mentalizing, **129**
Meta-representation, 139, 147
MGS, 249
Mimicking, 59, 264–265
Mind-blindness, 146–148
Mind reading, 279
Mirror neurons, 6–7, **65**, 65–66, 67, 133,
    136–137, 153
Mirror self-recognition, **111**, 268–269
MNI template, **38**
Modern Racism Scale, 214–215
Modularity, **5**, 5–6, 141
Money, 63
Montane voles, **190**, 190–191
Mood, **71**
Moral dilemmas, 227–229, 232–
    233, 237
Moral disgust, 87, **230**
Moral emotions, **79**, 230–231
Moral grammar, 230, 233
Moral intuition, 229
Moral norms, **227**, 235–237
Moral reasoning, 229

Morality, 229–239
  absence of, 237–238
  cognitive versus emotional processes, 229–233
  universal morals, 234–235
  variability, 235–237
Motherese, **267**
Motivation, **72**, 88–98
Motor mimicry, 264–265
MT, **112**
Mu suppression, **150**, 152
Mu waves, **151**, 151–152
Multi-cell recordings, 13, 14, 26
Multisensory neurons, 64–65
Myelination, **277**

N170, **29**
Name letter effect (NLE), **214**
Nash equilibrium, 166
Naturally occurring lesions, 14, 40–42
Nature–nurture debate, 53, **257**
Near-infrared spectroscopy (NIRS), **263**
Neural tube, **258**
Neurochemical lesions, 43
Neurodegenerative disorders, 41
Neuro-economics, **166**, 166–167
Neuroethics, **252**
Neuronal recycling, 62
Neurons, **20**, 20–22
Neuropsychology, **40**, 40–42
Neuroses, 74
Neurosurgery, 41
Neurotransmitters, **20**
  attachment, 189–193
  love, 183
Non-affective empathy, 136
Nucleus, 20
Number systems, 62–63

Obedience, **164**, 164–165
Observational measures, 16–17
Occipital face area (OFA), **104**, 104–105
Ontogenetic development, **257**
Opioids, 196
Orbitofrontal cortex
  acquired sociopathy/ASPD, 249–251
  adolescence, 278
  anger, 244
  architecture and function, 77
  attachment, 188
  morality, 237
  motivational value of rewards, 91–93
  mutual cooperation, 168
  psychopathy, 247–248
Orbitofrontal gyrus, 231
Organ Donor Dilemma, 229
Out-of-body experiences, **206**, 207
Outgroup prejudice, 214–220
Oxytocin, **190**, 190–193

Pain, 134, 194–196
Pair-bonding, **189**, 189–190

Papez circuit, **76**
Parasympathetic system, **22**, 24
Parietal lobe, 232
Parkinson's disease, 97, 114
Payoff matrix, **167**, 168
Pe, 220
Peacock's tail, 159
Peer influences, 275
Performance-based measures, 14–16
Periaqueductal gray, 196, 197, **244**
Peripheral nervous system, 22
Person identity node, **102**
Personal distress, **131**, 137
Personality, **120**
  Big Five, **122**, 122–123
  traits, **120**, 120–125, 210
Perspective taking, **130**
Phobia, **84**
Phylogenetic development, **257**
Physical attractiveness, 181–182
Physiognomy, 120
Piaget, Jean, 257
Pity, **131**
Pleasantness, 93
Pointing, 118–120
Polygraph, 22
Positron emission tomography (PET), 13, 14, 33, 34, 35–36
Post-synaptic neuron, 20
Posterior, 30
Posterior cingulate cortex, 232
Prairie voles, **189**, 189–192
Prayer, 222
Preferential looking, **16**, 16–17
Prefrontal cortex
  adolescence, 277, 278
  controlled behavior, 252
  empathy, 137
Prefrontal cortex (dorsolateral), 171–172, 232
Prefrontal cortex (lateral)
  attachment, 188
  cognitive control of emotions, 90
  inputs, 91
  personality judgments, 210
Prefrontal cortex (medial)
  adolescence, 279
  empathy, 132, 137
  morality, 231, 232
  personality judgments, 210
  temporal lobe connection, 91
  theory of mind, 143–144
  thinking about self, 213
Prefrontal cortex (ventrolateral), 196
Prefrontal cortex (ventromedial)
  adolescence, 278
  anger, 244
  cognitive control of emotions, 90
  culture and social identity, 212
  morality, 237
  motivational value of rewards, 91, 93
  mutual cooperation, 168
  personality judgments, 210
Prejudice, **202**, **213**, 214–220

Premotor cortex, 66, 132, 134
Pre-processing, **38**
Pre-synaptic neuron, 20
Pretend play, 270
Pretense, 139
Primary reinforcers, **89**
Prisoner's Dilemma, 166, **167**, 167–170
Proprioception, **206**
Prosocial behavior, 222, 223, 272–274
Prosopagnosia, **103**, 106, 109, 125
Proto-declarative pointing, **119**, 265–266
Proto-imperative pointing, **119**
Psychopathy, **236**, 237, 247–248
Psychopathy Checklist, 247
Public Goods Game, **172**, 172–175
Punishment, **71**, 88, 89, 238–239
Pupilometry, 22–23
Putamen, 32, 96

Questionnaires, 18–19

Racism, 214–220, 236
Random field theory, 39
Rate coding, **26**
Reaching, 118–120
Reactive aggression, **236**
Reading, 62
'Reading the mind in the eyes' test, 132
Receptive field, **65**
Reciprocal altruism (reciprocity), **158**, 159
Reductionism, **7**
Regret, **93**
Reinforcer, 89
Reliability, **18**
Religion, 221–223
Reputation, 160, 175
Response conflict, **94**
Response times, 14–15
Responsibility, **252**, 252–254
Reversal learning, **92**
Reverse inference, **8**
Reversible lesions, 43, 45
Reward, **71**, 85, 88, 89, 91–93, 95–98
Right and wrong, 227, 229–233
Right brain hypothesis, 73
Risk-taking, 278, 279–280
Romanian orphanages, 193
Rostral, 30

Safety issues, 40, 46
Sally–Anne task, 139, 140
Savant skills, 148
Schema, **142**
Schizophrenia, 206
Second-order intentionality, **140**, 140–141
Secondary reinforcers, **89**
Securely attached, **186**
Self-awareness, **203**, 268–269
Self-concept, 202–213
  agency, **204**, 204–206
  embodiment, **204**, 206
  memory, 207–210

Self-memory system, 207–208
Self-recognition, 111, 268–269
Self-stimulation, **95**
Semantic dementia, **143**
Separation, 194–196
Sexual selection, **159**
Sibling influences, 275
Simulation theory, 130–139, 141, 206
Single-sell recording, 13, 14, **26**, 26–27
Skin conductance response (SCR), 22, **23**, 25, 109
Slavery, 9
Smoothing, **39**
Social bonds, **179**, 179–180
Social brain, 5–7
Social brain hypothesis, 50
Social cognition, 5
Social cognitive neuroscience, 4
Social decision making, 166–176
Social dominance, 240, 241–242
Social events, 144
Social exclusion, 194–196
Social group size, 52
Social identity, 211–212
Social information transfer, 56–57
Social intelligence, **50**
Social intelligence hypothesis, **50**, 50–55
Social learning, **50**
  imitation, 59, 60–61
  infants, 259–268
  non-human species, 52, 57, 59–60
  violence, 242–243
Social neuroscience, emergence, 4–5
Social norms of fairness, 175
Social psychology, **4**
Social referencing, **89**, 89–90, 262
Social support, 179, 192
Sociopathy, **246**, 246–247, 249–251
Soma, 20
Somatic nervous systems, **22**
Spatial resolution, 13
Speed–accuracy trade-off, **15**, 16
Spiking rate, 22
Stanford Prison Experiment, 201
Stereotactic normalization, **38**
Stereotyping, **213**
Stimulation theory, **108**, 108–109
Stimulus enhancement, **60**
Stone tools, 59
Strange Situation Test, **185**, 185–186
Striatum, **96**; see also Dorsal striatum;
      Ventral striatum
Strokes, **40**
Stroop test, **94**, 95
Structural imaging, 33, 263
Subcortex, 30, 32–33

Subliminal perception, 17–18
Sulci, **31**
Super-ego, **74**
Superior, 30
Superior temporal sulcus
  adolescence, 279
  emotional body language, 114
  face perception, 105, 106–107
  joint attention, 116, 119
  morality, 231, 232
Survey measures, 18–19
Sweating, 23
Sympathetic system, **22**, 24
Sympathy, **131**
Synapse, **20**
Synaptic learning, 208–209
Synaptic potential, 20–21
Synaptogenesis, 258–259
Syntax, **54**
Systemizing, 149, 150

Talairach coordinates, **38**
Technology, 64–68
Temporal coding, **26**, 26–27
Temporal lobes, 222
Temporal poles, 142–143
Temporal resolution, **13**
Temporo-parietal junction (TPJ)
  altruism, 162–163
  empathy, 137
  loneliness, 197
  out-of-body experiences, 206
  self-recognition, 268
  theory of mind, 144–145
  tool use, 67
Tesla (T), 34
Testosterone, 183, **244**, 244–246
Thalamus, **33**
Theory of mind, 130, 139–145
  autism, 146–148
  development, 270–272, 279
  domain-general versus
      domain-specific accounts,
      141–142
  language function, 142
  neural substrates, 142–145
  religious beliefs, 222–223
  simulation theory, 141
Theory-theory, **141**
Third-order intentionality, 141
Threat detection, 84
Tit-for-tat, **168**
Tool use, 52, 58, 59, 64–68
Tools, **64**
Tradition, **57**
Traits, **120**, 120–125, 210

Transcranial magnetic stimulation (TMS),
      13, 14, 44–46, 109, 264
Transection, 43
Traumatic head injuries, 41
Triadic interactions, **262**
Triangular theory of love, **180**, 180, 181
Trier Social Stress Test (TSST), 192
Trolley Dilemma, 227–229, 232–233
Trust, **157**
Trust Game, **173**, 173–175
Trustworthiness, 125
Tumors, 41
Turn-taking, **265**
Type I error, **19**
Type II error, **19**

Ultimatum Game, 166, **170**, 170–172
Unconscious, measuring, 17–18
Unconscious emotional motivations, 74
Utilitarian, **228**

V5, **112**
Value, 166–167
Vasopressin, **190**, 190–193
Ventral, 30
Ventral striatum, **96**
  adolescence, 278
  anger, 87, 244
  architecture and function, 78
  loneliness, 197
  mutual cooperation, 168
  reward, 96–98
Ventral tegmental area, **96**
Ventral visual stream, **103**
Violence, see Aggression
Viral infections, 41
Virtual lesions, 45
Visual word form area (VWFA), **62**
Voices
  recognition, 259
  social cues, 115
Voxel, **34**

Wagering, 18
Walking upright, 58–59
Weak central coherence, 148–149
Westermarck effect, **235**
White matter, **30**, 277
Working memory, 252
Working self, 207
Writing, 62

Yawning, 129

Zero-order intentionality, 140